The Skeletal System

COLD SPRING HARBOR MONOGRAPH SERIES

The Bacteriophage Lambda
The Molecular Biology of Tumour Viruses
Ribosomes
RNA Phages
RNA Polymerase
The Operon
The Single-Stranded DNA Phages
Transfer RNA:
    Structure, Properties, and Recognition
    Biological Aspects
Molecular Biology of Tumor Viruses, Second Edition:
    DNA Tumor Viruses
    RNA Tumor Viruses
The Molecular Biology of the Yeast Saccharomyces:
    Life Cycle and Inheritance
    Metabolism and Gene Expression
Mitochondrial Genes
Lambda II
Nucleases
Gene Function in Prokaryotes
Microbial Development
The Nematode Caenorhabditis elegans
Oncogenes and the Molecular Origins of Cancer
Stress Proteins in Biology and Medicine
DNA Topology and Its Biological Effects
The Molecular and Cellular Biology of the Yeast Saccharomyces:
    Genome Dynamics, Protein Synthesis, and Energetics
    Gene Expression
    Cell Cycle and Cell Biology
Transcriptional Regulation
Reverse Transcriptase
The RNA World
Nucleases, Second Edition
The Biology of Heat Shock Proteins and Molecular Chaperones
Arabidopsis
Cellular Receptors for Animal Viruses
Telomeres
Translational Control
DNA Replication in Eukaryotic Cells
Epigenetic Mechanisms of Gene Regulation
C. elegans II
Oxidative Stress and the Molecular Biology of Antioxidant Defenses
RNA Structure and Function
The Development of Human Gene Therapy
The RNA World, Second Edition
Prion Biology and Diseases
Translational Control of Gene Expression
Stem Cell Biology
Prion Biology and Diseases, Second Edition
Cell Growth: Control of Cell Size
The RNA World, Third Edition
The Dog and Its Genome
Telomeres, Second Edition
Genomes
DNA Replication and Human Disease
Translational Control in Biology and Medicine
Invertebrate Neurobiology
The TGF-β Family
Molecular Biology of Aging
Adult Neurogenesis

The Skeletal System

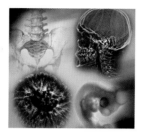

EDITED BY

Olivier Pourquié

Howard Hughes Medical Institute
Stowers Institute for Medical Research
Kansas City, Missouri

COLD SPRING HARBOR LABORATORY PRESS
Cold Spring Harbor, New York • www.cshlpress.com

The Skeletal System

Monograph 53
© 2009 by Cold Spring Harbor Laboratory Press, Cold Spring Harbor, New York
Printed in China

Publisher	John Inglis
Acquisition Editor	Alexander Gann
Development Director	Jan Argentine
Project Coordinator	Joan Ebert
Production Manager	Denise Weiss
Production Editor	Kaaren Kockenmeister
Desktop Editor	Susan Schaefer
Marketing Manager	Ingrid Benirschke
Sales Manager	Elizabeth Powers
Cover Designer	Michael Albano

Front cover artwork: Different perspectives of the skeletal system. *Top left,* engraving of a human lower spine and pelvis, ventral view, from W. Smellie (1754, London, NLM collection); *top right,* CT scan of an adult human skull, right view (courtesy of Dr. J. Chal); *bottom left,* Tissue section of an epithelial somite, stained for the actin cytoskeleton; *bottom right,* 4-day-old chicken embryo illustrating the segmented spine precursors and limb buds. Images provided by Jérome Chal; cover concept by Silvia Esteban.

Library of Congress Cataloging-in-Publication Data

The skeletal system / edited by Olivier Pourquié.
    p. cm. -- (Cold Spring Harbor monograph series)
  Includes index.
  ISBN 978-0-87969-825-6 (cloth : alk. paper)
 1. Human skeleton. 2. Human skeleton--Growth. 3. Bone. I. Pourquié,
Olivier.
  QM101.S547 2009
  611'.71--dc22

                                                          2008054989

10 9 8 7 6 5 4 3 2 1

All Cold Spring Harbor Laboratory Press publications may be ordered directly from Cold Spring Harbor Laboratory Press, 500 Sunnyside Boulevard, Woodbury, New York 11797-2924. Phone: 1-800-843-4388 in Continental U.S. and Canada. All other locations: (516) 422-4100. FAX: (516) 422-4097. E-mail: cshpress@cshl.edu. For a complete catalog of Cold Spring Harbor Laboratory Press publications, visit our World Wide Website http://www.cshlpress.com/.

Contents

Preface

Our understanding of skeletal biology has taken tremendous strides during the past few years. On the one hand, the spectacular recent breakthroughs in developmental biology have led to an understanding of the global rules shaping and positioning the cartilage and bone primordia in the vertebrate embryo. On the other hand, the discovery of key master regulators of the chondrocyte and bone lineage, such as Sox9 and Runx2, as well as the signaling pathways involved in the regulation of the differentiation of these lineages, has provided a much better understanding of these processes. This knowledge led to the elucidation of the molecular etiology of a majority of bone and cartilage genetic diseases.

The goal of this book is to provide a comprehensive and up-to-date summary of the field of skeletal biology. This is a large field and due to space limitations, some areas might be covered more extensively than others. However, an attempt was made to cover all stages of skeletal development and patterning, as well as differentiation of cartilage and bone cells. The complex area of bone physiology is discussed in some of the chapters, but is not addressed extensively. This book covers essentially three major themes.

The first theme relates to the development and patterning of bone in vertebrates. Several chapters deal with classical model systems that have been used to study bone and cartilage patterning in the vertebrate embryo. Specifically, the limb bud and the rules governing the formation of bone primordia are addressed, as well as the development of long bones, which provides a classical model to study the endochondral mode of bone formation. A chapter is devoted to cranial skeleton development, which is derived largely from the neural crest and is formed mostly by membrane bones. In addition, the process of segmental patterning and differentiation of the spine is discussed, which involves very different regulatory mechanisms. Also, the evolutionary origin of bone, which appeared early in the vertebrate lineage, constitutes the topic of an independent chapter.

The second theme explores the molecular mechanisms underlying the specification and differentiation of the cartilage and bone lineages. Four chapters address the major aspects of the control of differentiation of these lineages by transcription factors and signaling molecules. This area has recently exploded with the discovery of master regulators that are required for the differentiation of all cartilage and bone cells. The spatio-temporal deployment of these transcriptional regulators is exquisitely controlled by a handful of signaling pathways whose role is discussed extensively in two chapters. This theme also covers the genetic diseases resulting from mutations affecting bone and cartilage differentiation. A number of classical bone diseases, which have recently been associated with defects in the regulatory systems, are discussed as well.

The third and last theme covers major aspects of bone and cartilage terminal differentiation and remodeling. Bone and cartilage are supporting tissues that provide the rigid frame on which our body is built. The characteristic physical properties of these tissues result from the production of highly specialized extracellular matrix, which can become mineralized. Several chapters survey our current understanding of these different processes.

While this book is expected to meet the expectations of specialists, it is intended also to provide a survey of the field to newcomers. We hope that this book will take its place as a useful tool for those working in the ever-growing field of skeletal biology.

I am indebted to all the authors for their excellent contributions. I also thank Alex Gann at Cold Spring Harbor Laboratory for his initial encouragement with this volume, Kaaren Kockenmeister, and, in particular, I thank Joan Ebert for her continuous help and support. I am also grateful to Joanne Chatfield for her most valuable editorial assistance, and to Silvia Esteban for the cover illustration.

Olivier Pourquié

1

Evolutionary Origin of Bone and Cartilage in Vertebrates

Kinya G. Ota and Shigeru Kuratani
Laboratory for Evolutionary Morphology
Center for Developmental Biology, RIKEN
2-2-3 Minatojima-minami, Chuo, Kobe
Hyogo 650-0047, JAPAN

Most living vertebrates are characterized by possession of bones and cartilage as major components of the skeletal system, and the origin and evolution of these tissues remain intriguing questions. Classically, Geoffroy Saint-Hilaire (1818) tried to compare arthropods and vertebrates by two types of "inversions." One is the dorsoventral inversion that brings the ventral nervous system of arthropods to the dorsal side, as seen in vertebrates. The other is the inside-out inversion to transform the arthropod exoskeleton into the vertebrate-type endoskeleton. However, it is misleading to regard the endoskeleton as the major skeleton in vertebrates. Compared with the exoskeleton in arthropods, vertebrates also have exoskeletal elements. Thus, the vertebrate exoskeleton has been compared directly with that of arthropods, as suggested by Patten (1912), Gaskell (1908), and others. However, those ideas have now been refuted by new evidence of phylogenetic relationships among animal phyla, by improved knowledge of skeletal histology and cytology, and by molecular developmental evidence for skeletogenesis.

In the modern evolutionary scenario, which is largely based on phylogenetic trees constructed on molecular sequence data, it is generally accepted that vertebrates belong to the deuterostomes together with echinoderms, hemichordates, urochordates, and cephalochordates (amphioxus). This comprises a sister group to the protostomes, consisting of lophotrochozoans and ecdysozoans (Aguinaldo et al. 1997). It is along this phylogenetic tree that the origins of the vertebrate skeleton

The Skeletal System ©2009 Cold Spring Harbor Laboratory Press 978-087969-825-6

should be sought, by integrating the fossil evidence (Halstead 1974; Donoghue and Sansom 2002; Hall 2005). Comparative analyses of skeletal development in various living organisms will help guide us to identify which features of skeletogenesis would have been shared by which ancestral animals. In this chapter, we consider the evolutionary origins of skeletal tissues in vertebrates in the context of Evolutionary Developmental Biology, mainly focusing on the origin of cartilage.

SKELETAL SYSTEM AND SKELETAL TISSUES

Vertebrate bone is a vascularized tissue consisting of osteocytes with multiple interconnecting processes, embedded in an extracellular matrix (ECM), mineralized with hydroxyapatite, and containing type I collagen. Cartilage, on the other hand, is an avascular tissue containing chondrocytes that are usually round and separate from one another. The cartilage ECM is primarily composed of glycosaminoglycans and contains type II collagen (for review, see Halstead 1974; Hall 2005).

As noted, anatomically, the vertebrate skeleton consists of exo- and endoskeletons. The endoskeleton is further divided into the axial skeleton or vertebral column, the cranium (see p. 4), and the appendicular elements including the pectoral and pelvic girdles (for review, see Goodrich 1909, 1930; Romer and Parsons 1977; Donoghue and Sansom 2002). The endoskeleton either remains cartilaginous—sometimes becoming calcified—or undergoes ossification at late developmental stages. By contrast, the exoskeleton consists only of various mineralized structures, including bones, except for secondary cartilages associated with some taxa. Evolutionarily, bone arose as one type of various mineralized exoskeletal elements (for the spectrum of mineralized tissues, see Goodrich 1909; Halstead 1974; Romer and Parsons 1977; Donoghue and Sansom 2002; Sire and Huysseune 2003; Hall 2005). The exoskeletal bones are collectively called dermal bones, including all the bony elements developing in the dermis, such as those in the skull roof and facial skeleton, osteoderms and gastralia in some amniotes, and the plastron in turtles. A major part of the turtle carapace, on the other hand, is derived developmentally from the endoskeletal rib that has secondarily shifted its position dorsally and extended its growth into the dermis. Thus, it is misleading to call it "dermal" simply from its location (see Gilbert et al. 2001; Nagashima et al. 2007; see also Suzuki 1963 and below).

Related to the above classification, the terms for skeletal and ossification types are often confused and misused. A typical confusion is seen with "membrane bones" and "dermal bones." These are not synonyms and do not belong to the same context. According to Starck (1979), the term "membranous" refers to a type of ossification process, not a type of bone. Thus, ossification taking place in membranous tissue is always called "membranous ossification," which is seen in both dermal and cartilage bones (as seen in the perichondral ossification of long bones and late development of the turtle carapace). In an extreme case, endoskeletal elements can ossify purely membranously (Bellairs and Gans 1983). The term "dermal" should refer to the morphological property or evolutionary origin. As the exoskeleton develops primarily in the dermis, every dermal bone ossifies membranously, but membranous ossification does not always gives rise to dermal bones. Similarly, the term "endochondral" refers to a specific mode of ossification in which primordial cartilage is replaced by osteocytes. The term is used in a histogenetic context as every ossifying endochondral bone undergoes this process, and it does not necessarily refer to the morphological identity of that element.

The evolutionary sequence of each type of vertebrate skeleton remains enigmatic. Thus, no bony tissues are found in cyclostomes, but assuming the monophyly of cyclostomes (mainly based on molecular evolutionary data; for review, see Kuratani and Ota 2008), bone appears to have appeared first in (jawless) stem gnathostomes known as ostracoderms (for review, see Halstead 1974; Janvier 1996; Donoghue and Sansom 2002). The endoskeleton in living sharks and rays is predominantly cartilaginous. It often calcifies (forming calcified cartilages) and the exoskeleton in these taxa is devoid of true bones. However, this represents a secondary degenerative condition, as bony exoskeletons and endochondrally ossified vertebra already existed in the fossil placoderms, the sister group of all jawed vertebrates (for review, see Donoghue and Sansom 2002). The formation of ossified vertebrae appears to be a synapomorphy of crown gnathostomes (jawed vertebrates including the placoderms).

Vertebrae

The vertebrae are exclusively of somitic mesodermal origin and so are the ribs. In gnathostomes, the ventromedial portion of the epithelial somite is delaminated to form a mesenchymal cell population, called the sclerotome. This developmental process is induced by a signal derived from the notochord. In gnathostomes, the induced sclerotome expresses *Pax1/9* genes, which function together in condensation and differentiation of sclerotomes, further allowing the expression of *Sox9* and collagen type II for chondrification of centra and proximal ribs (Peters et al. 1999; for review, see Hall 2005; for early evolution of gnathostome centra, see

Goodrich 1909). The rest of the somite remains as the dermomyotome, which differentiates into skeletal muscles and dermis.

Among cyclostomes, only the adult lamprey shows vertebrae. These cartilages correspond to neural arches whose gnathostome counterparts can develop in the absence of *Pax1/9* functions, whereas *Pax1/9*-dependent centra are entirely absent from the lamprey. Notably, the sclerotome is very unclear in the lamprey embryo and no *Pax1/9* expressions have been reported so far in their somitic derivatives, although the genes themselves have been identified (Ogasawara et al. 1999, 2000; for the sclerotome of cyclostomes, see Hertwig 1906; Kusakabe and Kuratani 2005; Kuratani 2008).

In the adult hagfish (another cyclostome group), no skeletal elements are found in the trunk. However, Neumayer (1938) found vestigial cartilage nodules in the rostral part of the axis in the embryos. This is not so surprising, because in the hagfish (and the lamprey), the parachordals of the cranium chondrify from the paraxial mesoderm, implying that the vertebra-like mechanism for axial skeletal induction exists in these animals and involves signals from the notochord (see below; Kuratani 2008; see also Kuratani et al. 2004). It appears very likely that the loss of vertebrae represents a secondary condition in the hagfish, like the other "lost" organs such as eyes and a lateral line system (Wicht and Northcutt 1995). This is also consistent with the monophyly of cyclostomes (see p. 10). It has to be noted that there are no living vertebrate groups that completely lack a cartilaginous axial skeleton from their entire life history. This is consistent with the finding of cartilaginous vertebrae in the Cambrian fossil *Haikouichthys*, apparently belonging to a lineage basal to cyclostomes (Shu et al. 2003). Thus, the ancestral vertebrate initially appears to have possessed cartilaginous elements forming a splanchnocranium, a neurocranium, and vertebrae, and it is difficult to determine in which sequence these components arose.

Neural Crest and Cranium

Invention of the neural crest seems to have had profound significance for the evolution of vertebrates. Delaminated from the neural ridge epithelium and migrating into the various sites of the embryo, this ectodermal cell lineage not only gives rise to the peripheral nervous system or pigment cells, but also to the ectomesenchyme that differentiates into the skeletal system, mainly in the head (Fig. 1) (for review, see Gans and Northcutt 1983; Hall and Hörstadius 1988; Hall 1999; for the neural crest's contribution to craniogenesis, see also Platt 1893; Le Douarin 1982; Noden 1983, 1988; Meulemans and Bronner-Fraser 2007). This cell

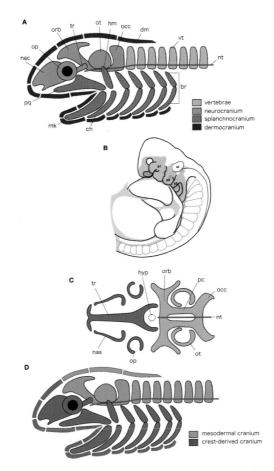

Figure 1. Schematic representation of the cranial morphological and developmental plans of gnathostomes. (*A*) Dorsal view of the generalized and simplified neurocranium in gnathostomes at an early embryonic state. Cartilaginous elements colored gray are derived from the neural crest and mesodermal elements are colored white. Note that the interface between the crest- and mesoderm-derived parts corresponds to the rostral tip of the notochord (*nt*). The scheme is based on Couly et al. (1993). (*B*) Morphological plan of the gnathostome cranium. The cranium is divided into the dorsal neurocranium that encapsulates the brain, the ventral splanchnocranium that supports the pharynx and mouth, and the dermocranium that serves as an external shield. (*C*) Developmental plan of the gnathostome cranium. The distribution of the paraxial mesoderm coincides with the notochord (chordal cranium). Thus, the prechordal region of the neurocranium is derived from the crest-derived ectomesenchyme (gray) and is called the prechordal cranium. Note that the dermocranium is also divided into the anterior and posterior moieties, derived from crest and mesoderm, respectively. (br) Branchial arch skeleton, (ch) ceratohyal, (dm) dermocranial elements, (occ) occipital cartilage (derived from rostral somites), (hm) hyomandibula, (hyp) hypophysis, (mk) Meckel's cartilage, (nas) nasal capsule, (nt) notochord, (op) optic capsule, (orb) orbital cartilage, (ot) otic capsule, (pc) parachordal cartilage, (pq) palatoquadrate, (tr) trabecula (primordia for nasal and interorbital septa), (vt) vertebrae.

lineage might have been the first mesenchymal lineage to differentiate into the skeleton in vertebrate evolution. Although in tunicate larvae migratory pigment cell precursors also express a gene expression repertoire reminiscent of the neural crest (Jeffery et al. 2004), they never differentiate into any skeletal tissues (tunicates do not develop bone or cartilage). In the amphioxus, which shows a striking resemblance to the vertebrate body plan, neural crest cells do not appear in development, nor do these animals possess the genetic cascades for crest specification (Fig. 2) (Meulemans and Bronner-Fraser 2007; see also Sauka-Spengler et al. 2007). The neural crest as the migratory cell lineage destined to form cephalic ectomesenchyme appears to be a synapomorphy that defines the vertebrates, including hagfish (Ota et al. 2007).

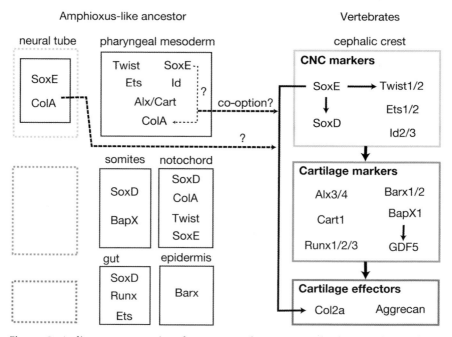

Figure 2. A diagram comparing the gene regulatory networks for vertebrate chondrogenic crest cells and several cell lineages in amphioxus. Bronner-Fraser and her colleagues have proposed a step-wise regulation of gene repertoires activated for the crest cells (*right*), which are not integrated in amphioxus lacking a crest cell lineage (*left*). In deuterostomes, a SoxE-Col cascade appears to represent a plesiomorphic trait. If nonvertebrate cartilage in the amphioxus-like ancestor evolved into the crest-derived cartilage in vertebrates, there are two possible candidates functioning in the living amphioxus for the origin of co-opted collagen gene regulatory networks (thick arrows). Discussion by Rychel and Swalla (2007) has been taken into consideration. (Modified from Meulemans and Bronner-Fraser 2007.)

According to traditional morphology based on the anatomical distributions and functions of skeletal elements in adult gnathostomes, the vertebrate cranium is divided into the dorsally located neurocranium that encapsulates the central nervous system, and the visceral cranium (splanchnocranium) supporting the pharynx and mouth (Fig. 1). The dermocranium must be added to these (de Beer 1937; Portmann 1969). However, viewed from developmental patterns, the neurocranium should be divided into rostral or crest-derived, and posterior or mesodermal, halves.

Using avian chimeric embryos, Couly et al. (1993) have found that the boundary between these two neurocranial parts corresponds to the rostral end of the notochord or the hypophyseal foramen. Namely, the chordal cranium is derived from the head mesoderm (forming parachordal and orbital cartilage derivatives) and rostral somites (forming occipitals). The prechordal cranium consists of trabecular derivatives and sensory capsules that are derived from the premandibular ectomesenchyme (crest cells occupying regions rostral to the mandibular arch) (Fig. 1C) (see Kuratani 2005). This distinction is consistent with the observation that the mesodermal mesenchyme requires the presence of a notochord to chondrify, whereas the neural crest-derived ectomesenchyme does not (Couly et al. 1993). Thus, the vertebrate skull consists of a crest-derived part (viscerocranium plus prechordal neurocranium) and the mesodermal part (posterior neurocranium) (Fig. 1D). The morphology of the prechordal cranium reflects the early distribution pattern of the ectomesenchyme in the pharyngula (Fig. 1B). In classical embryology, the trabecula of the gnathostomes were regarded as a remnant premandibular arch (Sewertzoff 1911; for review, see de Beer 1937; Jarvik 1980). Although this branchiomeric theory is not supported today, the trabecula does arise from the ectomesenchyme rostral to the mandibular arch (Cerny et al. 2004; see also Shigetani et al. 2000; for review, see Kuratani et al. 2001).

The basic scheme of crest/mesoderm contribution to vertebrate craniogenesis as described above appears to be shared, at least in part, by development in the lamprey (for the two types of mesenchyme in the lamprey, see Koltzoff 1901; Damas 1944; Horigome et al. 1999; for neural crest development in cyclostomes, see Horigome et al. 1999; Kuratani and Ota 2008; for contribution of the crest to the cyclostome viscerocranium, see Newth 1951, 1956; Langille and Hall 1988; McCauley and Bronner-Fraser 2003, 2006). In the lamprey, parachordal and trabecular cartilages are also recognized in the larval head (de Beer 1937; Johnels 1948; Kuratani et al. 2004). A recent labeling experiment has revealed that the lamprey trabecula first arises from the paraxial mesoderm and is of mandibular mesodermal origin. This is consistent with observations that most of the premandibular ectomesenchyme in the lamprey is used to form the upper lip of the larva (Johnels 1948; Kuratani et al. 2004; for relationships between oral patterning and trabecular evolution, see also Shigetani et al. 2002, 2005; Kuratani 2005). From the above, it appears most likely that the ancestral neurocranium involved both crest-derived ectomesenchyme and mesoderm in its formation. Thus, the real trabecula is most likely to be a gnathostome synapomorphy.

Such a dual cell lineage origin is also found in the dermal bones. Using transgenic mouse models, the frontal and more rostrally located dermal elements in the murine vault have been found to be of crest origin, whereas the parietal and more posterior dermal elements are mesodermal (Morriss-Kay 2001). In the chicken, the boundary between the crest and mesodermal territories is found in the middle of the frontal bone (Le Lièvre 1978; Evans and Noden 2006). Thus, the boundary between the two cell lineages in the dermal skull roof may vary between animal groups. It is misleading to conclude that all dermal elements must develop from the neural crest. In addition, the two cell lineages do not correspond to any of the various skeletal categorizations such as neuro- versus splanchnocranium, or endo- versus exoskeleton. Instead, they are associated with the distribution pattern of mesenchymal cells that was established developmentally before the pharyngula, which is rather conserved throughout vertebrates—possibly including the latest common ancestor of vertebrates (for review, see Kuratani et al. 2001).

Skeletal elements in each pharyngeal arch develop as serial homologs in the gnathostome splanchnocranium (Fig. 1). Positional specification of the splanchnocranial elements along the anteroposterior axis (development of the first arch into the mandibular identity and the second arch into the hyoid) is mediated by the nested pattern of homeobox genes, called the Hox code, as seen in vertebrae (for review, see Kessel 1992). The mandibular arch ectomesenchyme does not express any Hox genes, which provides the default state of the Hox code as a prerequisite for patterning of the jaw (Rijli et al. 1993; Couly et al. 1998). Besides, Hoxa-2 is expressed in second and posterior arches to specify the hyoid arch identity in the second arch (Rijli et al. 1993; Couly et al. 1998). Such a Hox code-dependent specification is partly shared by cyclostomes, whose oral patterning apparently depends on the Hox code-default state in the premandibular and mandibular ectomesenchyme (Takio et al. 2004, 2007; Shigetani et al. 2005). Although a similar Hox code appears in amphioxus development, it does not extend into the pharynx. The Hox code in the pharyngeal ectomesenchyme is thus a vertebrate-specific trait.

Dorsoventral specification of the splanchnocranium depends on the nested expression of another group of homeobox genes, the *Dlx* family. The ventral part of the arch ectomesenchyme specifically expresses both *Dlx5* and *Dlx6* to acquire a ventral identity, whereas *Dlx1* and *Dlx2* are expressed in the entire ectomesenchyme (Depew et al. 2002). However, this Dlx code appears to be a gnathostome synapomorphy and cyclostomes do not appear to possess the nested expression pattern of this gene family—although lamprey *Dlx* genes are expressed in the ectomesenchyme (Neidert et al. 2001). Evolution of the dorsoventrally articulated jaw may have depended on the establishment of the Dlx code. Although the above Cartesian grid pattern of Hox/Dlx codes provides morphological identities for the skeletal primordia, it is the crest cells per se that carry information for species-specific shape in skeletogenesis (Helms and Schneider 2003). Apparently, the autonomous cellular differentiation of the ectomesenchyme is partly mediated by modulation of the embryonic environment from epithelial–ectomesenchymal interactions.

EVOLUTION OF FIBRILLAR COLLAGENS IN VERTEBRATES

It is not easy to determine which of the mesodermal and crest-cell lineages first acquired the ability to chondrify (Hall 1999). Recent molecular studies have shown that it may not be the cell types or lineages, but rather a molecular cascade that marked the innovative evolution of the skeleton among the ancestral animals that preceded the origin of vertebrates (see below).

Collagen is one of the major ECM components for animals. It contains two highly conserved domains: collagen- and C-propeptide domains. A number of genes belong to this family among various taxa, ranging from sponges to humans (Exposito et al. 1990, 1992, 1993, 1995; Cluzel et al. 2000; Boot-Handford and Tuckwell 2003; Aouacheria et al. 2004, 2006). Among these, the fibrillar collagen gene subfamily contains important genes, *Col2A* and *Col1A* genes, encoding the main components of the vertebrate skeleton. *Col2A* gene encodes type II collagen for gnathostome cartilage and *Col1A* genes encode type I collagen for bone (see Hall 2005).

Col gene homologs have recently been isolated and their expressions observed in cyclostomes. For interpretation of the data, establishing the phylogenetic relationships between hagfishes, lampreys, and gnathostomes is central (Ota and Kuratani 2006). For example, the presence of *Col2A1* in gnathostomes is explicable based on the assumption that hagfish should be placed basal to the other vertebrates (lampreys and gnathostomes; Donoghue et al. 2000). Identification of a unique ECM component, lamprin in the cartilage of *Petromyzon marinus* and of myxinin in the Atlantic hagfish *Myxine glutinosa* by Glenda Wright and her colleagues (1984, 1988, 2001), appears consistent with this scenario. The amino acid sequences of lamprins and myxinin show no homologies with any collagen molecules, but are chemically similar to elastin (Robson et al. 1993, 2000; McBurney et al. 1996; see also Hall 2005). However, interpretations of recent molecular data have favored the monophyly of cyclostomes (Kuraku et al. 1999; Takezaki et al. 2003; Kuraku and Kuratani 2006) and the scenario for the evolution of the collagen genes is now being revised.

So far, two different *Col2A* genes, *Col2A1a* and *Col2A1b*, have been isolated from *P. marinus*, and their expressions have been detected. Interestingly, *Col2A* and *Sox9* are both expressed during chondrogenesis (Zhang et al. 2006). These results suggest that shared mechanisms for collagen gene expression and chondrogenesis would have already been established in the vertebrate common ancestor (Zhang et al. 2006), consistent with the expression patterns of chondrogenic genes in the pharyngeal arch crest cells that are conserved between lampreys and gnathostomes (McCauley and Bronner-Fraser 2006; Sauka-Spengler et al. 2007).

In the hagfish, classical histological studies have demonstrated that this animal uniquely possesses two types of cartilage: soft and hard (not identical to those in ammocoete larvae of the lamprey)(Cole 1905). Recent immunohistochemical studies have implied that the soft cartilages of *M. glutinosa* contain collagen-like proteins (Zhang and Cohn 2006). Involvement of two *Col1A* genes has also been implied in *Eptatretus burgeri* (Kimura and Matsui 1990), although no direct evidence has been provided showing the expression of these genes in the chondrocytes.

Molecular phylogeny has grouped the hagfish *Col2A1* and lamprey *Col2A1a* in the same clade, with the lamprey *Col2A1b* located on the root (Zhang and Cohn 2006), showing that *ColA2* was duplicated in the cyclostome lineage; no *Col2A1b* counterpart has been identified from the hagfish. Together with the morphological homology of cartilage elements between lamprey and hagfish, functionalization of *Col2A* genes remains an intriguing question for understanding of early skeletal evolution in vertebrates. Assuming monophyly of the cyclostomes, we now have two mutually exclusive hypotheses: (1) The cartilages of the common ancestor already contained Col2A1 and the noncollagenous components (lamprin and myxinin) appeared secondarily and independently in cyclostome lineages (Fig. 3); or (2) the ancestral cartilage would have con-

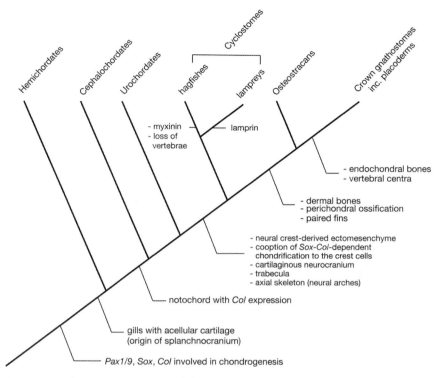

Figure 3. Evolutionary sequence of the vertebrate skeletal system. This phylogenetic tree is mainly based on molecular information that supports the monophyly of cyclostomes and places urochordates as the sister group of vertebrates. Note that the gnathostome lineage includes jawless fossil groups as its stem. Evolutionary changes to the skeletogenic program are indicated along the stem lineage towards the gnathostomes. (Modified from Donoghue and Sansom 2002, and Rychel and Swalla 2007.)

tained various types of noncollagenous ECM components, including lamprin or myxinin, which were lost secondarily from the gnathostome lineages. As we discus next, recent studies on invertebrate cartilage types seem to favor the former view.

CARTILAGE IN NONVERTEBRATE DEUTEROSTOMES

Gills are not synapomorphic for vertebrates, but are shared by cephalo-chordates, urochordates, hemichordates, and possibly echinoderms (Rychel et al. 2006 and references therein). These gills are presumably homologous to those in vertebrates. Apparently, their slits are all established with the function of *Pax1/9* genes (Ogasawara et al. 1999). Thus, endodermally expressed *Pax1/9* genes are of primary importance in the formation of the pharyngeal arches in all the deuterostomes with gills (Ogasawara et al. 1999; Rychel and Swalla 2007). In the pharyngeal arch cartilage of basal vertebrates, *Sox9* also plays a role directly upstream to regulate *Col2A* genes (see below) (McCauley and Bronner-Fraser 2002; Zhang et al. 2006).

Recent observations have revealed that acellular cartilage is distributed in the gills of hemichordates and amphioxus, in which there are no neural crest-derived ectomesenchyme elements such as those found in vertebrates (Rychel et al. 2006; Rychel and Swalla 2007; for cartilages in invertebrates, see Hall 2005). Moreover, as in the vertebrate cartilage, these invertebrate gill cartilages contain a fibrillar collagen protein similar to the vertebrate type II collagen. In two species of adult amphioxus, Rychel and Swalla (2007) have found that *SoxE*, the homolog of *Sox9* and the collagen gene itself, are both expressed in the pharyngeal endoderm, pharyngeal ectoderm, and mesodermal cells within the gill arches. In hemichordates as well, both the gene cognates are expressed in the pharyngeal endoderm, a tissue surrounding the acellular cartilage. Thus, the same molecular cascade and site of cartilage formation are conserved in deuterostomes, whereas cell lineages to express the genes and produce the cartilage seem to have changed. Interestingly, the endodermal cells to express *SoxE* and *Col* in amphioxus and hemichordates are the same cells to express *Pax1/9*. However, it is unknown if there might once have been an ancestral cascade established between *Pax1/9* and *SoxE*, and secondarily lost in vertebrates.

The curious phenomena discussed could be explained by two different scenarios. One is to deny the homology of pharyngeal arch cartilages (or gill arches themselves) and to assume the entirely independent origins of cartilages in nonvertebrate deuterostomes and vertebrates. This explanation will further assume the coincidental recruitment of the same molecular cascade—which might have been used in other unknown cartilaginous structures—to different cell lineages in the homologous organ in adapting to the same selective pressure. Alternatively, the gill cartilages might be homologous and this homologous molecular cascade was co-opted to control a different cell lineage in vertebrates, namely the neural crest cells. The latter possibility assumes that natural selection strongly favored the conserved position of the formed cartilage and the use of collagen, as well as evolutionary flexibility in the gene regulatory mechanism. A peculiar form of cartilage, called mucocartilage in the lamprey larva, might represent an intermediate state lacking chondrocytes. However, this cartilage does not contain collagen, but its main component is the lamprin noted previously (for review, see Gaskell 1908; Hall 2005).

Note that the neural crest per se might have arisen through co-option of the gene regulatory network (Fig. 2) (for review, see Meulemans and Bronner-Fraser 2007). What types of genomic changes could have been behind this co-option are not understood. The latter questions require an understanding of the evolutionary origin of cartilage at the genomic level.

SUMMARY AND CONCLUSIONS

Appearance of the neural crest and mesoderm-derived extensive mesenchyme were of fundamental importance for the evolution of the skeletal system in vertebrates. These are the cells that provided vertebrate-specific cellular cartilaginous tissues and dermal bones in the craniofacial domain, as well as massive axial skeleton in the trunk, especially in the gnathostomes. How the vertebrates have acquired these exo- and endoskeletal systems is, however, not straightforward. Although a prototype of the gene cascade employed in vertebrate skeletogenesis appears to be present in nonvertebrate chordates or even in deuterostomes, it is not always functioning in comparative cell lineages. Sub- and neounctionalizations of the duplicated genes, paucity of the fossil evidence, as well as co-option of the gene regulatory networks, would be the main reason for such an apparently complicated distribution of molecules involved in chondrogenesis. For better understanding of vertebrate skeletal tissue, a genome-wide search for skeletogenic genes in the cyclostomes and extensive expression and functional analyses of these genes in their embryos will be key.

ACKNOWLEDGMENTS

We are grateful to Hiroshi Wada and David McCauley for their critical reading of the manuscript.

REFERENCES

Aguinaldo, A.M., Turbeville, J.M., Linford, L.S., Rivera, M.C., Garey, J.R., Raff, R.A., and Lake, J.A. 1997. Evidence for a clade of nematodes, arthropods and other moulting animals. *Nature* **387:** 489–493.

Aouacheria, A., Cluzel, C., Lethias, C., Gouy, M., Garrone, R., and Exposito, J.Y. 2004. Invertebrate data predict an early emergence of vertebrate fibrillar collagen clades and an anti-incest model. *J. Biol. Chem.* **279:** 47711–47719.

Aouacheria, A., Geourjon, C., Aghajari, N., Navratil, V., Deleage, G., Lethias, C., and Exposito, J.Y. 2006. Insights into early extracellular matrix evolution: Spongin short chain collagen-related proteins are homologous to basement membrane type IV collagens and form a novel family widely distributed in invertebrates. *Mol. Biol. Evol.* **23:** 2288–2302.

Bellairs, A.d'A. and Gans, C. 1983. A reinterpretation of the amphisbaenian orbitosphenoid. *Nature* **302:** 243–244.

Boot-Handford, R.P. and Tuckwell, D.S. 2003. Fibrillar collagen: The key to vertebrate evolution? A tale of molecular incest. *Bioessays* **25:** 142–151.

Cerny, R., Lwigale, P., Ericsson, R., Meulemans, D., Epperlein, H.H., and Bronner-Fraser, M. 2004. Developmental origins and evolution of jaws: New interpretation of "maxillary" and "mandibular". *Dev. Biol.* **276:** 225–236.

Cluzel, C., Lethias, C., Garrone, R., and Exposito, J.Y. 2000. Sea urchin fibrillar collagen 2α chain participates in heterotrimeric molecules of (1α)(2)2α stoichiometry. *Matrix Biol.* **19:** 545–547.

Cole, F.J. 1905. A monograph on the general morphology of the Myxinoid fishes based on the study of *Myxine*. Part I. The anatomy of the skeleton. *Trans. R. Soc. Edinb.* **41:** 749–788.

Couly, G.F., Coltey, P.M., and Le Douarin, N.M. 1993. The triple origin of skull in higher vertebrates: A study in quail-chick chimeras. *Development* **117:** 409–429.

Couly, G., Grapin-Botton, A., Coltey, P., Ruhin, B., and Le Douarin, N.M. 1998. Determination of the identity of the derivatives of the cephalic neural crest: Incompatibility between *Hox* gene expression and lower jaw development. *Development* **125:** 3445–3459.

Damas, H. 1944. Recherches sur le développment de *Lampetra fluviatilis* L—Contribution à l'étude de la cephalogénèse des vertébrés. *Arch. Biol.* **55:** 1–289.

de Beer, G.R. 1937. *The development of the vertebrate skull.* Oxford University Press, London.

Depew, M.J., Lufkin, T., and Rubenstein, J.L. 2002. Specification of jaw subdivisions by *Dlx* genes. *Science* **298:** 371–373.

Donoghue, P.C. and Sansom, I.J. 2002. Origin and early evolution of vertebrate skeletonization. *Microsc. Res. Tech.* **59:** 352–372.

Donoghue, P.C., Forey, P.L., and Aldridge, RJ. 2000. Conodont affinity and chordate phylogeny. *Biol. Rev. Camb. Philos. Soc.* **75:** 191–251.

Evans, D.J. and Noden, D.M. 2006. Spatial relations between avian craniofacial neural crest and paraxial mesoderm cells. *Dev. Dyn.* **235:** 1310–2135.

Exposito, J.Y., Ouazana, R., and Garrone, R. 1990. Cloning and sequencing of a *Porifera* partial cDNA coding for a short-chain collagen. *Eur. J. Biochem.* **190:** 401–406.

Exposito, J.Y., D'Alessio, M., Solursh, M., and Ramirez, F. 1992. Sea urchin collagen evolutionarily homologous to vertebrate pro-α 2(I) collagen. *J. Biol. Chem.* **267:** 15559–15562.

Exposito, J.Y., D'Alessio, M., Di Liberto, M., and Ramirez, F. 1993. Complete primary structure of a sea urchin type IV collagen α chain and analysis of the 5′ end of its gene. *J. Biol. Chem.* **268:** 5249–5254.

Exposito, J.Y., Boute, N., Deleage, G., and Garrone, R. 1995. Characterization of two genes coding for a similar four-cysteine motif of the amino-terminal propeptide of a sea urchin fibrillar collagen. *Eur. J. Biochem.* **234:** 59–65.

Gans, C. and Northcutt, R.G. 1983. Neural crest and the origin of vertebrates: A new head. *Science* **220:** 268–274.

Gaskell, W.H. 1908. *On the origin of vertebrates.* Longmans, Green, London.

Geoffroy Saint-Hilaire, E. 1818. *Philosophie anatomique,* 1st ed. (see citation in Le Guyader 1998). J.B. Baillière, Paris.

Gilbert, S.F., Loredo, G.A., Brukman, A., and Burke, A.C. 2001. Morphogenesis of the turtle shell: The development of a novel structure in tetrapod evolution. *Evol. Dev.* **3:** 47–58.

Goodrich, E.S. 1909. Vertebrata craniata (First fascicle: Cyclostomes and fishes). In *A treatise on zoology* (ed. R. Lankester), Vol. 9. Adam and Charles Black, London.

Goodrich, E.S. 1930. *Studies on the structure and development of vertebrates.* Macmillan, London.

Hall, B.K. 1999. *The neural crest in development and evolution.* Springer Verlag, New York.

Hall, B.K. 2005. *Bones and cartilage: Developmental and evolutionary skeletal biology.* Elsevier, Amsterdam.

Hall, B.K. and Hörstadius, S. 1988. *The neural crest.* Oxford University Press, New York.

Halstead, L.B. 1974. *Vertebrate hard tissues.* Wykeham, London.

Helms, J.A. and Schneider, R.A. 2003. Cranial skeletal biology. *Nature* **423:** 326–331.

Hertwig, O., Ed. 1906. *Handbuch der vergleichenden und experimentellen Entwicklungslehre der Wirbeltiere.* Verlag von Gustav Fischer, Jena, Germany.

Horigome, N., Myojin, M., Hirano, S., Ueki, T., Aizawa, S., and Kuratani, S. 1999. Development of cephalic neural crest cells in embryos of *Lampetra japonica*, with special reference to the evolution of the jaw. *Dev. Biol.* **207:** 287–308.

Janvier, P. 1996. *Early vertebrates.* Oxford Scientific, New York.

Jarvik, E. 1980. *Basic structure and evolution of vertebrates,* vol. 2. Academic, New York.

Jeffery, W.R., Strickler, A.G., and Yamamoto, Y. 2004. Migratory neural crest-like cells form body pigmentation in a urochordate embryo. *Nature* **431:** 696–699.

Johnels, A.G. 1948. On the development and morphology of the skeleton of the head of *Petromyzon. Acta Zool.* **29:** 139–279.

Kessel, M. 1992. Respecification of vertebral identities by retinoic acid. *Development* **115:** 487–501.

Kimura, S. and Matsui, R. 1990. Characteriztion of two genetically distinct type I-like collagens from hagfish (*Eptatretus burgeri*). *Comp. Biochem. Physiol.* **95B:** 137–143.

Koltzoff, N.K. 1901. Entwicklungsgeschichte des Kopfes von *Petromyzon planeri. Bull. Soc. Nat. Moscou* **15:** 259–289.

Kuraku, S. and Kuratani, S. 2006. Timescale for cyclostome evolution inferred with a phylogenetic diagnosis of hagfish and lamprey cDNA sequences. *Zool. Sci.* **23:** 1053–1064.

Kuraku, S., Hoshiyama, D., Katoh, K., Suga, H., and Miyata, T. 1999. Monophyly of lampreys and hagfishes supported by nuclear DNA-coded genes. *J. Mol. Evol.* **49:** 729–735.

Kuratani, S. 2005. Cephalic crest cells and evolution of the craniofacial structures in vertebrates: Morphological and embryological significance of the premandibular-mandibular boundary. *Zoology* **108:** 13–26.

Kuratani, S. 2008. Development of agnathans. In *Development of nonteleostean fishes* (ed. Y. Kunz-Ramsay). Science Publishers, Enfield, New Hampshire. (In press.)

Kuratani, S. and Ota, G.K. 2008. Primitive versus derived traits in the developmental program of the vertebrate head: Views from cyclostome developmental studies. *J. Exp. Zool. B Mol. Dev. Evol.* **310:** 294–314.

Kuratani, S., Nobusada, Y., Horigome, N., and Shigetani, Y. 2001. Embryology of the lamprey and evolution of the vertebrate jaw: Insights from molecular and developmental perspectives. *Philos. Trans. R. Soc. Lond. B Biol. Sci.* **356:** 1615–1632.

Kuratani, S., Murakami, Y., Nobusada, Y., Kusakabe, R., and Hirano, S. 2004. Developmental fate of the mandibular mesoderm in the lamprey, *Lethenteron japonicum:* Comparative morphology and development of the gnathostome jaw with special reference to the nature of trabecula cranii. *J. Exp. Zool. B Mol. Dev. Evol.* **302**: 458–468.

Kusakabe, R. and Kuratani, S. 2005. Evolution and developmental patterning of the vertebrate skeletal muscles: Perspectives from the lamprey. *Dev. Dyn.* **234**: 824–834.

Langille, R.M. and Hall, B.K. 1988. Role of the neural crest in development of the trabeculae and branchial arches in embryonic sea lamprey, *Petromyzon marinus* (L). *Development* **102**: 301–310.

Le Douarin, N.M. 1982. *The neural crest.* Cambridge University Press, United Kingdom.

Le Guyader, H. 1998. *Étienne Geoffroy Saint-Hilaire (1772–1884): Un naturaliste visionnaire.* Belin, Paris.

Le Lièvre, C.S. 1978. Participation of neural crest-derived cells in the genesis of the skull in birds. *J. Embryol. Exp. Morphol.* **47**: 17–37.

McBurney, K.M., Keeley, F.W., Kibenge, F.S., and Wright, G.M. 1996. Spatial and temporal distribution of lamprin mRNA during chondrogenesis of trabecular cartilage in the sea lamprey. *Anat. Embryol.* **193**: 419–426.

McCauley, D.W. and Bronner-Fraser, M. 2002. Conservation of *Pax* gene expression in ectodermal placodes of the lamprey. *Gene* **287**: 129–139.

McCauley, D. and Bronner-Fraser, M. 2003. Neural crest contributions to the lamprey head. *Development* **130**: 2317–2327.

McCauley, D. and Bronner-Fraser, M. 2006. Importance of SoxE in neural crest development and the evolution of the pharynx. *Nature* **441**: 750–752.

Meulemans, D. and Bronner-Fraser, M. 2007. Insights from amphioxus into the evolution of vertebrate cartilage. *PLoS One* **2**: e787.

Morriss-Kay, G.M. 2001. Derivation of the mammalian skull vault. *J. Anat.* **199**: 143–151.

Neidert, A.H., Virupannavar, V., Hooker, G.W., and Langeland, J.A. 2001. Lamprey *Dlx* genes and early vertebrate evolution. *Proc. Natl. Acad. Sci.* **98**: 1665–1670.

Nagashima, H., Kuraku, S., Uchida, K., Ohya, Y.K., Narita, Y., and Kuratani, S. 2007. On the carapacial ridge in turtle embryos: Its developmental origin, function and the chelonian body plan. *Development* **134**: 2219–2226.

Neumayer, L. 1938. Die Entwicklung des Kopfskelettes von Bdellostoma. St. L. *Arch. Ital. Anat. Embryol.* (suppl.) **40**: 1–222.

Newth, D.R. 1951. Experiments on the neural crest of the lamprey embryo. *J. Exp. Biol.* **28**: 247–260.

Newth, D.R. 1956. On the neural crest of the lamprey embryo. *J. Embryol. Exp. Morphol.* **4**: 358–375.

Noden, D.M. 1983. The role of the neural crest in patterning of avian cranial skeletal, connective, and muscle tissues. *Dev. Biol.* **96**: 144–165.

Noden, D.M. 1988. Interactions and fates of avian craniofacial mesenchyme. *Development* (suppl.) **103**: 121–140.

Ogasawara, M., Wada, H., Peters, H., and Satoh, N. 1999. Developmental expression of *Pax1/9* genes in urochordate and hemichordate gills: Insight into function and evolution of the pharyngeal epithelium. *Development* **126**: 2539–2550.

Ogasawara, M., Shigetani, Y., Hirano, S., Satoh, N., and Kuratani, S. 2000. *Pax1/Pax9*-related genes in an agnathan vertebrate, *Lampetra japonica*: Expression pattern of *LjPax9* implies sequential evolutionary events towards the gnathostome body plan. *Dev. Biol.* **223**: 399–410.

Ota, K.G. and Kuratani, S. 2006. The history of scientific endeavors towards understanding hagfish embryology. *Zool. Sci.* **23**: 403–418.

Ota, K.G., Kuraku, S., and Kuratani, S. 2007. Hagfish embryology with reference to the evolution of the neural crest. *Nature* **446:** 672–675.

Patten, W.M. 1912. *The evolution of the vertebrates and their kin.* Blakiston, Philadelphia.

Peters, H., Wilm, B., Sakai, N., Imai, K., Maas, R., and Balling, R. 1999. *Pax1* and *Pax9* synergistically regulate vertebral column development. *Development* **126:** 5399–5408.

Platt, J.B. 1893. Ectodermic origin of the cartilages of the head. *Anat. Anz.* **8:** 506–509.

Portmann, A. 1969. *Einführung in die vergleichende Morphologie der Wirbeltiere.* Schwabe, Basel, Switzerland.

Rijli, F.M., Mark, M., Lakkaraju, S., Dierich, A., Dollé, P., and Chambon, P. 1993. Homeotic transformation is generated in the rostral branchial region of the head by disruption of *Hoxa-2,* which acts as a selector gene. *Cell* **75:** 1333–1349.

Robson, P., Wright, G.M., Sitarz, E., Maiti, A., Rawat, M., Youson, J.H., and Keeley, F.W. 1993. Characterization of lamprin, an unusual matrix protein from lamprey cartilage. Implications for evolution, structure, and assembly of elastin and other fibrillar proteins. *J. Biol. Chem.* **268:** 1440–1447.

Robson, P., Wright, G.M., Youson, J.H., and Keeley, F.W. 2000. The structure and organization of lamprin genes: Multiple-copy genes with alternative splicing and convergent evolution with insect structural proteins. *Mol. Biol. Evol.* **17:** 1739–1752.

Romer, A.S. and Parsons, T.S. 1977. *The vertebrate body,* 5th ed. Saunders, Philadelphia.

Rychel, A.L. and Swalla, B.J. 2007. Development and evolution of chordate cartilage. *J. Exp. Zool. B Mol. Dev. Evol.* **308:** 325–335.

Rychel, A.L., Smith, S.E., Shimamoto, H.T., and Swalla, B.J. 2006. Evolution and development of the chordates: Collagen and pharyngeal cartilage. *Mol. Biol. Evol.* **23:** 541–549.

Sauka-Spengler, T., Meulemans, D., Jones, M., and Bronner-Fraser, M. 2007. Ancient evolutionary origin of the neural crest gene regulatory network. *Dev. Cell* **13:** 405–420.

Sewertzoff, A.N. 1911. Die Kiemenbogennerven der Fische. *Anat. Anz.* **38:** 487–495.

Shigetani, Y., Nobusada, Y., and Kuratani, S. 2000. Ectodermally-derived FGF8 defines the maxillomandibular region in the early chick embryo: Epithelial–mesenchymal interactions in the specification of the craniofacial ectomesenchyme. *Dev. Biol.* **228:** 73–85.

Shigetani, Y., Sugahara, F., Kawakami, Y., Murakami, Y., Hirano, S., and Kuratani, S. 2002. Heterotopic shift of epithelial-mesenchymal interactions for vertebrate jaw evolution. *Science* **296:** 1319–1321.

Shigetani, Y., Sugahara, F., and Kuratani, S. 2005. Evolutionary scenario of the vertebrate jaw: The heterotopy theory from the perspectives of comparative and molecular embryology. *Bioessays* **27:** 331–338.

Shu, D.G., Morris, S.C., Han, J., Zhang, Z.F., Yasui, K., Janvier, P., Chen, L., Zhang, X.L., Liu, J.N., Li, Y., and Liu, H.Q. 2003. Head and backbone of the early Cambrian vertebrate *Haikouichthys. Nature* **421:** 526–529.

Sire, J.Y. and Huysseune, A. 2003. Formation of dermal skeletal and dental tissues in fish: A comparative and evolutionary approach. *Biol. Rev.* **78:** 219–249.

Starck, D. 1979. *Vergleichende Anatomie der Wirbeltiere. II.* Springer, Berlin.

Suzuki, H.K. 1963. Studies on the osseous system of the slider turtle. *Ann. N.Y. Acad. Sci.* **109:** 351–410.

Takezaki, N., Figueroa, F., Zaleska-Rutczynska, Z., and Klein, J. 2003. Molecular phylogeny of early vertebrates: Monophyly of the agnathans as revealed by sequences of 35 genes. *Mol. Biol. Evol.* **20:** 287–292.

Takio, Y., Pasqualetti, M., Kuraku, S., Hirano, S., Rijli, F.M., and Kuratani, S. 2004. Evolutionary biology: Lamprey *Hox* genes and the evolution of jaws. *Nature* **429:** 1 p. following 262.

Takio, Y., Kuraku, S., Kusakabe, R., Murakami, Y., Pasqualetti, M., Rijli, F.M., Narita, Y., Kuratani, S., and Kusakabe, R. 2007. *Hox* gene expression patterns in *Lethenteron japonicum* embryos insights into the evolution of the vertebrate Hox code. *Dev. Biol.* **308:** 606–620.

Wicht, H. and Northcutt, R.G. 1995. Ontogeny of the head of the Pacific hagfish (*Eptatretus stouti*, Myxinoidea): Development of the lateral line system. *Phil. Trans. R. Soc. Lond. B Biol. Sci.* **349:** 119–134.

Wright, G.M., Keeley, F.W., Youson, J.H., and Babineau, D.L. 1984. Cartilage in the Atlantic hagfish, *Myxine glutinosa*. *Am. J. Anat.* **169:** 407–424.

Wright, G.M., Armstrong, L.A., Jacques, A.M., and Youson, J.H. 1988. Trabecular, nasal, branchial, and pericardial cartilages in the sea lamprey, *Petromyzon marinus*: Fine structure and immunohistochemical detection of elastin. *Am. J. Anat.* **182:** 1–15.

Wright, G.M., Keeley, F.W., and Robson, P. 2001. The unusual cartilaginous tissues of jawless craniates, cephalochordates, and invertebrates. *Cell Tissue Res.* **304:** 165–174.

Zhang, G. and Cohn, M.J. 2006. Hagfish and lancelet fibrillar collagens reveal that type II collagen-based cartilage evolved in stem vertebrates. *Proc. Natl. Acad. Sci.* **103:** 16829–16833.

Zhang, G., Miyamoto, M.M., and Cohn, M.J. 2006. Lamprey type II collagen and Sox9 reveal an ancient origin of the vertebrate collagenous skeleton. *Proc. Natl. Acad. Sci.* **103:** 3180–3185.

2

Developmental Patterning of the Limb Skeleton

Kimberly L. Cooper and Clifford J. Tabin

Department of Genetics
Harvard Medical School
Boston, Massachusetts 02115

THE VERTEBRATE SKELETON IS COMPOSED of approximately 200 bones, ranging in shape and size from the delicate bones of the mammalian inner ear to the robust femur. Each individual bone forms in a precise location and orientation with respect to its neighbors and in relation to force generating and transmitting tissues—the muscles, tendons, and ligaments. The appropriate structure of the bones is essential for function of the skeleton to support and move the body, and depends on an array of molecular cues that pattern their formation early in development.

Our knowledge of developmental mechanism patterning all tissues and organs of the body, including the skeleton, is largely derived from experiments using two model systems—chick and mouse embryos. While aspects of patterning the craniofacial and axial skeletal elements have been elucidated, development of the bones of the limbs is particularly well understood. The limbs are easily accessible for embryological manipulation and are expendable for the survival of prenatal animals, allowing for analysis of late developmental phenotypes after genetic or surgical perturbation. The developing limb bud has therefore become an important model for the investigation of cellular and molecular mechanisms that pattern the tissues that give rise to bones.

The tetrapod limb is of additional interest from an evolutionary perspective because it is a conserved but malleable structure whose adaptive variations in form increase an animal's fitness in different ecological niches—by promoting mobility, aiding in the acquisition of food, fighting against or escaping from predators, and assisting in reproduction. Thus, the differences in limb development of diverse species are increasingly attracting attention. At the same time, the evolutionary conservation of molecular mechanisms that generate the basic underlying pattern of the limb has made studies in diverse animals comparable.

The mesodermal progenitors of the limb skeleton originate as the embryo gastrulates to form the germ layers that give rise to all of the organ systems of the body. The mesodermal layer is subdivided into three distinct mediolateral compartments: the paraxial mesoderm (PM), the intermediate mesoderm (IM), and the lateral plate mesoderm (LPM). It is the LPM that gives rise to the skeleton and connective tissue of the limbs, as well as the heart, circulatory system, and the body wall. The outgrowth of LPM that forms the limb bud is ensheathed in a layer of ectoderm that plays a vital role in its patterning and outgrowth. Patterning of the limb is orchestrated along all three axes. In the familiar human arm, for example, the proximal–distal axis runs from the shoulder to the tips of the fingers. This axis can be subdivided into three distinct regions: the stylopod (humerus), the zeugopod (radius and ulna), and the autopod (wrist and digits of the hand). A line from the thumb to the little finger defines the anterior–posterior axis, and the dorsal–ventral axis extends from the back of the hand to the palm. These axes are established very early in the developing limb bud, and the signaling pathways interact with each other so that all three axes are inter-regulated and are established and maintained in concert. The primordia for the skeletal elements of the limb are laid down in the context of these axes.

INITIATION

Experimental evidence suggests that the entire thoracic and lumbar LPM is competent to initiate formation of a limb bud (Cohn et al. 1995). A bead of fibroblast growth factor (FGF) protein implanted into the flank along the dorsal–ventral boundary of a developing chick embryo induces the formation of an ectopic limb bud that will develop into a fully formed limb, though missing the girdle structures that do not have a limb bud origin (Cohn et al. 1995; Vogel et al. 1995; Ohuchi et al. 1997). Beads placed more anteriorly induce wings, and beads placed more posteriorly result in the formation of legs. These experiments raise the question of why only the mesoderm at specific somite levels gives rise to limb buds in an unmanipulated embryo. While the precise somite number associated with the developing fore and hind limb varies between species, limb position is consistently in register with expression of specific sets of *Hox* genes within the somites (Burke et al. 1995). Since the *Hox* genes form a code for the morphology of the vertebrae formed from the somites (Kessel and Gruss 1991; Burke et al. 1995), this assures that the limbs will be in correct register with the support structures of the pelvic and pectoral girdles. At limb bud initiation stages in the chick, *Hoxd9* expression shifts in the LPM so that its anterior boundary is at the anterior limit of the wing, while the limits of *Hoxb9* and *Hoxc9* are restricted to the posterior wing boundary (Cohn et al. 1997). *Hoxb9* expression is strong in the flank but decreases at the anterior boundary of the hind limb, such that it is expressed less intensely in the presumptive leg while *Hoxc9* and *Hoxd9* remain strong. Application of FGF to the flank shifts the anterior and posterior expression boundaries of these *Hox* genes to reflect the positional identity of the ectopic limb. It is thus likely that expression of these and other *Hox* genes play key roles in establishing the location and morphologies of the limbs. The shift in *Hox* gene expression in these experiments is limited to the LPM and does not affect axial identity consistent with a large body of work, suggesting that *Hox* expression is independently regulated in the neural tube, somitic, and LPM (Oberg and Eichele 1999; Nowicki and Burke 2000).

While limb bud mesenchyme originates from the LPM, there is evidence suggesting that its induction may also depend on more axial structures. Embryological experiments placing an impenetrable foil barrier in the longitudinal axis between the LPM and the IM block the outgrowth and subsequent patterning of the limb bud. These experiments indicate that axial signals to the LPM are required for patterning and outgrowth of the limb (Stephens and McNulty 1981). Since *Fgf8* is expressed in the IM, and an FGF8-soaked bead is sufficient to induce ectopic limb bud formation, *Fgf8* became a favorite candidate for the inductive signal from the IM (Crossley et al. 1996). However, loss of *Fgf8* expression in the IM results in limbs that form normally, indicating that *Fgf8* may not be the limb-initiating signal from the IM (Boulet et al. 2004). Furthermore, loss of the *Fgf8* expressing structure, the mesonephros, fails to impact limb development (Fernandez-Teran et al. 1997). An alternative explanation for the foil barrier experiments may lie in the overlying ectoderm. Lineage analyses of grafted tissue in chick–quail chimeras indicate that the ectoderm overlying the prospective ventral LPM gives rise to ventral limb ectoderm. Surprisingly, the ectoderm overlying dorsal LPM converges to form the distal-most limb ectoderm, the AER, while the dorsal limb ectoderm is derived from the ectoderm that overlies the IM at early stages (Michaud et al. 1997). Consistent with these experiments, the expression of several genes that ultimately come to be expressed in the

AER, such as *Radical Fringe* and *Wnt3a*, are initially expressed throughout the dorsal LPM ectoderm (Laufer et al. 1997; Rodriguez-Esteban et al. 1997; Kengaku et al. 1998). Therefore, it is plausible that the limb defects observed in foil barrier experiments may be attributed to effects of these manipulations on the morphogenesis and reorganization of the overlying ectoderm, rather than to the obstruction of axial signals.

The ability of beads of FGF protein to induce ectopic limbs in the flank reflects the role of endogenous *Fgf* signaling. Before the morphological appearance of a limb bud in the chick and mouse, *Fgf10* is expressed broadly in the IM and LPM and becomes restricted to the LPM of the prospective limb field (Ohuchi et al. 1997; Min et al. 1998; Sekine et al. 1999). Furthermore, implantation of an FGF8-soaked bead induces expression of *Fgf10* in the LPM prior to limb bud induction but approximately 17 hours after bead placement, indicating a second factor is involved, while similar placement of an FGF10-soaked bead results in immediate outgrowth (Ohuchi et al. 1997; Isaac et al. 2000). In the chick, *Wnt2b* is expressed in the IM and LPM at the level of the forelimb, and *Wnt8c* is expressed in the LPM at the hind limb level. Application of either of these WNTs, like application of FGF8, induces formation of ectopic limbs in the flank by first inducing expression of *Fgf10*. Application of an activated form of β-*catenin* or a negative regulator of β-*catenin* signaling indicates that a canonical *Wnt* signaling pathway is likely involved in endogenous limb bud induction (Kawakami et al. 2001).

PROXIMAL–DISTAL AXIS AND THE AER

Growth in the proximal–distal axis is tied to the morphological transformation of the most distal ectoderm into a thickened epithelium—the apical ectodermal ridge (AER)—extending along the anterior–posterior axis. The AER is established as a critical signaling center early in the developing limb by signals emanating from the mesenchyme. As previously discussed, *Wnt/β-catenin* signaling from the flank induces expression of *Fgf10* in the region of the LPM that will form the mesenchyme of the limb bud. *Fgf10* activity in turn leads to formation of a morphological AER and likely induces and maintains expression of other *Fgfs* within the AER (Isaac et al. 2000). Indeed, *Fgf10* is essential for AER initiation in mouse embryos (Chen et al. 1998; Sekine et al. 1999). Experimentally, ectopic activation of *Fgf8* in the AER occurs approximately 17 hours after FGF10 bead placement (Isaac et al. 2000), which suggests that there may be an intermediate signal. A candidate for this signal in the chick is *Wnt3a* (*Wnt3* in mice). *Wnt3a* is expressed in the ectoderm at the dorsal–ventral border that in turn activates the AER (Kengaku et al. 1998; Kawakami et al. 2001). Ectopic expression of *Wnt3a* induces ectopic AER formation and ectopic *Fgf8* expression in a β-*catenin* and *Lef/Tcf*-dependent manner (Kengaku et al. 1998). In the mouse, *Wnt3* acts very similarly to chick *Wnt3a*, although it is expressed more broadly throughout limb ectoderm. Mice deficient for *Wnt3* or β-*catenin* in the AER fail to initiate *Fgf8* expression in the AER and do not form limbs (Barrow et al. 2003). Additionally, the *Lef/Tcf* mutant is very similar to the *Wnt3* limb phenotype (Galceran et al. 1999; Kawakami et al. 2001).

Once the AER is formed, it plays a critical role in limb patterning. Embryological manipulations performed 60 years ago determined that removal of the AER results in truncations of distal skeletal elements (Saunders 1948). Removal of the AER from an early limb bud results in loss of the stylopod, zeugopod, and autopod, while removal at later times results in successively smaller truncations such that removal at late bud stages results in loss of the distal autopod (Summerbell 1974; Rowe and Fallon 1982). Transplanting the AER between wings and legs (Zwilling 1959), and between early and late limb buds, indicates that the AER plays a permissive role in the patterning of distal structures but does not itself instruct formation of particular skeletal elements in the proximal–distal axis (Rubin and Saunders 1972). Candidates for the permissive factor produced by the AER came from the discovery that several members of the *Fgf* family of molecules, including *Fgfs 4, 8, 9*, and *17*, are expressed in the AER (Martin 1998; Sun et al. 2000) and confer proliferative and antiapoptotic activity to the underlying mesoderm. Removal of the AER and replacement by a bead soaked in FGF protein is sufficient to restore the outgrowth of the limb, and resulting skeletal elements develop normally (Niswander and Martin 1993; Fallon et al. 1994; Niswander et al. 1994). Roles for these factors in maintaining limb outgrowth are further supported by genetic studies. For example, genetic removal of *Fgf4* and *Fgf8* in the mouse indicates that these two *Fgfs* are required for maintenance, but not for initiation of the morphological AER. Loss of *Fgf8* alone results in loss of proximal-most structures (humerus/femur) and reduction of anterior structures, likely due to redundancy with other *Fgfs* in the posterior AER (Lewandoski et al. 2000; Moon and Capecchi 2000). Consistent with this, loss of function of both *Fgf8* and *Fgf4* results in loss of all limb structures (Sun et al. 2002; Boulet et al. 2004).

Since the AER had been shown not to be instructive in specifying different proximodistal limb elements, the mechanism by which this is achieved must lie within the mesenchyme itself. To conceptualize this mechanism, the Progress Zone Model for limb development (Summerbell

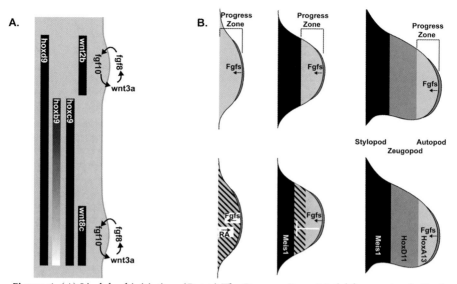

Figure 1. (*A*) Limb bud initiation. (*B, top*) The Progress Zone Model for proximal–distal outgrowth and (*bottom*) dynamic changes in proximal–distal gene expression.

et al. 1973) was theorized, initially based on evidence largely derived from the results of AER extirpation experiments described previously and later supported by X-irradiation-induced phocomelia (Wolpert et al. 1979). This model states that mesenchymal cells underlying the AER (the progress zone) are maintained in an uncommitted state by signals from the AER as long as they are within the progress zone, but their fate becomes fixed once they leave the zone. Cells in the progress zone of the earliest limb bud are fated for proximal-most limb structures, but over time they autonomously acquire increasingly more distal fates as long as they remain within range of the permissive AER signal. However, the cells are also dividing and hence the more proximal progress zone cells are continually being displaced from the range of AER influence. Thus, cells that exit the progress zone early form proximal structures, and cells that stay in the progress zone, continue their autonomous respecification for a longer time and as a consequence become fated to form more distal structures. The AER removal experiments are explained by the hypothesis that cells in the progress zone cease to become progressively distalized in fate when the permissive signal from the AER is removed and are therefore only capable of differentiating as proximal structures. While the Progress Zone Model has provided a compelling explanation for a number of classical experiments, it suffers from the lack of a clear molecular mechanism for "counting" the time spent in the progress zone, and more importantly, from the lack of any molecular markers turning on in successive waves within the progress zone as time progresses.

More recent work shows that removal of the AER results in cell death that occurs in an approximately 200 μm-deep domain of cells underlying the AER (Dudley et al. 2002). This result led to an alternative proposal to the Progress Zone Model, suggesting that cells may have a prepattern that is established relatively early in limb development. The model hypothesizes that prepattern is realized as structural identity as the limb bud "telescopes" out during growth, but these domains are very small in the early limb bud and die without the growth-promoting influence of the AER. According to this view, the role played by the AER may be to keep the zone of prepatterned cells alive and undifferentiated until they proliferate and expand to form a sufficient precursor population for the appropriate structures. Similar to the Progress Zone Model, a major problem for the "early specification" model is that, to date, no expression patterns have been described for genes that would clearly demarcate domains of future limb "segments" in the very early limb bud. It now seems likely that a new model is needed to most parsimoniously explain proximal–distal specification (Tabin and Wolpert 2007). While any new model must be consistent with the earlier classical experiments, there is now a wealth of molecular information on which such hypotheses can be based. In particular, there is a collection of genes that are dynamically expressed in prospective segments that at minimum give an indication of the way proximal–distal specification progresses in the limb, and may in fact be involved in actually determining the skeletal elements in the proximal–distal axis. Specifically, *Meis1* is expressed in the proximal limb bud under control of retinoic acid produced in the flank (Mercader et al. 2000). Response to retinoic acid is repressed by distal expression of *Fgf* that functions to restrict *Meis1* expression to the proximal limb bud. *Fgf* activity also positively influences, directly or indirectly, the expression of more distally expressed genes, including the Hox genes. As the limb grows distally, the domains of RA and *Fgf* responsiveness are further apart, and a new domain demarcated by the expression of *HoxA11* and *D11* appears. Thus, the progressive emergence of proximal–distal pattern in the limb appears to be attributable, at least in part, to the competitive influence of signals from the flank and tip of the limb bud (Mercader et al. 2000; Tabin and Wolpert 2007). Appropriate limb patterning may result from a combination of the influence of *Fgfs* that keep cells in an undifferentiated state and the dynamically changing patterns of gene expression that influence a cell's segmental identity as it leaves the progress zone. Cells that leave the progress zone early, when *Meis1* expression is the most pronounced, will differentiate as stylopod; cells exiting when the *Hox 11* paralogues are expressed in the domain out of reach of *Fgf* will become zeugopod; and the last cells to leave the progress zone express *Hox12* and *Hox13* genes, and become the autopod.

THE HOX GENES

Hox genes play multiple interconnected roles to regulate limb development. The chick and mouse each have four *Hox* clusters (A–D) and a total of 39 genes. Of paramount importance to the developing limb are the 5′ Group A and D genes that play roles in both proximal–distal segment specification and anterior–posterior patterning. An internal cluster deletion of *HoxD8-11* combined with a complete loss of the *HoxA* cluster results in a variable proximal truncation of the stylopod and loss of all distal elements, indicating that *Hox* genes play a vital role in the early phase of limb outgrowth (Tarchini et al. 2006). This function is likely through initiation and maintenance of mesenchymal *Fgf10* expression (Zákány et al. 2007). *Hox*-dependent initiation and maintenance of the AER also requires early low levels of *HoxD12* and *HoxD13* expression in the mesenchyme.

As the limb grows out, different paralogous groups play prominent roles in each of the proximal–distal limb segments. For example, the

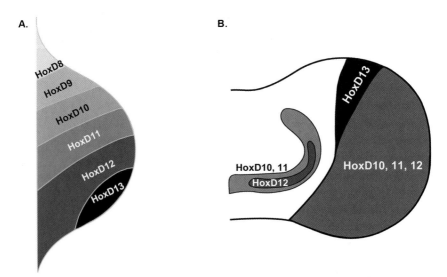

Figure 2. (*A*) Early nested expression of the Group D Hox genes. (*B*) Late expression of Group D Hox genes at the autopod stage.

growth of the zeugopod is dependent upon the Group 11 paralogues (*HoxA11* and *HoxD11* in the forelimb; *HoxA11*, *HoxC11*, and *HoxD11* in the hind limb). Loss of function of both *HoxA11* and *D11* results in normal development of the forelimb stylopod and autopod but a significant abrogation of the zeugopod (Davis et al. 1995; Boulet and Capecchi 2004). Of particular note, the early zeugopod cartilage elements form normally but fail to grow, indicating in this example that the *Hox* genes do not specify the limb segment per se. The role of the *Hox11* paralogues in the autopod is obscured by the strong expression of *HoxD12* and *D13* genes that exert a phenotypic dominance, known as posterior prevalence, over the more anterior genes. Deletion of *HoxD12* and *D13* from the limbs results in up-regulation of *HoxD11* expression and polydactyly, indicating *HoxD11* can also promote chondrogenesis and digit formation (Fromental-Ramain et al. 1996; Kmita et al. 2005)

The autopod is a tetrapod-specific adaptation that appears to result from acquisition of novel autopod-specific regulatory elements that control expression of the most 5′ *Hox* genes (Spitz et al. 2003; Lehoczky et al. 2004; Gonzalez et al. 2007), although recent studies of fin formation in cartilaginous fish have raised the possibility that such regulatory elements are a basal feature that was subsequently lost in the teleost lineage (Davis et al. 2007; Freitas et al. 2007). A combined loss of *HoxA13* and *HoxD13* results in a complete absence of the autopod, reflecting their requirement for its establishment (Fromental-Ramain et al. 1996). In addition to their role in regulating formation and growth of the proximal–distal segments, *Hox* genes also play an important role in establishing the signaling center controlling the anterior–posterior axis.

POLARIZATION AND THE ZPA

In the 1960s, John Saunders and Mary Gasseling discovered that when a section of the posterior limb bud was transplanted to the anterior limb bud of a host embryo, ectopic digits developed in the anterior limb (Saunders and Gasseling 1968). Moreover, these extra digits had morphologies that were a mirror image of the endogenous digits, the ectopic set having a reversed polarity relative to the endogenous digits, and the posterior mesodermal tissue with this inductive capacity was thus termed the "zone of polarizing activity" (ZPA). Further grafting experiments where defined numbers of ZPA cells were placed in the anterior limb bud determined that a larger number of ZPA cells induce more anterior digit identities, while fewer cells are only capable of inducing posterior identities (Tickle 1981). These results provided experimental evidence in support of the hypothesis that the ZPA activity might reflect the production of a secreted morphogen that acts to pattern the identities of the digits in a dose-dependent manner (Wolpert 1969; Tickle et al. 1975).

It was not until 1993 that this "posteriorizing" signal was identified as a vertebrate homolog of the *Drosophila* patterning molecule hedgehog, called *Sonic hedgehog* (*Shh*). This secreted molecule is strongly expressed in the most posterior aspect of the tetrapod limb bud, coinciding precisely with the domain possessing polarizing activity and making it a perfect candidate for the posteriorizing factor. Cells expressing *Shh* were grafted into the anterior mesoderm of host embryos and induced mirror-image duplications identical to ZPA grafts (Riddle et al. 1993), indicating that *Shh* was likely the signal in the ZPA that polarizes the anterior–posterior axis of the limb. The evidence showing that *Shh* can not only induce ectopic digits but is also in fact required for normal patterning of the developing limb came in the form of a loss-of-function allele generated in mice. The limb phenotype of $Shh^{-/-}$ mice indicates that *Shh* expression is required for formation of all posterior limb structures: the ulna/fibula in the zeugopod and all digits of the autopod, except for digit 1 (Chiang et al. 1996).

For years, the prevailing theory remained that *SHH* acts as a morphogen gradient, carrying out its patterning role in a concentration-dependent manner, and indeed a graded distribution of SHH protein can be detected across the posterior limb bud (Lewis et al. 2001). Moreover, there is a graded response in the expression of target genes such as *Ptc1* and *Gli1* (Marigo and Tabin 1996; Marigo et al. 1996). However, it was also shown that cells fated to become anterior digits can be "promoted" to posterior digits with extended exposure to SHH, suggesting a possible temporal component as well (Yang et al. 1997). Recent data have given further support to this idea. Transgenic mice carrying the Cre recombinase under control of *Shh* regulatory elements were generated and crossed to a lacZ reporter mouse that would mark every cell and every descendent of a cell that had at any point in its ancestry expressed *Shh*. The resulting limbs expressed the lacZ reporter throughout digit 5, digit 4, and half of digit 3. This result indicated that a great amount of cell division occurs to expand the population of posterior limb bud cells across the digit-forming region. Furthermore, *Shh* need not function only as a morphogen from a very long distance if cells acquire their identities in part through expression of *Shh* for a defined amount of time—digit 3 for the shortest, and digit 5 for the longest (Harfe et al. 2004). Since the *Shh* knockout determined that the development of the anterior-most digit 1 is independent of *Shh* (Chiang et al. 1996), the only digit to require *Shh* that does not itself ever express it is digit 2.

For appropriate anterior–posterior patterning of the limb bud, it is critical that *Shh* expression be restricted to the correct posterior, distal, and marginal location. Each modular part of this expression pattern is regulated by a different signaling system. The marginal restriction observed in wild-type embryos and embryos with ectopic anterior expression domains appears to be downstream of a signal emanating from the nonAER dorsal–ventral border ectoderm, since transplantation of this tissue to the dorsal surface expands *Shh* expression anteriorly (Nissim et al. 2007). However, in normal limb development, *Shh* is only expressed at the posterior margin, indicating a second level of regulation in the anterior–posterior axis.

A number of positive and negative factors regulate *Shh* expression along the AP axis, promoting its expression in the posterior and repressing it in the anterior. *Hand2* (formerly known as *dHand*) is expressed in the posterior limb bud in a domain preceding and encompassing the expression of *Shh*. Chick and mouse embryos that express *Hand2* ectopically in the anterior limb bud develop anterior duplications of digits, reflecting the phenotype of ZPA transplants and *Shh* misexpression (Charité et al. 2000; Fernandez-Teran et al. 2000). Mice homozygous for a loss-of-function allele of *Hand2* have a phenotype reminiscent of the *Shh* mutant and fail to initiate expression of *Shh* in the posterior limb bud (Charité et al. 2000). Additionally, *Hand2* expression is down-regulated in *Shh* mutant limbs, indicating that a positive feedback loop leads to the robustness of the posterior expression domain (Charité et al. 2000; Fernandez-Teran et al. 2000).

Early posterior-restricted expression of *Hand2* is complementary to the anterior expression of the transcription factor *Gli3*. Normally full length GLI3 activator (GLI3A) protein is cleaved into a repressor form (GLI3R) in the anterior limb bud (Wang et al. 2000). The mouse mutant *Extratoes(Xt)*, which is deficient in *Gli3* signaling, expresses ectopic *Hand2* anteriorly, indicating that *Gli3* normally represses *Hand2* expression in the anterior. Analysis of early expression patterns in both *Xt* and *Hand2* mutants indicates that *Gli3* is required for the initial restriction of *Hand2* expression and that *Hand2* feeds back to repress *Gli3* expression in the posterior (te Welscher et al. 2002). *Shh* completes the regulatory interactions of the anterior–posterior axis and patterns the digits by inhibiting the cleavage of full-length GLI3A to GLI3R. Loss of function of *Shh* results in loss of all but the anterior-most digit, while loss of *Gli3* results in the formation of many ectopic digits. A combined loss of function of both *Shh* and *Gli3* results in limbs that are identical to loss of *Gli3* alone, indicating that *Gli3* is the only member of the *Gli* family responding to *Shh* signaling in the limb (Litingtung et al. 2002; te Welscher et al. 2002).

The *Hox* genes also regulate and are in turn regulated by *Shh* signaling. The posterior *HoxA* and *HoxD* genes (10–13) are expressed in a collinear fashion in the limb bud so that *HoxD10* is expressed most broadly, and *HoxD13* is restricted to the most posterior aspect of the limb (Zákány et al. 2004). Combinations of deletions of these two *Hox* clusters result in loss of *Shh* expression, indicating both a qualitative and quantitative requirement for *Hox* genes to establish *Shh* expression in the ZPA, and proximal–distal regions of *Shh* expression appear to be controlled by expression of increasingly distally restricted *Hox* genes (Tarchini et al. 2006). Analysis of *HoxD11–13* expression in the chick *oligozeugodactyly* mutant, a limb-specific loss of *Shh* expression, reveals that early establishment of *Hox* expression is *Shh* independent, but continued high-level expression of *Hox* genes in their initial domains, as well as in the later distal phase of expression, requires functional *Shh* (Ros et al. 2003). A number of other transcription factors have been identified that are also involved in the posterior restriction of *Shh*. These include the posteriorly expressed, positively acting *Hoxb8* and the anteriorly expressed, negatively acting *Alx4* and *Twist* (Stratford et al. 1997; Takahashi et al. 1998; O'Rourke et al. 2002). Finally, the distal restriction of *Shh* expression involves cross interactions with the AER that will be discussed below.

DORSAL–VENTRAL SPECIFICATION

The dorsal–ventral axis of the developing limb is characterized by the expression of *Wnt7a* in the dorsal ectoderm, which is initiated in the dorsal flank prior to limb outgrowth (Dealy et al. 1993; Parr et al. 1993). Downstream of this signal, *Lmx1* (*Lmx1b* in the mouse) is expressed in the dorsal limb mesoderm (Riddle et al. 1995; Vogel et al. 1995). The ventral ectoderm and ventral part of the AER express the transcription factor *Engrailed1* (*En1*) (Davis et al. 1991; Gardner and Barald 1992; Loomis et al. 1996). Loss of *En1* in developing mouse limbs results in radial nails and loss of paw pads at the digit tips, indicating that *En1* is essential for the formation of ventral structures (Loomis et al. 1996; Cygan et al. 1997), and in the *En1* mutant, *Wnt7a* expression spreads to the ventral ectoderm. Conversely, loss of function of *Lmx1b* (Chen et al. 1998) results in loss of nails and radialization of ventral structures. *Wnt7a* appears to regulate the dorsal expression of *Lmx1*, since removal of the ectoderm from chick wings results in loss of *Lmx1* expression, and *Lmx1* is induced by ectopic *Wnt7a*. Furthermore, ectopic *Lmx1* is sufficient to rescue dorsal structures in the absence of *Wnt7a* (Riddle et al. 1995; Vogel et al. 1995).

Bmp signaling is also critical for the establishment of dorsal–ventral polarity. The activated downstream component of the *Bmp* pathway, phospho-SMAD1, is detected throughout the ventral ectoderm as well as in the mesoderm of wild-type limbs. When the *Bmp* receptor, *Bmp1ra*, is deleted from limb bud ectoderm, the resulting limbs are severely malformed and missing the ventral flexor tendons. The skeletal phenotype is preceded by the expansion of *Wnt7a* and *Lmx1b* into the ventral territory, and a near complete loss of the ventral expression of *En1* (Ahn et al. 2001). All of these genetic and experimental manipulations primarily affect the distal limb bud. The control of dorsal–ventral polarity in the proximal limb bud is much less well understood.

AXIS INTEGRATION

In addition to the many feedback loops that exist to specify each individual axis, there is significant regulation between the axes. While molecular aspects of all three axes seem superficially distinct, there is a

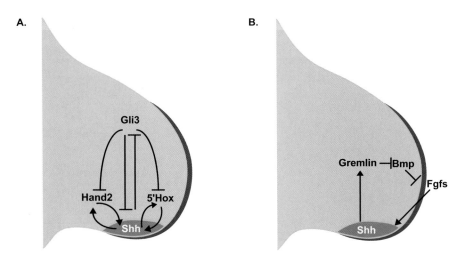

Figure 3. (*A*) Regulation of the proximal–distal axis. (*B*) Inter-regulation of the anterior–posterior and proximal–distal axes.

remarkable amount of integration through loops that keep the axes intricately connected. The best characterized of these is the signaling loop between the AER and the ZPA that connects the proximal–distal and anterior–posterior axes (Laufer et al. 1994; Niswander et al. 1994). *Bmp2* and *Bmp4* are expressed in the mesoderm, as well as strongly throughout the AER (Akita et al. 1996). Their activity, if left unchecked, would down-regulate the expression of *Fgfs* in the AER and terminate limb development. However, mesodermal expression of the *Bmp* antagonist, *Gremlin*, which is induced by posterior expression of *Shh* in the ZPA, inhibits this *Bmp* activity and maintains the AER during limb outgrowth (Capdevila et al. 1999; Zuniga et al. 1999). *Gremlin* antagonism of *Bmp* signaling is required for expression of *Fgf4*, *-9*, and *-17*. *Fgf8* is indirectly regulated by this pathway because *Gremlin* expression is required to maintain the AER. Expression of *Shh* is in turn positively regulated by expression of *Fgf4* and *Fgf9*, completing the regulatory loop between the anterior–posterior and proximal–distal axes. It is this dependence on *Fgf* activity that keeps *Shh* expressed distally within the limb mesenchyme. The mouse *limb deformity* mutant, defective for expression of *Gremlin*, reflects an early breakdown in this loop that disrupts patterning in the anterior–posterior axis. Expression of *Shh*, *Fgf4*, and *-8* are down-regulated prematurely, resulting in both distal truncations and anterior–posterior defects in digit specification (Khokha et al. 2003).

Positive regulation between the two axes is propagated as the limb grows distally and terminates as the digit plate expands in the anterior and posterior axis. Concomitant with the expansion of the autopod is the down-regulation of *Shh* and termination of the feedback loop. Limb mesenchyme cells that have expressed *Shh* are unable to express *Gremlin*, and it is this key characteristic that breaks the loop between the axes and terminates growth of the limb (Scherz et al. 2004). As anterior cells are distanced further from the range of *Shh* signaling, they down-regulate *Gremlin* expression, allowing *Bmps* to down-regulate expression of *Fgfs* in the AER. Loss of distal *Fgf* expression results in loss of *Shh* expression in the ZPA, completing the breakdown of the feedback loop. A second level of integration between the anterior–posterior and proximal–distal axes is attained by shared requirement for both *Shh* and *Fgf* signaling to activate expression of certain key targets. Thus, while *Ptc1* and *Gli1* are activated by *Shh* alone, and *Msx1* is activated primarily by *Fgfs*, the induction of *Bmp2* or *HoxD13* requires simultaneous expression of both.

A connection between the anterior–posterior and dorsal–ventral axes lies in the regulation of *Shh*, which, in addition to its obvious restriction to the distal, posterior margin of the limb bud, is also biased towards the dorsal mesenchyme. This is due to an additional partial dependence on *Wnt7a* signaling, linking the dorsal–ventral and anterior–posterior axes (Yang and Niswander 1995). The fact that some *Shh* expression remains in the absence of *Wnt7a* likely reflects partial redundancy with other *Wnt* genes, such as *Wnt7b* expression throughout the limb bud ectoderm (Parr et al. 1993).

SUMMARY AND PERSPECTIVE

While the past decades have resulted in great advances in the understanding of patterning mechanisms in the developing limb, a number of unanswered questions remain. As the seemingly homogenous population of mesenchymal limb bud cells grows distally from the body wall, subsets of these cells condense and undergo morphological changes, leading ultimately to their differentiation into chondrocytes that will form the scaffold for endochondral ossification of the skeletal elements. The morphological and gene expression changes that occur upon condensation are becoming clear, but the signal(s) that establish the timing and position of condensation remain a mystery. A striking characteristic of the structure of the limb skeleton is the reiterative branching pattern that occurs in the proximal–distal direction of outgrowth. The limb initiates condensations proximally with a single skeletal element in the stylopod, the humerus. This single element branches to form a Y-shaped structure in the zeugopod that becomes the radius and ulna. The branch-point of the Y subsequently cavitates to form the hinge joint of the elbow. While the location of the branch is established by the upstream molecular cascades that subdivide the limb bud into stylopod, zeugopod, and autopod, the cause of branch formation is entirely unknown. Progressing more distally, the pattern becomes increasingly branched into the multitude of elements making up the wrist and digits.

It is interesting to note that while natural selection can act to drive morphological diversity in forelimbs and hind limbs concurrently (e.g., the reduction of digits in horses), a phenomenon known as "concerted evolution," it can also alter their morphology independently, as is the case for the modification of bat forelimbs into wings or the specific hind limb adaptations of kangaroos. The vast majority of the genes that regulate limb patterning appear to be required in both fore and hind limbs equitably. In addition, of the handful of genes that are expressed in only one pair of limbs, only one, *Pitx1*, is required for establishing limb-type identity, and only to a partial extent (Minguillon et al. 2005). However, many of the studies of limb development have focused on complete functional knockouts of individual genes and not on their *cis*-regulation. There may be an as of yet untapped wealth of more subtle regulation of the limb patterning genes that subdivides their expression and function in the fore and hind limb so that variations in form can be subtly altered independently in each paired appendage. Inter-species comparisons of gene regulation and a continued investigation of the mechanisms that further sculpt the preformed limb during subsequent growth and endochondral ossification will likely elucidate mechanisms for adaptive variation that give the limb skeleton its unique malleability.

ACKNOWLEDGMENTS

We appreciate the assistance of Dr. Jessica L. Whited with preparation of the figures. We apologize to the authors whose work we were unable to include due to space limitations. Our work on limb patterning is supported by the National Institutes of Health grant R37 HD32443. K.L.C. is supported by NRSA postdoctoral fellowship F32 HD052349 from the NIH.

REFERENCES

Ahn, K., Mishina, Y., Hanks, M.C., Behringer, R.R., and Crenshaw III, E.B. 2001. BMPR-IA signaling is required for the formation of the apical ectodermal ridge and dorsal-ventral patterning of the limb. *Development* **128:** 4449–4461.

Akita, K., Francis-West, P., and Vargesson, N. 1996. The ectodermal control in chick limb development: *Wnt-7a, Shh, Bmp-2* and *Bmp-4* expression and the effect of FGF-4 on gene expression. *Mech. Dev.* **60:** 127–137.

Barrow, J.R., Thomas, K.R., Boussadia-Zahui, O., Moore, R., Kemler, R., Capecchi, M.R., and McMahon, A.P. 2003. Ectodermal *Wnt3/β*-catenin signaling is required for the establishment and maintenance of the apical ectodermal ridge. *Genes Dev.* **17:** 394–409.

Boulet, A.M. and Capecchi, M.R. 2004. Multiple roles of *Hoxa11* and *Hoxd11* in the formation of the mammalian forelimb zeugopod. *Development* **131:** 299–309.

Boulet, A.M., Moon, A.M., Arenkiel, B.R., and Capecchi, M.R. 2004. The roles of *Fgf4* and *Fgf8* in limb bud initiation and outgrowth. *Dev. Biol.* **273:** 361–372.

Burke, A.C., Nelson, C.E., Morgan, B.A., and Tabin, C. 1995. *Hox* genes and the evolution of vertebrate axial morphology. *Development* **121:** 333–346.

Capdevila, J., Tsukui, T., Rodriquez Esteban, C., Zappavigna, V., and Izpisúa Belmonte, J.C. 1999. Control of vertebrate limb outgrowth by the proximal factor *Meis2* and distal antagonism of BMPs by Gremlin. *Mol. Cell* **4:** 839–849.

Charité, J., McFadden, D.G., and Olson, E.N. 2000. The bHLH transcription factor dHAND controls *Sonic hedgehog* expression and establishment of the zone of polarizing activity during limb development. *Development* **127:** 2461–2470.

Chen, H., Lun, Y., Ovchinnikov, D., Kokubo, H., Oberg, K.C., Pepicelli, C.V., Gan, L., Lee, B., and Johnson, R.L. 1998. Limb and kidney defects in *Lmx1b* mutant mice suggest an involvement of *LMX1B* in human nail patella syndrome. *Nat. Genet.* **19**: 51–55.

Chiang, C., Litingtung, Y., Lee, E., Young, K.E., Corden, J.L., Westphal, H., and Beachy, P.A. 1996. Cyclopia and defective axial patterning in mice lacking *Sonic hedgehog* gene function. *Nature* **383**: 407–413.

Cohn, M.J., Izpisúa-Belmonte, J.C., Abud, H., Heath, J.K., and Tickle, C. 1995. Fibroblast growth factors induce additional limb development from the flank of chick embryos. *Cell* **80**: 739–746.

Cohn, M.J., Patel, K., Krumlauf, R., Wilkinson, D.G., Clarke, J.D., and Tickle, C. 1997. Hox9 genes and vertebrate limb specification. *Nature* **387**: 97–101.

Crossley, P.H., Minowada, G., MacArthur, C.A., and Martin, G.R. 1996. Roles for FGF8 in the induction, initiation, and maintenance of chick limb development. *Cell* **84**: 127–136.

Cygan, J.A., Johnson, R.L., and McMahon, A.P. 1997. Novel regulatory interactions revealed by studies of murine limb pattern in *Wnt-7a* and *En-1* mutants. *Development* **124**: 5021–5032.

Davis, C.A., Holmyard, D.P., Millen, K.J., and Joyner, A.L. 1991. Examining pattern formation in mouse, chicken and frog embryos with an *En*-specific antiserum. *Development* **111**: 287–298.

Davis, A.P., Witte, D.P., Hsieh-Li, H.M., Potter, S.S., and Capecchi, M.R. 1995. Absence of radius and ulna in mice lacking *hoxa-11* and *hoxd-11*. *Nature* **375**: 791–795.

Davis, M.C., Dahn, R.D., and Shubin, N.H. 2007. An autopodial-like pattern of Hox expression in the fins of a basal actinopterygian fish. *Nature* **447**: 473–476.

Dealy, C.N., Roth, A., Ferrari, D., Brown, A.M., and Kosher, R.A. 1993. Wnt-5a and Wnt-7a are expressed in the developing chick limb bud in a manner suggesting roles in pattern formation along the proximodistal and dorsoventral axes. *Mech. Dev.* **43**: 175–186.

Dudley, A.T., Ros, M.A., and Tabin, C.J. 2002. A re-examination of proximodistal patterning during vertebrate limb development. *Nature* **418**: 539–544.

Fallon, J.F., Lopez, A., Ros, M.A., Savage, M.P., Olwin, B.B., and Simandl, B.K. 1994. FGF-2: Apical ectodermal ridge growth signal for chick limb development. *Science* **264**: 104–107.

Fernandez-Teran, M., Piedra, M.E., Simandl, B.K., Fallon, J.F., and Ros, M.A. 1997. Limb initiation and development is normal in the absence of the mesonephros. *Dev. Biol.* **189**: 246–255.

Fernandez-Teran, M., Piedra, M.E., Kathiriya, I.S., Srivastava, D., Rodriguez-Rey, J.C., and Ros, M.A. 2000. Role of dHAND in the anterior-posterior polarization of the limb bud: Implications for the Sonic hedgehog pathway. *Development* **127**: 2133–2142.

Freitas, R., Zhang, G., and Cohn, M.J. 2007. Biphasic *Hoxd* gene expression in shark paired fins reveals an ancient origin of the distal limb domain. *PLoS ONE* **2**: e754.

Fromental-Ramain, C., Warot, X., Messadecq, N., LeMeur, M., Dollé, P., and Chambon, P. 1996. *Hoxa-13* and *Hoxd-13* play a crucial role in the patterning of the limb autopod. *Development* **122**: 2997–3011.

Galceran, J., Farinas, I., Depew, M.J., Clevers, H., and Grosschedl, R. 1999. *Wnt3a*$^{-/-}$-like phenotype and limb deficiency in *Lef1*$^{-/-}$*Tcf1*$^{-/-}$ mice. *Genes Dev.* **13**: 709–717.

Gardner, C.A. and Barald, K.F. 1992. Expression patterns of *engrailed*-like proteins in the chick embryo. *Dev. Dyn.* **193**: 370–388.

Gonzalez, F., Duboule, D., and Spitz, F. 2007. Transgenic analysis of *Hoxd* gene regulation during digit development. *Dev. Biol.* **306:** 847–859.

Harfe, B.D., Scherz, P.J., Nissim, S., Tian, H., McMahon, A.P., and Tabin, C.J. 2004. Evidence for an expansion-based temporal Shh gradient in specifying vertebrate digit identities. *Cell* **118:** 517–528.

Isaac, A., Cohn, M.J., Ashby, P., Ataliotis, P., Spicer, D.B., Cooke, J., and Tickle, C. 2000. FGF and genes encoding transcription factors in early limb specification. *Mech. Dev.* **93:** 41–48.

Kawakami, Y., Capdevila, J., Buscher, D., Itoh, T., Rodriguez Esteban, C., and Izpisúa Belmonte, J.C. 2001. WNT signals control FGF-dependent limb initiation and AER induction in the chick embryo. *Cell* **104:** 891–900.

Kengaku, M., Capdevila, J., Rodriguez-Esteban, C., De La Peña, J., Johnson, R.L., Izpisúa Belmonte, J.C., and Tabin, C.J. 1998. Distinct WNT pathways regulating AER formation and dorsoventral polarity in the chick limb bud. *Science* **280:** 1274–1277.

Kessel, M. and Gruss, P. 1991. Homeotic transformations of murine vertebrae and concomitant alteration of *Hox* codes induced by retinoic acid. *Cell* **67:** 89–104.

Khokha, M.K., Hsu, D., Brunet, L.J., Dionne, M.S., and Harland, R.M. 2003. Gremlin is the BMP antagonist required for maintenance of Shh and Fgf signals during limb patterning. *Nat. Genet.* **34:** 303–307.

Kmita, M., Tarchini, B., Zakany, J., Logan, M., Tabin, C.J., and Duboule, D. 2005. Early developmental arrest of mammalian limbs lacking *HoxA/HoxD* gene function. *Nature* **435:** 1113–1116.

Laufer, E., Nelson, C.E., Johnson, R.L., Morgan, B.A., and Tabin, C. 1994. *Sonic hedgehog* and *Fgf-4* act through a signaling cascade and feedback loop to integrate growth and patterning of the developing limb bud. *Cell* **79:** 993–1003.

Laufer, E., Dahn, R., Orozco, O.E., Yeo, C.Y., Pisenti, J., Henrique, D., Abbott, U.K., Fallon, J.F., and Tabin, C. 1997. Expression of *Radical fringe* in limb-bud ectoderm regulates apical ectodermal ridge formation. *Nature* **386:** 366–373.

Lehoczky, J.A., Williams, M.E., and Innis, J.W. 2004. Conserved expression domains for genes upstream and within the *HoxA* and *HoxD* clusters suggests a long-range enhancer existed before cluster duplication. *Evol. Dev.* **6:** 423–430.

Lewandoski, M., Sun, X., and Martin, G.R. 2000. Fgf8 signaling from the AER is essential for normal limb development. *Nat. Genet.* **26:** 460–463.

Lewis, P.M., Dunn, M.P., McMahon, J.A., Logan, M., Martin, J.F., St-Jacques, B., and McMahon, A.P. 2001. Cholesterol modification of sonic hedgehog is required for long-range signaling activity and effective modulation of signaling by Ptc1. *Cell* **105:** 599–612.

Litingtung, Y., Dahn, R.D., Li, Y., Fallon, J.F., and Chiang, C. 2002. *Shh* and *Gli3* are dispensable for limb skeleton formation but regulate digit number and identity. *Nature* **418:** 979–983.

Loomis, C.A., Harris, E., Michaud, J., Wurst, W., Hanks, M., and Joyner, A.L. 1996. The mouse *Engrailed-1* gene and ventral limb patterning. *Nature* **382:** 360–363.

Marigo, V. and Tabin, C.J. 1996. Regulation of patched by sonic hedgehog in the developing neural tube. *Proc. Natl. Acad. Sci.* **93:** 9346–9351.

Marigo, V., Johnson, R.L., Vortkamp, A., and Tabin, C.J. 1996. Sonic hedgehog differentially regulates expression of *GLI* and *GLI3* during limb development. *Dev. Biol.* **180:** 273–283.

Martin, G.R. 1998. The roles of FGFs in the early development of vertebrate limbs. *Genes Dev.* **12:** 1571–1586.

Mercader, N., Leonardo, E., Piedra, M.E., Martinez, A.C., Ros, M.A., and Torres, M. 2000. Opposing RA and FGF signals control proximodistal vertebrate limb development through regulation of Meis genes. *Development* **127:** 3961–3970.

Michaud, J.L., Lapointe, F., and Le Douarin, N.M. 1997. The dorsoventral polarity of the presumptive limb is determined by signals produced by the somites and by the lateral somatopleure. *Development* **124:** 1453–1463.

Min, H., Danilenko, D.M., Scully, S.A., Bolon, B., Ring, B.D., Tarpley, J.E., DeRose, M., and Simonet, W.S. 1998. *Fgf-10* is required for both limb and lung development and exhibits striking functional similarity to *Drosophila branchless*. *Genes Dev.* **12:** 3156–3161.

Minguillon, C., Del Buono, J., and Logan, M.P. 2005. *Tbx5* and *Tbx4* are not sufficient to determine limb-specific morphologies but have common roles in initiating limb outgrowth. *Dev. Cell* **8:** 75–84.

Moon, A.M. and Capecchi, M.R. 2000. Fgf8 is required for outgrowth and patterning of the limbs. *Nat. Genet.* **26:** 455–459.

Nissim, S., Allard, P., Bandyopadhyay, A., Harfe, B.D., and Tabin, C.J. 2007. Characterization of a novel ectodermal signaling center regulating Tbx2 and Shh in the vertebrate limb. *Dev. Biol.* **304:** 9–21.

Niswander, L. and Martin, G.R. 1993. FGF-4 and BMP-2 have opposite effects on limb growth. *Nature* **361:** 68–71.

Niswander, L., Jeffrey, S., Martin, G.R., and Tickle, C. 1994. A positive feedback loop coordinates growth and patterning in the vertebrate limb. *Nature* **371:** 609–612.

Nowicki, J.L. and Burke, A.C. 2000. *Hox* genes and morphological identity: Axial versus lateral patterning in the vertebrate mesoderm. *Development* **127:** 4265–4275.

O'Rourke, M.P., Soo, K., Behringer, R.R., Hui, C.C., and Tam, P.P. 2002. *Twist* plays an essential role in FGF and SHH signal transduction during mouse limb development. *Dev. Biol.* **248:** 143–156.

Oberg, K.C. and Eichele, G. 1999. *Hox* gene expression and regulation in the presumptive wing region of the chick lateral plate mesoderm. *Dev. Biol.* **210:** 151–180.

Ohuchi, H., Nakagawa, T., Yamamoto, A., Araga, A., Ohata, T., Ishimaru, Y., Yoshioka, H., Kuwana, T., Nohno, T., Yamasaki, M., Itoh, N., and Noji, S. 1997. The mesenchymal factor, FGF10, initiates and maintains the outgrowth of the chick limb bud through interaction with FGF8, an apical ectodermal factor. *Development* **124:** 2235–2244.

Parr, B.A., Shea, M.J., Vassileva, G., and McMahon, A.P. 1993. Mouse *Wnt* genes exhibit discrete domains of expression in the early embryonic CNS and limb buds. *Development* **119:** 247–261.

Riddle, R.D., Johnson, R.L., Laufer, E., and Tabin, C. 1993. *Sonic hedgehog* mediates the polarizing activity of the ZPA. *Cell* **75:** 1401–1416.

Riddle, R.D., Ensini, M., Nelson, C., Tsuchida, T., Jessell, T.M., and Tabin, C. 1995. Induction of the LIM homeobox gene *Lmx1* by WNT7a establishes dorsoventral pattern in the vertebrate limb. *Cell* **83:** 631–640.

Rodriguez-Esteban, C., Schwabe, J.W., De La Peña, J., Foys, B., Eshelman, B., and Izpisúa Belmonte, J.C. 1997. *Radical fringe* positions the apical ectodermal ridge at the dorsoventral boundary of the vertebrate limb. *Nature* **386:** 360–366.

Ros, M.A., Dahn, R.D., Fernandez-Teran, M., Rashka, K., Caruccio, N.C., Hasso, S.M., Bitgood, J.J., Lancman, J.J., and Fallon, J.F. 2003. The chick *oligozeugodactyly* (*ozd*) mutant lacks sonic hedgehog function in the limb. *Development* **130:** 527–537.

Rowe, D.A. and Fallon, J.F. 1982. The proximodistal determination of skeletal parts in the developing chick leg. *J. Embryol. Exp. Morphol.* **68:** 1–7.

Rubin, L. and Saunders Jr., J.W. 1972. Ectodermal-mesodermal interactions in the growth of limb buds in the chick embryo: Constancy and temporal limits of the ectodermal induction. *Dev. Biol.* **28:** 94–112.

Saunders Jr., J.W. 1948. The proximo-distal sequence of origin of the parts of the chick wing and the role of the ectoderm. *J. Exp. Zool.* **282:** 628–668.

Saunders, J.W. and Gasseling, M.T. 1968. Ectodermal and mesenchymal interactions in the origin of limb symmetry. In *Ectodermal and mesenchymal interactions: 18th Hahnemann Symposium* (ed. R. Fleischmajer and R.E. Billingham), pp. 78–97. Williams and Wilkins, Baltimore.

Scherz, P.J., Harfe, B.D., McMahon, A.P., and Tabin, C.J. 2004. The limb bud Shh-Fgf feedback loop is terminated by expansion of former ZPA cells. *Science* **305:** 396–399.

Sekine, K., Ohuchi, H., Fujiwara, M., Yamasaki, M., Yoshizawa, T., Sato, T., Yagishita, N., Matsui, D., Koga, Y., Itoh, N., and Kato, S. 1999. Fgf10 is essential for limb and lung formation. *Nat. Genet.* **21:** 138–141.

Spitz, F., Gonzalez, F., and Duboule, D. 2003. A global control region defines a chromosomal regulatory landscape containing the *HoxD* cluster. *Cell* **113:** 405–417.

Stephens, T.D. and McNulty, T.R. 1981. Evidence for a metameric pattern in the development of the chick humerus. *J. Embryol. Exp. Morphol.* **61:** 191–205.

Stratford, T.H., Kostakopoulou, K., and Maden, M. 1997. *Hoxb-8* has a role in establishing early anterior-posterior polarity in chick forelimb but not hindlimb. *Development* **124:** 4225–4234.

Summerbell, D. 1974. A quantitative analysis of the effect of excision of the AER from the chick limb-bud. *J. Embryol. Exp. Morphol.* **32:** 651–660.

Summerbell, D., Lewis, J.H., and Wolpert, L. 1973. Positional information in chick limb morphogenesis. *Nature* **244:** 492–496.

Sun, X., Lewandoski, M., Meyers, E.N., Liu, Y.H., Maxson Jr., R.E., and Martin, G.R. 2000. Conditional inactivation of *Fgf4* reveals complexity of signalling during limb bud development. *Nat. Genet.* **25:** 83–86.

Sun, X., Mariani, F.V., and Martin, G.R. 2002. Functions of FGF signalling from the apical ectodermal ridge in limb development. *Nature* **418:** 501–508.

Tabin, C. and Wolpert, L. 2007. Rethinking the proximodistal axis of the vertebrate limb in the molecular era. *Genes Dev.* **21:** 1433–1442.

Takahashi, M., Tamura, K., Buscher, D., Masuya, H., Yonei-Tamura, S., Matsumoto, K., Naitoh-Matsuo, M., Takeuchi, J., Ogura, K., Shiroishi, T., et al. 1998. The role of *Alx-4* in the establishment of anteroposterior polarity during vertebrate limb development. *Development* **125:** 4417–4425.

Tarchini, B., Duboule, D., and Kmita, M. 2006. Regulatory constraints in the evolution of the tetrapod limb anterior-posterior polarity. *Nature* **443:** 985–988.

te Welscher, P., Fernandez-Teran, M., Ros, M.A., and Zeller, R. 2002. Mutual genetic antagonism involving GLI3 and dHAND prepatterns the vertebrate limb bud mesenchyme prior to SHH signaling. *Genes Dev.* **16:** 421–426.

Tickle, C. 1981. The number of polarizing region cells required to specify additional digits in the developing chick wing. *Nature* **289:** 295–298.

Tickle, C., Summerbell, D., and Wolpert, L. 1975. Positional signaling and specification of digits in chick limb morphogenesis. *Nature* **254:** 199–202.

Vogel, A., Rodriguez, C., Warnken, W., and Izpisúa Belmonte, J.C. 1995. Dorsal cell fate specified by chick Lmx1 during vertebrate limb development. *Nature* **378:** 716–720.

Wang, B., Fallon, J.F., and Beachy, P.A. 2000. Hedgehog-regulated processing of Gli3 produces an anterior/posterior repressor gradient in the developing vertebrate limb. *Cell* **100:** 423–434.

Wolpert, L. 1969. Positional information and the spatial pattern of cellular differentiation. *J. Theor. Biol.* **25:** 1–47.

Wolpert, L., Tickle C., and Sampford, M. 1979. The effect of cell killing by x-irradiation on pattern formation in the chick limb. *J. Embryol. Exp. Morphol.* **50:** 175–193.

Yang, Y. and Niswander, L. 1995. Interaction between the signaling molecules WNT7a and SHH during vertebrate limb development: Dorsal signals regulate anteroposterior patterning. *Cell* **80:** 939–947.

Yang, Y., Drossopoulou, G., Chuang, P.T., Duprez, D., Marti, E., Bumcrot, D., Vargesson, N., Clarke, J., Niswander, L., McMahon, A., et al. 1997. Relationship between dose, distance and time in *Sonic Hedgehog*-mediated regulation of anteroposterior polarity in the chick limb. *Development* **124:** 4393–4404.

Zákány, J., Kmita, M., and Duboule, D. 2004. A dual role for *Hox* genes in limb anterior-posterior asymmetry. *Science* **304:** 1669–1672.

Zákány, J., Zacchetti, G., and Duboule, D. 2007. Interactions between *HOXD* and *Gli3* genes control the limb apical ectodermal ridge via *Fgf10*. *Dev. Biol.* **306:** 883–893.

Zuniga, A., Haramis, A.P., McMahon, A.P., and Zeller, R. 1999. Signal relay by BMP antagonism controls the SHH/FGF4 feedback loop in vertebrate limb buds. *Nature* **401:** 598–602.

Zwilling, E. 1959. Interaction between ectoderm and mesoderm in duck-chicken limb bud chimaeras. *J. Exp. Zool.* **142:** 521–532.

3

Patterning and Differentiation of the Vertebrate Spine

Jérome Chal[1] and Olivier Pourquié[1,2,3]
[1]Stowers Institute for Medical Research
[2]Howard Hughes Medical Institute
[3]Department of Anatomy and Cell Biology
The University of Kansas School of Medicine
Kansas City, Kansas 66103

ONE OF THE MOST STRIKING FEATURES of the human spine is its periodic organization. This so-called "segmental" arrangement of the vertebrae along the anteroposterior body axis is established during embryonic development. Structures called somites, which contain the precursors of the vertebrae, form in a rhythmic fashion at the posterior end of the embryo during the process of somitogenesis. Somites are sequentially added to the growing axis, thus establishing the characteristic periodic pattern of the future vertebral column. The primary segmentation of the vertebrate embryo displayed by somitic organization also underlies much of the segmental organization of the body, including muscles, nerves, and blood vessels. In amniotes, somites are the major component of the paraxial mesoderm that form bilaterally along the nerve cord as a result of primitive streak and tail bud regression during body axis formation. Somites bud off from the anterior presomitic mesoderm (PSM) as epithelial spheres surrounding a core of mesenchymal cells called the somitocoele. The dorsal portion of the somite remains epithelial and forms the dermomyotome, which differentiates into muscle and dermis while its ventral moiety undergoes an epithelio-mesenchymal transition, leading to the formation of the sclerotome. The sclerotome gives rise to the skeletal elements of the vertebral column: the vertebrae, ribs, intervertebral disks, and tendons. Most of our understanding of amniote somitogenesis at the morphogenetic and molecular levels results from studies involv-

The Skeletal System ©2009 Cold Spring Harbor Laboratory Press 978-087969-825-6

ing the chicken (*Gallus gallus*) and the mouse (*Mus musculus*). In this chapter, we essentially focus on the patterning and development of the spine in amniote species such as chickens, mice, and humans.

SEGMENTAL PATTERNING OF THE VERTEBRAL COLUMN: THE CLOCK AND WAVEFRONT SYSTEM

Origin of the Vertebral Precursors: The Paraxial Mesoderm

Together with the head mesoderm, somites form the paraxial mesoderm, which appears as bilateral strips of tissue that flank the neural tube and notochord, and are bound laterally by the intermediate and lateral plate mesoderm. In the chicken, mouse, and human embryos, the first somite lies immediately posterior to the otic vesicle (Huang et al. 1997; Spörle and Schughart 1997; O'Rahilly and Müller 2003). Anterior to this somite, the paraxial mesoderm is referred to as the head or cephalic mesoderm, and it contributes to the skeletal muscles and bones of the head (see Chapter 4 by Le Douarin and Creuzet). The paraxial mesoderm is generated when cells located in a defined region of the epiblast (the superficial layer of the embryo) ingress into the primitive streak during gastrulation (Waddington 1952; Rosenquist 1966; Nicolet 1971; Tam and Beddington 1987; Schoenwolf et al. 1992; Hatada and Stern 1994; Tam and Trainor 1994; Psychoyos and Stern 1996; Pourquié 2004; Iimura et al. 2007).

At the beginning of gastrulation, the presumptive territory of the paraxial mesoderm in the epiblast is located bilaterally to the forming primitive streak that defines the future anteroposterior axis of the embryo (Fig. 1A). During primitive streak formation, these territories converge toward the streak and begin to ingress while undergoing an epithelio-mesenchymal transition. The first paraxial mesoderm precursors to ingress form the head mesoderm, which lies at the anterior tip of the embryo. Then, the primitive streak begins to shrink and regress, and its anterior tip, which corresponds to the Spemann organizer of amniotes called Hensen's node or the node, moves posteriorly. This regression lays in its wake the forming body axis and progressively forms more posterior levels of the paraxial mesoderm (Fig. 1B). Fate maps of the early embryo show that the paraxial mesoderm precursors are located in the anterior portion of the streak and in the adjacent epiblast, as well as in the node region where the notochord precursors are located (Fig. 1A,B) (Waddington 1952; Rosenquist 1966; Nicolet 1971; Tam and Beddington 1987; Selleck and Stern 1991; Schoenwolf et al. 1992; Hatada and Stern

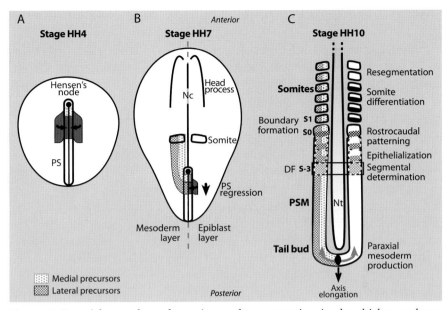

Figure 1. Paraxial mesoderm formation and segmentation in the chicken embryo. (*A*) Hamburger and Hamilton Stage 4 (Stage HH4) chicken embryo (Hamburger and Hamilton 1992). At this stage, the primitive streak (PS), which corresponds to the amniote blastopore, has reached its maximal length. Presumptive territories of somites are located in the superficial epiblast just below Hensen's node (medial precursor population, green) and in two symmetrical domains located on both sides of the PS (lateral precursor population, purple). These cells are ingressing (arrows) through the PS to form the paraxial mesoderm. (*B*) Stage HH7 chicken embryo. The PS and node have begun their posterior regression (arrowhead), leaving in their wake the embryonic axis comprising the head process anteriorly and the notochord (Nc) axially. Epiblast cells (purple) continue to ingress in the PS (arrow) and join the descendents of a population of resident stem cells located in the anterior primitive streak (green) to generate the paraxial mesoderm that forms two strips of tissue bilaterally to the notochord. The mesodermal layer is represented on the *left* side without the superficial epiblast. Medial and lateral precursors are derived from the streak stem cells and the epiblast layer, and contribute to the medial and lateral compartments of the forming somites, respectively. (*C*) Stage HH10 chicken embryo. Somitogenesis progresses posteriorly in concert with axis elongation (arrow). Paraxial mesoderm cells are produced at the tail bud level and undergo a maturation process in the presomitic mesoderm (PSM), leading to the periodic formation of new pairs of somites (S0). Dorsal views, anterior to the top. (Nt) neural tube, (DF) determination front. (Presumptive somite nomenclature according to Pourquié and Tam 2001).

1994; Tam and Trainor 1994; Psychoyos and Stern 1996; Iimura et al. 2007).

Whereas the primitive streak contributes to the formation of the most anterior somites, more posteriorly, the paraxial mesoderm is formed from the tail bud (see Fig. 1C). In this structure, gastrulation movements, similar to those observed in the primitive streak, are still observed (Pasteels 1937; Gont et al. 1993; Catala et al. 1995; Kanki and Ho 1997; Knezevic et al. 1998; Davis and Kirschner 2000; Cambray and Wilson 2002, 2007; Stern 2006). In the tail bud, precursors of the paraxial mesoderm are located immediately posterior to the chordo-neural hinge region (Charrier et al. 1999; Cambray and Wilson 2007; McGrew et al. 2008). During axis elongation, newly ingressed paraxial mesoderm cells are deposited bilaterally to the node or chordo-neural hinge and the axial structures, resulting in the formation of the two strips of paraxial mesoderm tissue (Fig. 1C) (Selleck and Stern 1991; Schoenwolf et al. 1992; Hatada and Stern 1994; Psychoyos and Stern 1996).

The origin of the paraxial mesoderm during gastrulation can be traced to two distinct populations of precursors (Rosenquist 1966; Nicolet 1970; Bellairs 1986; Selleck and Stern 1991; Iimura et al. 2007). These two populations exhibit different fates, forming the future medial and lateral somitic compartments (Fig. 1, green and purple, respectively). The medial compartment gives rise to the epaxial muscles, the vertebral column, and the dermis of the back; whereas the lateral compartment produces essentially the ribs and the hypaxial muscles that include intercostals and limb muscles (Ordahl and Le Douarin 1992; Olivera-Martinez et al. 2000). In the chicken embryo, medial somites derive from a population of precursors that exhibit a stem-cell-like behavior and are located in the anteriormost primitive streak and Hensen's node, while the lateral somite precursors derive from the epiblast adjacent to the anterior primitive streak (Fig. 1) (Bellairs 1986; Selleck and Stern 1991; Iimura et al. 2007). A stem-cell-like population, located in the primitive streak and contributing to somites, has also been identified in the mouse (Nicolas et al. 1996; Eloy-Trinquet and Nicolas 2002). The self-renewal capacity of these stem cells has been revealed by lineage analysis using various cell-labeling strategies and by serial transplants in chicken and mouse embryos (Selleck and Stern 1991; Nicolas et al. 1996; Cambray and Wilson 2002, 2007; McGrew et al. 2008). The precursors of the medial somites control the segmentation process and impose their segmental pattern on the lateral somite (Freitas et al. 2001). This dual origin of paraxial mesoderm precursors is likely conserved across vertebrates (Iimura et al. 2007).

A Molecular Oscillator Involved in Establishing the Periodicity of Vertebrae

In all vertebrate species, pairs of somites bud off periodically at the anterior tip of the PSM at a specific pace (for example, every 90 minutes in chickens, 120 minutes in mice, and between 4–8 hours in humans) (Romanoff 1960; Tam 1981; Palmeirim et al. 1997; William et al. 2007; Aulehla et al. 2008). Microsurgical experiments in the chicken embryo have revealed that segmentation of the PSM is a tissue-autonomous process and is independent of the surrounding embryonic tissues (Menkes and Sandor 1969; Christ et al. 1974; Packard 1978; Palmeirim et al. 1998; Jouve et al. 2000). The rhythmic production of somites from the PSM has inspired theoretical models such as the "clock and wavefront" model (Cooke and Zeeman 1976). This model proposed that the periodicity of somites results from the action of a molecular oscillator (called the clock) traveling along the embryonic axis. In this model, the periodic segment formation is triggered during a defined (permissive) phase of the oscillation, while the oscillator is constantly displaced posteriorly by a wave of maturation (called the wavefront), hence, ensuring the spacing of the response to the oscillator. A number of subsequent theoretical models have been proposed, many of which have also relied upon the conversion of a temporal oscillation into a spatial periodic pattern (for reviews, see Dale and Pourquié 2000; Kulesa et al. 2007).

The first evidence of the existence of an oscillator coupled to somitogenesis was provided by the periodic expression of the transcription factor *hairy1* mRNA in the chicken embryo PSM (Palmeirim et al. 1997). During the formation of each somite, the PSM is swiped by a dynamic wave of *hairy1* mRNA expression. During each somitogenesis cycle, *hairy1* is first activated in the posterior PSM and then progressively in more anteriorly located cells, giving the illusion of a traveling wave of gene expression (Fig. 2A–D). These *hairy1* transcriptional oscillations were proposed to reflect the existence of a molecular oscillator called the segmentation clock (Palmeirim et al. 1997). Subsequently, it has been shown in the fish, frog, chicken, and mouse that several members of the *Hes/Her/Hairy* family of the basic helix-loop-helix (bHLH) transcriptional repressors exhibit such a cyclic behavior, indicating that the oscillator is conserved in vertebrates (Palmeirim et al. 1997; Holley et al. 2000; Jiang et al. 2000; Jouve et al. 2000; Bessho et al. 2001; Dunwoodie et al. 2002; Li et al. 2003b). In the chicken, the genes *hairy1*, *hairy2*, and *Hey2*, and in the mouse, the genes *Hes1*, *Hes5*, *Hey1*, and *Hes7* oscillate in the PSM (Palmeirim et al. 1997; Jouve et al. 2000; Leimeister et al. 2000;

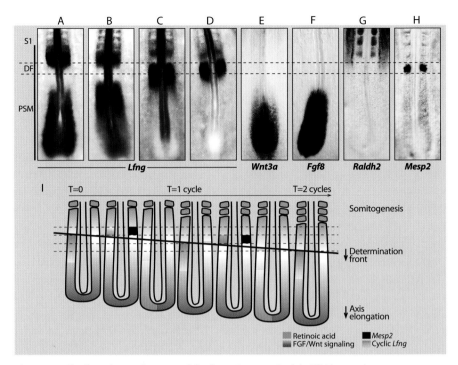

Figure 2. Clock and wavefront model of somitogenesis. (*A–H*) Expression pattern comparison of key components of the clock and wavefront system during somitogenesis in chicken embryos. (*A–D*) Wave of expression of the cyclic gene *Lfng* during one somite formation. All embryos have 18 somites and have been hybridized with the *Lfng* probe. A similar expression pattern dynamic is observed for the *hairy1* gene. (*E*) *Wnt3a*, (*F*) *Fgf8*, (*G*) *Raldh2*, (*H*) *Mesp2* expression by in situ hybridization. (S1) last formed somite, (DF) determination front (dotted lines). (*I*) Clock and wavefront segmentation model. Antagonistic gradients of Fgf/Wnt signaling (purple) and retinoic acid signaling (green) position the determination front (thick, black line). The periodic signal of the segmentation clock is shown in orange (represented on the *left* side only). As the embryo extends posteriorly, the determination front moves caudally. Cells that reach the determination front are exposed to the periodic clock signal, initiating the segmentation program and activating gene expression, such as *Mesp2* (black stripes, represented on the *right* side only), in a stripe that prefigures the future segment. This establishes the segmental pattern of the presumptive somites. Dorsal views, anterior to the top.

Bessho et al. 2001; Dunwoodie et al. 2002; Dequeant et al. 2006). In several developmental processes, the *Hes/Her/Hairy* genes are targets of the Notch pathway, but in the PSM, their oscillatory expression does not seem to absolutely require Notch signaling (Kageyama et al. 2007; Niwa et al. 2007; Ozbudak and Lewis 2008). In amniotes, other classes of cyclic genes have been identified based on their rhythmic expression pattern in the PSM (Fig. 2A–D). These cyclic genes belong, in their vast majority, to the Notch, Wnt, and fibroblast growth factor (Fgf) signaling pathways (Dequeant and Pourquié 2008). Their dynamic expression sequence has now been imaged in real time in live mouse embryos or in isolated PSM cells using luciferase or green fluorescent protein-based methods (Masamizu et al. 2006; Aulehla et al. 2008). Cyclic expression at the protein level has been demonstrated only for a subset of cyclic genes that includes *Hes7* and *Lunatic fringe* (*Lfng*) (Bessho et al. 2003; Dale et al. 2003). As discussed below, analysis of the mutation and misexpression of the characterized cyclic genes demonstrate that many are important for proper somitogenesis.

Periodic Notch Activation Is Associated with Somite Formation

Notch signaling has been shown to play an essential role at different stages of somitogenesis (for review, see Pourquié 2001; Weinmaster and Kintner 2003; Rida et al. 2004; Gridley 2006; Lewis and Ozbudak 2007; Ozbudak and Pourquié 2008). Notch receptors (Notch 1–4 in mammals) are single-pass transmembrane proteins that undergo regulated, sequential proteolysis that is required for proper membrane presentation, as well as for signal transduction (Fig. 3A). Three cleavages have been identified and are mediated sequentially: first, by a Furin-like protease in the Golgi complex; next, at the cell surface by members of the ADAM (A disintegrin and metalloprotease) family of metalloproteases (such as ADAM17/TACE and ADAM10/Kuz); and finally, by the Presenilin/γ-secretase complex (cleavage S3) upon ligand binding to release the Notch intracellular domain (NICD) (Kopan and Ilagan 2004; Bray 2006). The originality of the Notch signaling pathway resides in the absence of a secondary messenger, since NICD acts directly as a transcription factor (Fig. 3A). Notch receptors are activated by transmembrane ligands of the Delta/Serrate/Lag-2 (DSL) family, namely, Delta-like 1, 3, and 4 (Dll1, -3, and -4) and Jagged/Serrate 1 and 2 (Jag1 and -2) in mammals. Notch signaling is mediated by the CBF1/RBP-Jk, Suppressor of Hairless Su(H)/Lag1 (CSL) transcription complex that associates with NICD. In the absence of NICD, CSL acts as a repressor. Upon nuclear translocation of NICD and binding to CSL, histone acetylases (HAT) and other components, such as Mastermind-like proteins (MAML1-3), are recruited and lead to transcriptional activation of specific target genes (Fig. 3A) (Baron 2003; Fryer et al. 2004).

Notch signaling has been implicated in a large array of developmental processes, where it can act as a switch that controls cell fate or regu-

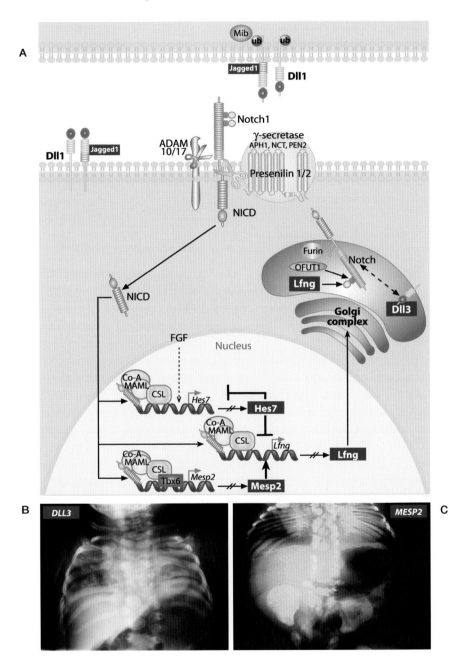

Figure 3. (*See facing page for legend.*)

lates tissue patterning (Artavanis-Tsakonas et al. 1999). In the mouse PSM, the activation of the Notch1 receptor is periodic and parallels the rhythm of somite production. Rhythmic cleavage of NICD can be detected using a specific antibody that recognizes the cleaved form of the Notch1 receptor (Huppert et al. 2005; Morimoto et al. 2005). In mice and zebrafish, mutation of *Notch1* (*Notch1* in mice and *Notch1a/des* in zebrafish) results in defects in somite formation, suggesting that it acts as the major Notch receptor involved in somite segmentation (Conlon et al. 1995; Holley et al. 2002). While *Notch2* is also expressed in the mouse PSM, the *Notch2* null embryo does not exhibit segmentation defects (Hamada et al. 1999). However, the double *Notch1* and *-2* null embryo exhibits more severe segmentation defects than the single *Notch1* mutant, suggesting some functional redundancy between the two receptors (Huppert et al. 2005). Mice mutant for *Presenilins* or for the other components of the γ-secretase complex (e.g., *Nicastrin* [*Nct*] or *Aph-1a*) exhibit a Notch-like segmentation defect, and double *PS1* and *-2* mutants do not form somites (Fig. 3A) (Shen et al. 1997; Wong et al. 1997; Donoviel et al. 1999; Herreman et al. 2000; Koizumi et al. 2001; Li et al. 2003a; Ma et al. 2005). Moreover, in the chicken and zebrafish, pharmacological inhibition of γ-secretase to block Notch signaling results in segmentation defects (Geling et al. 2002; Dale et al. 2003; Mara et al. 2007; Riedel-Kruse et al. 2007; Ozbudak and

---

Figure 3. Notch signaling pathway and human vertebral abnormalities. (*A*) The Notch signaling pathway. The transmembrane Notch receptor is activated by the binding of Delta and Jagged ligands from neighboring cells, leading to proteolytic cleavage and release of the Notch intracellular domain (NICD). Notch cleavage involves the γ-secretase complex. NICD translocates to the nucleus and interacts with the DNA-binding CSL protein, recruiting Mastermind (MAML) and leading to the activation of transcription of target genes, such as *Hes7*, *Lfng*, and *Mesp2*. The ubiquitination (Ub) by the ubiquitin ligase Mib controls the endocytosis of Notch ligands, which is necessary for proper signaling. During Notch receptor trafficking to the membrane, Notch undergoes sequential proteolytic cleavages by Furin and ADAMs protease. Notch undergoes sequential glycosylation by OFUT1 and Lfng. Dll3 is located principally in the Golgi complex where it interacts in *cis* to modulate Notch signaling. Notch pathway components found to be mutated in association with human vertebral segmentation abnormalities are shown in red boxes. (*B*) Radiograph of a patient with Spondylocostal dysostosis (SCDO1) showing the severe axial skeletal malformations associated with a mutation in the *DLL3* gene. (Radiograph provided by Peter Turnpenny, M.D.) (*C*) Radiograph of patient with spondylothoracic dysostosis (STD) illustrating severe vertebral and rib malformations associated with a mutation in the *MESP2* gene. (Radiograph provided by Albert Cornier, M.D., and reprinted from Cornier et al. 2008 [© Elsevier].) (Co-A) Coactivator.

Lewis 2008). After nuclear translocation, NICD activates transcription of target genes, such as *Lfng* or *Notch-regulated ankyrin repeat protein* (*Nrarp*) (see Fig. 3A) (Krebs et al. 2001; Lamar et al. 2001; Cole et al. 2002; Morales et al. 2002; Dale et al. 2003; Pirot et al. 2004). Periodic activation of transcription downstream of Notch is indicated by the rhythmic waves of *Nrarp* and *Lfng* expression in the PSM (see Fig. 2A–D) (Forsberg et al. 1998; McGrew et al. 1998; Aulehla and Johnson 1999; Dequeant et al. 2006). Mice mutant for *RBP-Jk* (Oka et al. 1995; del Barco Barrantes et al. 1999) and *Lfng* (Evrard et al. 1998; Zhang et al. 2002) exhibit marked defects in somitogenesis with few anterior somites formed and an overall disruption of posterior segmentation.

The rhythmic activation of Notch signaling in the PSM could be triggered by the periodic expression of the Notch ligands. In amniotes, *Dll1*, *Dll3*, and *Jag/Serrate1*, genes are expressed in the PSM (Bettenhausen et al. 1995; Henrique et al. 1995; Lindsell et al. 1995; Myat et al. 1996; Dunwoodie et al. 1997). Genes coding for Notch ligands exhibit cyclic expression in the mouse (*Dll1*) and zebrafish (*deltaC*) PSM (Jiang et al. 2000; Maruhashi et al. 2005). Moreover, mice mutant for *Dll1* and *Dll3* (*Pudgy*), and zebrafish mutant for *deltaD* (*aei*) and *deltaC* (*bea*) exhibit somitogenesis defects (Hrabě de Angelis et al. 1997; Kusumi et al. 1998; van Eeden et al. 1998; Barrantes et al. 1999; Jiang et al. 2000; Jouve et al. 2000; Dunwoodie et al. 2002; Jülich et al. 2005). In the mouse, Dll1 (which is expressed at the surface of PSM cells) activates Notch signaling; whereas, Dll3 (a highly divergent Notch ligand that is localized primarily in the Golgi complex) acts intracellularly through an unknown mechanism (see Fig. 3A) (Geffers et al. 2007). In vitro experiments and analysis of mice mutant for *Dll3* indicate that Dll3 modulates Notch signaling in a cell-autonomous fashion through a mechanism that is independent of Presenilin (Takahashi et al. 2003; Ladi et al. 2005). In addition, several ubiquitin ligases modulate Notch signaling by regulating Notch ligand endocytosis (Le Borgne 2006). In mice and zebrafish, mutations in genes coding for E3 ubiquitin ligases (e.g., *Mind bomb* [*Mib1*] and *Neuralized* [*Neur1/2*]) result in somitogenesis defects (see Fig. 3A) (Itoh et al. 2003; Chen and Casey Corliss 2004; Barsi et al. 2005; Koo et al. 2005; Zhang et al. 2007).

Alternatively, periodic Notch activation could also be triggered by intracellular negative feedback loops. *Hes7* codes for a bHLH transcriptional repressor that can act downstream of Notch and can repress its own transcription, as well as that of the negative feedback inhibitor *Lfng* (see Fig. 3A) (Bessho et al. 2001, 2003; Dale et al. 2003; Morimoto et al. 2005; Niwa et al. 2007). *Hes7* is periodically expressed in the mouse PSM, and in mutant mice, its inactivation results in an arrest of *Lfng* oscillations and in an up-regulation of *Hes7* transcription due to the lack of *Hes7* repressive activity, resulting in somitogenesis defects (see Fig. 3A) (Bessho et al. 2003). Thus, Hes7 was proposed to play a role in the control of cyclic gene oscillations in the PSM (Kageyama et al. 2007). In zebrafish, homologs of the *Hes7* gene—*Her1* and *Her7*—display a cyclic expression in the PSM, and their overexpression or their knockdown leads to segmentation defects (Takke and Campos-Ortega 1999; Holley et al. 2000, 2002; Henry et al. 2002; Oates and Ho 2002; Gajewski et al. 2003; Lewis 2003; Rida et al. 2004; Oates et al. 2005a; Giudicelli et al. 2007; Holley 2007; Ozbudak and Lewis 2008). It has been proposed that a simple *Her1/Her7*-based, delayed negative feedback mechanism generates oscillations and, thus, acts as the clock pacemaker (Lewis 2003). In zebrafish, it has been proposed that the role of Notch signaling is restricted to the synchronization of oscillations between individual PSM cells and, therefore, coordinates the waves of cyclic gene expression along the PSM (Jiang et al. 2000; Horikawa et al. 2006; Mara et al. 2007; Riedel-Kruse et al. 2007; Ozbudak and Lewis 2008). Evidence is still lacking for such a role for Notch signaling in amniotes.

A second negative feedback loop relying on Lfng and regulating Notch activation has been described in amniotes (Dale et al. 2003; Morimoto et al. 2005). Lfng is a glycosyltransferase that modifies the Notch extracellular domain during its trafficking to the cell surface (see Fig. 3A) (Haines and Irvine 2003). In the PSM, Lfng periodically inhibits Notch signaling and participates in a negative feedback loop involved in the control of Notch pathway oscillation and somite boundary positioning (Dale et al. 2003; Serth et al. 2003; Huppert et al. 2005; Morimoto et al. 2005; Shifley et al. 2008). Lfng-modification activity is dependent upon the priming of the Notch receptor by the O-fucosyltransferase (OFUT1), which is expressed ubiquitously and is essential for mesoderm segmentation in mice (see Fig. 3A) (Shi and Stanley 2003). In the PSM, *Lfng* cyclic expression is controlled at the transcriptional level and is observed only in amniotes (see Fig. 2A–D) (Forsberg et al. 1998; McGrew et al. 1998; Aulehla and Johnson 1999; Leve et al. 2001; Cole et al. 2002; Morales et al. 2002; Qiu et al. 2004).

Oscillations of Canonical Wnt and of Fgf Signaling in the Mouse Embryo

Wnt ligands are soluble proteins that bind to a receptor complex composed of Frizzled (Fz) and low-density-lipoprotein receptor-related protein 5/6 (LRP5/6). In the absence of Wnt signaling, the protein β-catenin is sequestered in a "destruction complex" that contains the scaffolding proteins Adenomatosis polyposis coli (APC), Disheveled (Dsh), Axin1 and -2, and the kinases Glycogen synthase kinase 3 (GSK3) and Casein kinase 1 (CK1) that phosphorylate β-catenin, targeting it for proteosomal degradation. Wnt binding to its receptor complex leads to the release of β-catenin from the destruction complex and, thus, prevents its phosphorylation (Fuerer et al. 2008; Klaus and Birchmeier 2008). This stops β-catenin degradation and allows it to translocate to the nucleus and to bind to the lymphoid enhancer factor/T-cell-specific factor (LEF/TCF) transcriptional complex. In the absence of Wnt signaling, LEF/TCF (*Lef1* and *Tcf1*, *-3*, and *-4* in mammals) binds Groucho-related proteins and acts as a repressor. Upon activation of the pathway, the complex, formed from LEF/TCF and β-catenin, is turned into a transcriptional activator that drives the expression of a specific set of target genes (Nusse 2005; Hoppler and Kavanagh 2007). *Axin2* is a classical target of the Wnt canonical pathway that has been shown to be expressed periodically in the mouse embryo PSM (Jho et al. 2002; Leung et al. 2002; Aulehla et al. 2003). This implicated the Wnt/β-catenin pathway as a novel player in the mouse segmentation clock (Aulehla et al. 2003). While mice mutant for the cyclic gene *Axin2* do not exhibit mesodermal defects, mice mutant for *Axin1* (*Fused*) exhibit segmentation defects (Zeng et al. 1997; Yu et al. 2005).

In addition to *Axin2*, several other cyclic Wnt targets have been identified in a microarray screen aimed at identifying all cyclic genes in the mouse PSM transcriptome (Dequeant et al. 2006). These genes include several negative feedback inhibitors of the pathway, including: *Dickkopf homolog 1 (Dkk1)*, *Dapper homolog 1 (Dact1)*, and *Naked cuticle 1 (Nkd1)* (Ishikawa et al. 2004; Dequeant et al. 2006; Suriben et al. 2006). Inactivation of several of these inhibitors, such as *Dkk1* and *Dact1*, results in segmentation defects (Mukhopadhyay et al. 2001; MacDonald et al. 2004). Wnt ligands are distributed in a graded but steady fashion along the PSM. Thus, as proposed for Notch oscillations, despite the presence of a constant (nonperiodic) Wnt signaling input in the PSM, oscillations of Wnt target transcription could be triggered by the periodic activation of unstable negative feedback inhibitors (Goldbeter and Pourquié 2008). However, as discussed further, periodic expression of cyclic genes, such as *Dkk1*, is still detected in mouse mutants that express a constitutively stable version of β-catenin in the PSM, thus, arguing against the role of periodic β-catenin destabilization in the clock pacemaker (Aulehla et al. 2008; Dunty et al. 2008).

The Fgf pathway constitutes a third signaling pathway activated periodically in the mouse and chicken PSM. Upon binding by Fgf ligands, the Fgf receptors (*FgfR1–4* in mammals) dimerize and through transphosphorylation, initiate several transduction cascades that include: (1) MAP kinase or ERK kinase (MEK), Extracellular-signal-regulated kinase/Mitogen-activated protein kinase (ERK/MAPK) cascade, (2) the PI3 Kinase cascade, and (3) the Phospholipase Cγ- (PLCγ) cascade, ultimately leading to the activation of specific target genes (Bottcher and Niehrs 2005). Cyclic expression of the Fgf targets *Snail homolog 1* and *2* (*Snail1* and *-2*), *Sprouty homolog 2* (*Spry2*), and *Dual specificity phosphatase 4* and *6* (*Dusp4* and *-6*) suggests that the Fgf signaling pathway is activated periodically in the posterior PSM (Dale et al. 2006; Dequeant et al. 2006; Niwa et al. 2007). This periodic regulation of Fgf signaling is further supported by the dynamic phosphorylation of ERK in the mouse PSM (Niwa et al. 2007). As for the Wnt and Notch pathways, many of the Fgf pathway cyclic genes are negative feedback inhibitors. Fgf signaling has been proposed to control the initiation of *Hes7* oscillations in the tail bud, while Notch signaling would maintain these oscillations in the PSM (see Fig. 3A) (Niwa et al. 2007). Thus, in the posterior PSM, Fgf-dependent *Hes7* expression controls the cyclic activation of the Fgf inhibitor *Dusp4*, hence, potentially driving the periodic inhibition of this pathway (Niwa et al. 2007).

Does the Cyclic Gene Network Act as the Clock Pacemaker?

The complex epistatic relationships and multiple cross talks between the Notch, Fgf, and Wnt signaling pathways in the PSM make the analysis of their respective contributions to the segmentation clock mechanism particularly challenging. Global analysis of cyclic gene expression reveals that Notch- and Fgf-related cyclic genes oscillate mostly in antiphase to Wnt-cyclic genes, suggesting a cross talk between these signaling pathways (Dequeant et al. 2006; Goldbeter and Pourquié 2008).

Oscillations of *Axin2* and *Spry2* are maintained in the mouse *RBP-Jk* mutants and in transgenic mouse embryos constitutively activating Notch in the PSM (Aulehla et al. 2003; Hirata et al. 2004; Dequeant et al. 2006; Feller et al. 2008). In contrast, *Lfng* and *Axin2* oscillations are disrupted in the mouse *Wnt3a* hypomorphic *vestigial tail* (*vt*) mutant (Aulehla et al. 2003). Together, these findings argue that Notch does not act as the clock pacemaker and that Wnt signaling acts upstream of the Notch oscillations.

The identification of many cyclic Wnt pathway negative feedback inhibitors raises the possibility that the periodic β-catenin destabilization, which should result from the action of these negative feedback loops, could act as the clock pacemaker that controls oscillations of the Notch and Fgf pathways. However, using a transgenic reporter mouse in which the *Lfng* promoter is fused to the yellow fluorescent protein Venus, *Lfng* has been shown to oscillate with the same periodicity in β-catenin gain-of-function mutants and in wild-type mice (Aulehla et al. 2008). Therefore, periodic β-catenin production is not required to control the rhythmicity of Notch activation in the PSM.

Conditional deletion of *FgfR1* in the PSM abolishes oscillations of *Hes7, Lfng, Axin2*, and *Spry2* (Niwa et al. 2007; Wahl et al. 2007). Similarly, in vitro treatment of mouse embryos with the Fgf inhibitor SU5402 leads to a rapid arrest of *Axin2* and *Spry2* oscillations; whereas, *Hes7* and *Lfng* oscillations cease only after a delay of more than one cycle (Niwa et al. 2007; Wahl et al. 2007). These results are consistent with a role for Fgf signaling in the control of the Notch and Wnt oscillations. However, introducing a constitutively stable version of β-catenin in a mutant mouse embryo, in which *FGFR1* is conditionally deleted in the PSM, restores the formation of the *Lfng* stripes of expression (Aulehla et al. 2008). These embryos show constitutive nuclear β-catenin expression and lack Fgf signaling in the PSM; yet, they appear to be capable of producing *Lfng* oscillations. Therefore, this result argues against a role for an Fgf-based negative feedback loop in the pacemaker of the segmentation clock.

Taken together, these data support the argument that none of the three signaling pathways periodically activated in the PSM individually appear to act as a global clock pacemaker. Hence, this raises doubts about the current models in which the periodic gene expression associated with the segmentation clock is presented as resulting from the dynamic properties of the cyclic gene network (Dequeant and Pourquié 2008). In amniotes, it is possible that each subnetwork has the capacity to generate its own oscillations independent of the oscillations of the other subnetworks—while coupling among the subnetworks entrain them to each other (Goldbeter and Pourquié 2008; Ozbudak and Pourquié 2008). Alternatively, it is possible that the network of cyclic genes that underlie the segmentation clock is entrained by an outside pacemaker that remains to be identified.

Defects in the Segmentation Clock Lead to Congenital Scoliosis in Humans

Congenital abnormalities in vertebral segmentation occur in a wide variety of rare but well-characterized disorders, encompassing many diverse and poorly understood phenotypic patterns (Turnpenny et al. 2007). Congenital forms of scoliosis involve structural malformations of the spine that are visible on radiographs and include segmental abnormalities such as hemivertebrae, wedge-shaped vertebrae, vertebral fusions, and bars. These observations in patients with congenital scoliosis are suggestive of various abnormalities that may occur during early developmental patterning (see Fig. 3B,C). A new nomenclature, developed by the International Consortium for Vertebral Anomalies and Scoliosis (ICVAS), has been created to better describe these conditions (Turnpenny et al. 2007).

Manifestations leading to a diagnosis of congenital scoliosis include: (1) generalized vertebral segmentation abnormalities as observed in patients with spondylocostal dysostosis (SCD) or spondylothoracic dysostosis (STD), (2) regionalized conditions, such as Klippel-Feil syndrome (which affects only the cervical region), and (3) conditions that involve only one or two vertebrae. Congenital scoliosis can also be associated with anomalies in other organ systems, most frequently involving the renal, cardiac, or neural systems. In some cases, patients with congenital scoliosis with abnormalities of the chest wall present a major surgical challenge. A better understanding and increased knowledge of the disease mechanism(s) will aid in improving the prediction of the clinical course of the disease, particularly in children.

Although most cases of congenital scoliosis were previously thought to be sporadic, recent evidence suggests that a considerable genetic component may be involved. Strikingly, thus far, the four mutations associated with congenital scoliosis in humans have been identified in genes associated with the segmentation clock mechanism (see Fig. 3A). These mutations result in monogenic autosomal recessive forms of SCD (Turnpenny et al. 2007; Sparrow et al. 2008). Homozygosity mapping and linkage analysis in consanguineous Arab-Israeli and Pakistani pedigrees with a particular form of SCD (SCDO1 [MIM 277300]) led to the discovery of multiple mutations in the Notch ligand *DLL3* (Bulman et al. 2000). These *DLL3* mutations result in abnormal vertebral segmentation throughout the entire spine with all vertebrae losing their normal form and regular three-dimensional shape (see Fig. 3B). A second, milder form of SCD (SCDO2 [MIM 608681]) has been associated with a 4-base-pair duplication in the *MESP2* gene that is essential for normal segmentation (discussed later in this chapter) (Whittock et al. 2004). A mutation in *LFNG* was identified in members of one family with SCD (SCDO3 [MIM 609813]) (Sparrow et al. 2006). A mutation in the *HES7* gene was also found to be associated with a case of SCD (SCDO4) (Sparrow et al. 2008). The identified mutation impairs HES7 protein to heterodimerize with the E47 cofactor. In addition, a recessive null mutation in the *MESP2* gene was also identified in patients with STD (Jarcho-Levin syndrome) (see Fig. 3C) (Cornier et al. 2008). Also, various mutations of the Notch ligand *JAGGED1* have been associated with the autosomal dominant Alagille syndrome in which misshaped vertebrae (butterfly vertebrae) are observed (AGS [MIM 118450]) (Li et al. 1997; Oda et al. 1997; Joutel and Tournier-Lasserve 1998). Together, these data strongly support the argument that human somitogenesis also relies on molecular mechanisms similar to those identified in mouse and chicken embryos.

The Wavefront: Translating the Periodic Signal into Repeated Segments

The Determination Front Controls PSM Maturation

The concept of a "wavefront," as originally proposed by Cooke and Zeeman, suggests that a transitional zone exists in the PSM where cells that reach this level undergo a "catastrophe" corresponding to a sudden change in cell properties and leading to somite formation (Cooke and Zeeman 1976). Several studies have demonstrated the existence of such a transitional region in the PSM, called the determination front (Dubrulle et al. 2001; Sawada et al. 2001). This region is located at approximately the S-3/S-2 level in chicken embryos (presumptive somite -3/2; nomenclature according to Pourquié and Tam [2001]) and corresponds to an important transition in gene regulation and cellular properties (see Figs. 1C and 2). Grafting experiments in chicken embryos reveal that the segmental pattern is established only anterior to the determination front (Dubrulle et al. 2001).

The determination front is positioned by antagonistic gradients of Fgf, Wnt, and retinoic acid (RA) signalings, and regresses posteriorly as the embryo elongates along the anteroposterior axis (see Fig. 2E–G,I) (Aulehla and Herrmann 2004; Dubrulle and Pourquié 2004a; Moreno and Kintner 2004). Mathematical modeling of this system led to the proposal that the determination front is associated with a bistability window defined by the mutually antagonistic gradients of Fgf and RA (Goldbeter et al. 2007; Dequeant and Pourquié 2008). In this window, cells are poised to switch abruptly from one steady state to the other in response to a triggering signal, such as the signaling pulse delivered by the segmentation clock. This abrupt transition could explain the sudden rhythmic appearance of stripes of gene expression at the determination front level. Molecularly, segmental genes (such as *mesoderm posterior 2* [*Mesp2*]) become derepressed at the determination front level (Delfini et al. 2005), and stripes of *Mesp2* gene expression form in a rhythmic sequence in response to the clock signal (see Fig. 2H,I) (Saga et al. 1997). As we shall see, the stripes of *Mesp2* expression provide the template from which the morphological segments—the somites—will form (Morimoto et al. 2005; Saga 2007).

Fgf/Wnt Signaling Gradients Control the Maturation of the PSM

Parallel posterior-to-anterior gradients of Fgf and Wnt signaling (evidenced by graded phosphorylated ERK and nuclear β-catenin, respectively) are established in response to the graded expression of secreted ligands, such as Fgf8 and Wnt3a, along the PSM (see Fig. 2E,F) (Dubrulle et al. 2001; Sawada et al. 2001; Aulehla et al. 2003; Dubrulle and Pourquié 2004b; Delfini et al. 2005; Aulehla et al. 2008). In the posterior PSM, cells are exposed to a high level of Fgf and Wnt activity, and are maintained in an immature, undifferentiated state (Dubrulle et al. 2001; Sawada et al. 2001; Aulehla et al. 2008; Dunty et al. 2008). The transition from the immature to the competent state of PSM cells can be visualized by the down-regulation of the *Mesogenin1* (*Msgn1*) gene at the level of the determination front (Buchberger et al. 2000; Yoon et al. 2000).

Molecular Mechanisms Generating the Signaling Gradients. The posterior gradients and thus the determination front move posteriorly in the wake of axis elongation. This posterior displacement of the *Fgf8* gradient has been studied and shown to rely on an original mechanism involving *Fgf8* mRNA decay (Dubrulle and Pourquié 2004b). Transcription of the *Fgf8* mRNA is restricted to the PSM precursors in the tail bud, and it ceases when their descendents enter the posterior PSM. Thus, as the axis elongates, cells gradually become located more anteriorly in the PSM, and their *Fgf8* mRNA content progressively decays (see Fig. 2F). This results in the establishment of an *Fgf8* mRNA gradient that is converted into a graded ligand distribution pattern and in graded Fgf activity along the PSM (Sawada et al. 2001; Dubrulle and Pourquié 2004b; Delfini et al. 2005). A similar mechanism is assumed to be responsible for establishing the Wnt gradient (Aulehla et al. 2003). As a result of the progressive decay of the *Fgf/Wnt* mRNA and proteins in PSM cells, the determination front is constantly displaced posteriorly (Goldbeter et al. 2007). In the zebrafish, chicken, mouse, and snake, the regression speed of the determination front (marked by the anterior boundary of the *Msgn1* expression domain) during somitogenesis is similar to the speed of somite formation (Gomez et al. 2008). Thus, during embryogenesis, this original gradient formation mechanism ensures a tight coupling between axis elongation and segmentation.

Fgf Signaling and Maturation of the PSM. The role of the Fgf signaling gradient in positioning the determination front was first demonstrated by experiments challenging the slope of the signaling gradient in chicken embryos by grafting FGF8-soaked beads next to the PSM or by overexpressing an *Fgf8*-expressing construct in the PSM by electroporation (Dubrulle et al. 2001). This resulted in an anterior extension of posterior PSM markers (such as *Brachyury*), in down-regulation of segmentation and differentiation markers (such as *Paraxis*, *Mesp2*, or *myogenic differentiation 1* [*MyoD*]), and in the formation of smaller somites (Dubrulle et al. 2001; Delfini et al. 2005). Conversely, inhibition of Fgf signaling was achieved by treating chicken embryos with pharmacological inhibitors (Dubrulle et al. 2001). This resulted in a posterior shift of the anterior boundary of genes associated with a posterior PSM identity (such as *Fgf8*), and in the formation of larger somites. Together, these data suggested that high levels of Fgf signaling are required to maintain the posterior identity of PSM cells (Dubrulle et al. 2001). This further led to the idea that the progressive decrease in Fgf signaling activity along the PSM defines a specific threshold below which the cells become competent to respond to the signaling pulse delivered by the segmentation clock. Changing the position of the threshold by acting on Fgf levels leads to a change in the position of the somite boundary position, resulting in smaller or larger somites. The position of this threshold was proposed to correspond to the determination front (Dubrulle et al. 2001).

Fgf8 and *FgfR1* have emerged as key players during embryonic axis elongation and somitogenesis, with conserved expression domains and function in the vertebrate PSM (see Fig. 2F) (Deng et al. 1994; Yamaguchi et al. 1994; Crossley and Martin 1995; Sun et al. 1999; Dubrulle et al. 2001; Sawada et al. 2001; Hoch and Soriano 2006; Niwa et al. 2007; Wahl et al. 2007). Fgf signaling in the PSM activates the ERK/MAPK and PI3K-Akt cascade (Yamaguchi et al. 1992, 1994; Deng et al. 1994; Crossley and Martin 1995; Ciruna et al. 1997; Sawada et al. 2001; Dubrulle and Pourquié 2004b; Delfini et al. 2005; Lunn et al. 2007; Niwa et al. 2007; Wahl et al. 2007). In mice mutant for *Fgf8* and *FgfR1*, loss of function of Fgf signaling results in severe gastrulation defects (Deng et al. 1994; Sun et al. 1999). Conditional deletion of *Fgf8* in the PSM does not lead to abnormal somitogenesis, suggesting that the *Fgf3*, *-4*, and *-17* ligands, which are also expressed in the mouse tail bud and PSM, probably act redundantly with *Fgf8* to pattern the paraxial mesoderm (Bottcher and Niehrs 2005; Perantoni et al. 2005; Wahl et al. 2007). In contrast, *FgfR1* is the only Fgf receptor expressed in the mouse PSM (Wahl et al. 2007). Conditional deletion of *FGFR1* in the PSM results in severe segmentation defects and axis truncation, suggesting that Fgf signaling in the PSM is mediated essentially by FgfR1 (Niwa et al. 2007; Wahl et al. 2007).

Wnt Signaling and Maturation of the PSM. In amniotes, Wnt signaling in the posterior PSM involves *Wnt3a* and *-5a*, and the *Frizzled1, -2,* and *-7* receptors (see Fig. 2E) (Takada et al. 1994; Yamaguchi et al. 1999a; Cauthen et al. 2001; Aulehla et al. 2003). Loss of function of Wnt signaling in mice mutant for *Wnt3a* or its downstream effectors *Lef1* and *Tcf1* causes severe axis truncation immediately following the forelimb level (Takada et al. 1994; Galceran et al. 1999). Similarly, axis truncations are observed when β-catenin function is deleted in the PSM (Aulehla et al. 2008; Dunty et al. 2008). Mice bearing the *Wnt3a* hypomorphic mutation *vt* are truncated at the tail level, whereas mice bearing the *vt* allele and the *Wnt3a* knockout allele exhibit a more severe truncation phenotype, and a variable number of lumbar (but not caudal) vertebrae are formed (Greco et al. 1996). Interestingly, mutations of *Wnt3a, Lef1/Tcf1,* or their targets *T* and *Tbx6,* also cause the posterior paraxial mesoderm to switch to a neural tube fate (Takada et al. 1994; Yoshikawa et al. 1997; Chapman and Papaioannou 1998). These results clearly indicate that Wnt3a, acting through the canonical Wnt signaling pathway, is required for production of the posterior paraxial mesoderm. In humans, abnormalities of this gradient system could result in congenital anomalies such as caudal agenesis, a rare condition that occurs sporadically but is more specific to pregnant women with diabetes mellitus (Catala 2002).

In mice, nuclear β-catenin forms a clear gradient parallel to the *Wnt3a* mRNA gradient in the posterior PSM (Aulehla et al. 2008). β-catenin conditional loss- and gain-of-function studies in mouse embryos demonstrate that the level of nuclear β-catenin controls PSM cell maturation (Aulehla et al. 2008; Dunty et al. 2008). Gain of function of Wnt/β-catenin signaling leads to an anterior expansion of the expression domain of posterior genes, such as *Brachyury, Msgn1,* or *Tbx6.* Furthermore, in these experiments, both segmentation and the onset of the differentiation program in the PSM are blocked (Aulehla et al. 2008; Dunty et al. 2008). The most striking effect in the gain-of-function mutant is an anteroposterior extension of the oscillatory domain that results in a multistripe, oscillatory expression pattern in the enlarged PSM (Aulehla et al. 2008; Dunty et al. 2008). These results imply that Wnt/β-catenin controls the size of the oscillatory domain and maintains the immature state of posterior cells and, thus, controls the onset of segmentation and differentiation in the PSM (Aulehla et al. 2008). Therefore, Wnt signaling also plays a major role in the positioning of the determination front. Whether Wnt acts in parallel or synergistically with Fgf in this process is not understood.

Wnt signaling is regulated at multiple levels in the PSM. Secreted WNT antagonists (such as Dkk1 and Secreted frizzled-related proteins [Sfrps]), are implicated in axis elongation and somitogenesis regulation (Hoang et al. 1998; Baranski et al. 2000; Ladher et al. 2000; Terry et al. 2000; Kawano and Kypta 2003). Dkk1 interacts with the coreceptor LRP6 by competitive binding with WNT ligands (Sëmenov et al. 2001; Satoh et al. 2006). Mice mutant for *Dkk1* and *LRP6* exhibit segmentation defects (Mukhopadhyay et al. 2001; Kokubu et al. 2004; MacDonald et al. 2004). Sfrps function by sequestering WNT ligands (Hsieh et al. 1999; Kawano and Kypta 2003; Satoh et al. 2006). Mice compound mutant for *Sfrp1,* *-2,* and *-5* exhibit defects in mesoderm segmentation and a shortening of the anteroposterior axis (Satoh et al. 2006, 2008).

Cross Talk between Fgf and Wnt Signaling. The interactions between Fgf and Wnt signaling in the PSM remain unclear. In mice, *Fgf8* expression is absent from the *Wnt3a* hypomorph *vt* mutants, suggesting that Wnt signaling acts upstream of Fgf signaling in the tail bud (Aulehla et al. 2003, 2008). However, gain of function of β-catenin in the PSM of mouse embryos results only in a partial gain of function of Fgf signaling, suggesting that Wnt signaling is necessary but insufficient for Fgf signaling in the PSM (Aulehla et al. 2008).

In addition, the Wnt and Fgf pathways have been shown to simultaneously control the expression of a number of transcription factors—*Brachyury/T, Tbx6, Gbx2, Cdx1, -2, and -4,* and *Msgn1* genes—that in turn control different aspects of PSM maturation (Pownall et al. 1996; Yamaguchi et al. 1999a,b; Arnold et al. 2000; Ciruna and Rossant 2001; Lohnes 2003; Bussen et al. 2004; Hofmann et al. 2004; Pilon et al. 2006, 2007; Wahl et al. 2007; Wittler et al. 2007; Wardle and Papaioannou 2008). Mutations of several of these genes cause similar truncation phenotypes as those observed in mutations in the Fgf or Wnt pathways (Herrmann et al. 1990; Schulte-Merker et al. 1994; Wilson et al. 1995; Chapman et al. 1996; Chapman and Papaioannou 1998; White et al. 2003).

How Fgf and Wnt signalings elicit a response, which is graded for some genes and cyclic for others, is not understood. Additionally, the control of the surface distribution of the Fgf and Wnt receptors (FgfR and Fz, respectively) allows spatial restriction of the signaling along the PSM. *Shisa* is a conserved gene family (*Shisa1-3*) that encodes endoplasmic reticulum-localized chaperone proteins (He 2005; Yamamoto et al. 2005; Filipe et al. 2006; Nagano et al. 2006; Furushima et al. 2007). *Shisa2* is expressed in the anterior PSM and acts as a negative feedback inhibitor of Fgf and Wnt signalings by regulating the chaperone machinery to prevent FgfR and Fz maturation and cell-surface expression (Yamamoto et al. 2005; Nagano et al. 2006). In Xenopus, *Shisa2* knockdown results in an anterior shift of the determination front, resulting in delayed PSM maturation (Nagano et al. 2006). Thus, several distinct molecular mechanisms can modulate the Fgf/Wnt signaling gradients in the PSM.

Other Signaling Pathways Involved in PSM Patterning. The noncanonical Wnt pathway has also been implicated in the control of paraxial mesoderm differentiation. In the mouse embryo, *Wnt5a* is expressed in a caudal-to-rostral gradient in the PSM, and mice mutant for *Wnt5a* exhibit a shorter PSM, skeletal defects, and axial truncation (Yamaguchi et al. 1999a; Schwabe et al. 2004). This phenotype is observed also in mice double mutant for *Dsh1* and *-2*, which are downstream transducers of Wnt signaling (Hamblet et al. 2002; Wang et al. 2006). In humans, disorders such as the Robinow syndrome (RRS [MIM 268310]), are associated with autosomal recessive mutations in the receptor tyrosine kinase orphan receptor *ROR2*, resulting in spine abnormalities that include hemivertebrae (Afzal et al. 2000; van Bokhoven et al. 2000; Patton and Afzal 2002). ROR2 has been shown to interact with Wnt5a (Oishi et al. 2003). Among the identified human *ROR2* mutations, some have been shown to affect specifically the targeting of the ROR2 receptor to the cell surface (Chen et al. 2005). In amniotes, *Ror2* is expressed in the PSM (Al-Shawi et al. 2001; Matsuda et al. 2001), and mice mutant for *Ror2* or double mutant for *Ror1* and *-2* exhibit several defects including vertebral malformations reminiscent of RRS (DeChiara et al. 2000; Takeuchi et al. 2000; Nomi et al. 2001; Schwabe et al. 2004; Raz et al. 2008).

Other evidence suggests that the transforming growth factor-β (TGF-β) signaling, in addition to being essential for mesoderm specification (Miura et al. 2006; Shen 2007), may be implicated in positioning the determination front. In the mouse, *Smad interacting protein1* (*SIP1*) is expressed segmentally in the anterior PSM, and mice mutant for *SIP1* exhibit a shortened axis and form only a few anterior somites (Maruhashi et al. 2005).

Progressive Epithelialization of the PSM Accompanies Somite Formation

The posterior PSM is a loose mesenchymal tissue. As the cells reach the anterior PSM, they undergo a progressive mesenchymal-to-epithelial transition that involves changes in cell shape, cell substrate, and cell–cell interactions (Fig. 4A–C) (Christ et al. 2007). As the PSM cells epithelial-

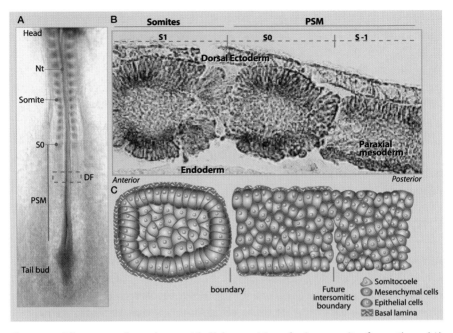

Figure 4. The mesenchymal-to-epithelial transition during somite formation. (*A*) Dorsal view of a two-day-old chicken embryo highlighting the distribution of the epithelial somite pairs, bilaterally to the neural tube (Nt). The posterior PSM is a loose mesenchymal tissue produced by the tail bud. The PSM initiates epithelialization around the level of the determination front (DF, boxed), and eventually forms an epithelial pair of somites (S0) at its anterior extremity. (*B*) Saggital section of the paraxial mesoderm showing the cellular polarization associated with the transition of PSM to somites. Tissue section stained for *N-cadherin* expression by in situ hybridization. (*C*) Corresponding schematic representation of the cellular epithelialization during somite boundary formation. Epithelialization is associated with the deposition of a basal lamina on the outer surface of the forming somite. Somitocoele cells do not epithelialize. (S-1) Next somite to form, (S0) forming somite, (S1) newly formed somite (according to Pourquié and Tam 2001).

ize, they adopt an elongated and polarized shape, and their basolateral side comes in contact with the nascent basal lamina on the outer surface of the forming somite while their apical domain establishes adherent junctions with somitocoele cells (Fig. 4B,C). In the chicken embryo, the first signs of epithelialization become evident around the determination front level, with the cell nuclei aligning dorsally and ventrally in the PSM (Duband et al. 1987).

The *Snail* gene family, coding for zinc-finger transcriptional repressors, has been implicated in the control of the mesenchymal state of several cell types by negatively regulating the expression of cadherins (Cano et al. 2000). In the mouse and chicken, these genes are expressed in a periodic fashion in the posterior mesenchymal PSM under the control of Fgf signaling (Dale et al. 2006). In chicken embryos, overexpression of *Snail2* blocks segmentation and epithelialization in the anterior PSM, suggesting that down-regulation of *Snail* genes is required in the anterior PSM for cells to become epithelial (Dale et al. 2006).

In the PSM, the expression of the conserved bHLH transcription factor *Paraxis* (*Tcf15*) begins around the level of the determination front and, hence, coincides with the onset of PSM epithelialization. *Paraxis* expression is maintained in epithelial somites, where it becomes restricted to the differentiating dermomyotome (Quertermous et al. 1994; Blanar et al. 1995; Burgess et al. 1995; Barnes et al. 1997; Shanmugalingam and Wilson 1998; Carpio et al. 2004; Tseng and Jamrich 2004; Wilson-Rawls et al. 2004). The dorsal ectoderm has been shown to be required for proper PSM epithelialization, possibly via *Paraxis* activation in the PSM (Sosic et al. 1997; Palmeirim et al. 1998; Correia and Conlon 2000; Schmidt et al. 2004; Linker et al. 2005). Mouse embryos mutant for *Paraxis* show specific defects in somite epithelialization, and newborn mutant mice exhibit vertebral fusions and chondrogenesis defects (Burgess et al. 1996; Takahashi et al. 2007a). Thus, *Paraxis* controls the morphological transition from mesenchyme to epithelium during somite formation, as well as later steps of spine development (Takahashi et al. 2007a). In addition to their role during somite differentiation (discussed later in this chapter), the transcription factors *Mesenchyme homeobox 1* and *2* (*Meox1* and *-2*) and *Paired box gene 3* (*Pax3*) regulate the epithelialization of the forming somite (Schubert et al. 2001; Mankoo et al. 2003).

The RA Signaling Gradient Antagonizes Fgf/Wnt Signaling

RA Signaling and Maturation of the PSM. RA is a Vitamin A derivative that exhibits pleiotropic effects during embryonic development (Niederreither and Dollé 2008). Signal transduction requires direct RA binding to its nuclear receptor, formed by a heterodimer of RA and Retinoid X receptors (RARs and RXRs). These receptors act as ligand-dependent transcriptional activators of genes that contain RA-response elements (RAREs) (Niederreither and Dollé 2008). RA signaling is regulated by controlling the amount of biologically active RA. While the retinal dehydrogenases RALDH1-4 and CYP1B1 enzymes synthesize RA from precursors, the CYP26 cytochrome p450 family of proteins (*Cyp26a1*, *-b1*, and *-c1*) is implicated in RA catabolism in the vertebrate embryo (Chambers et al. 2007; Vilhais-Neto and Pourquié 2008). Importantly, gene expression and loss-of-function studies demonstrate that *Raldh2* (*Aldh1a2*), which catalyzes the last step of RA biosynthesis, is expressed in somites and is the major source of RA during early embryogenesis (see Fig. 2G). Other enzymes, such as RDH10 and CYP1B1, have been implicated in RA production in the embryo (Chambers et al. 2007; Sandell et al. 2007), but their role in paraxial mesoderm patterning has not been established.

In amniotes, RA signaling forms a decreasing rostrocaudal gradient that is opposite to the Fgf and Wnt gradients in the PSM (Begemann and Meyer 2001; Begemann et al. 2001; Diez del Corral et al. 2003; Moreno and Kintner 2004; Sirbu and Duester 2006). In the trunk region, *Raldh2* is expressed in the anteriormost PSM and segmented region, and is excluded from the tail bud and posterior PSM (see Fig. 2G) (Niederreither et al. 2002b). Using a RARE-LacZ reporter mouse, RA signaling was found to be restricted to the anterior PSM and segmented region, and was absent from the posterior PSM and tail bud where *Cyp26a1* is expressed downstream of Fgf (Rossant et al. 1991; Abu-Abed et al. 2001; Vermot et al. 2005). Analysis of mice deficient for RA production (null for *Raldh2*) or of Vitamin A-deficient (VAD) quail embryos reveals that perturbation of RA signaling results in mesoderm segmentation defects (Niederreither et al. 1999; Diez del Corral et al. 2003; Vermot et al. 2005). Moreover, mice or zebrafish mutant for *Cyp26a1* exhibit axis truncation, suggesting that the regulation of the amount of RA in the embryo is critical for proper axis elongation (Abu-Abed et al. 2001; Sakai et al. 2001). Interestingly, mice mutant for *Cyp26a1* exhibit down-regulation of Wnt signaling targets such as *T*. Moreover, ectopic neural structures form in place of the paraxial mesoderm, a phenotype also observed in a number of Wnt signaling mutants (Sakai et al. 2001). Thus, in the PSM, increased RA signaling correlates with a loss of Wnt signaling.

In the chicken embryo, treatment of posterior PSM explants with RA agonists can down-regulate *Fgf8* expression, while a graft of an FGF8-soaked bead in the PSM represses *Raldh2* expression (Diez del Corral et al. 2003). Furthermore, in mice mutant for *Raldh2* and in chicken or quail embryos deprived of RA, the *Fgf8* domain is extended along the PSM (Diez del Corral et al. 2003; Vermot and Pourquié 2005). Experiments in chicken and Xenopus embryos indicate that RA can activate transcription of key segmentation genes, such as the *Mesp2* homologs, either directly or by counteracting Fgf signaling that represses their expression (Moreno and Kintner 2004; Delfini et al. 2005). Together, this led to the proposal that the mutual inhibition of the Fgf and RA gradients is involved in positioning the determination front (see Fig. 2). Nevertheless, somites do form in the mouse *Raldh2* mutant in which RA signaling is not detected (Niederreither et al. 2002a). Furthermore, in the *FgfR1* conditional knockout, no significant posterior shift of the RARE-lacZ domain is observed, suggesting that Fgf is not the only antagonist to the RA gradient (Wahl et al. 2007). Whether the posterior Wnt gradient, by itself, can antagonize RA signaling remains to be investigated.

RA Signaling Controls Somite Bilateral Symmetry. Whereas the spine is a symmetrical structure, obvious left–right asymmetries are observed in the positioning and structure of vertebrate internal organs, such as the heart and liver (Shiratori and Hamada 2006). Establishment of these asymmetries is downstream of a left–right signaling pathway that is active during gastrulation (Shiratori and Hamada 2006). The secreted factor Nodal acts on the left side of the embryo to activate the expression of specific genes, such as *Paired-like homeodomain transcription factor 1 (Pitx1)*, in the left lateral plate (Shiratori and Hamada 2006; Speder et al. 2007). These initial asymmetries control the subsequent and specific left–right development of the internal organs. Strikingly, in normal embryos, the asymmetric signals produced by the node cross the forming paraxial mesoderm without affecting the perfect symmetry of paraxial mesoderm patterning and development (Yokouchi et al. 1999). During early somitogenesis, RA signaling has been shown to play a role in maintaining the bilateral coordination of paraxial mesoderm development (Kawakami et al. 2005; Vermot et al. 2005; Vermot and Pourquié 2005; Sirbu and Duester 2006). Thus, mice mutant for *Raldh2* or chicken embryos treated with the compound Disulfiram (in which RA synthesis is blocked) exhibit asymmetric somite formation at the cervical level. This asymmetry of somite formation is downstream of the left–right machinery, as it can be reversed by changing the situs of the embryos (Vermot et al. 2005; Vermot and Pourquié 2005). Thus, RA signaling acts in the paraxial mesoderm by buffering the asymmetric signal generated by the left–right machinery. While this appears to be a conserved feature among vertebrates, the underlying molecular mechanisms remain unknown (Kawakami et al. 2005; Vermot et al. 2005; Vermot and Pourquié 2005; Sirbu and Duester 2006). In humans, a majority of patients with idiopathic scoliosis exhibit a spine curvature toward the right, suggesting an underlying defect in the left–right symmetry (Ahn et al. 2002). While the molecular mechanisms underlying these diseases have not been identified, pathways controlling the left–right symmetry of the spine during embryogenesis, such as the RA pathway, are indeed attractive candidates.

Segmental Patterning and Somite Boundary Specification

The segmentation process is initiated at the determination front level when a segment-wide domain acquires a distinct genetic identity that isolates it from the posterior PSM (see Figs. 1C and 2). This change can be visualized, by activation of the transcription of the *Mesp* gene family, as a striped domain that prefigures the future segment (see Fig. 2H). *Mesp* family genes (*Mesp1* and *-2* in the mouse and *Meso1* and *-2* in the chicken) code for bHLH transcription factors, which show a conserved expression pattern and function during somitogenesis (Saga et al. 1996, 1997; Buchberger et al. 1998, 2002; Oginuma et al. 2008). *Mesp1* and *Mesp2* are first expressed as a dynamic stripe located at the determination front level and then become restricted to the future rostral somitic compartment in more anterior regions of the PSM. *Mesp1* and *Mesp2* gene expression becomes down-regulated when the somite forms. While high Fgf signaling in the posterior PSM inhibits *Mesp2* expression, Tbx6 and Notch signaling synergize to induce *Mesp2* expression at the determination front level (see Figs. 3A and 5) (Takahashi et al. 2003; Delfini et al. 2005; Yasuhiko et al. 2006; Wahl et al. 2007). Furthermore, *Mesp2* gene activation requires the transcription factors *Forkhead box c1* (*Foxc1*) and *Forkhead box c2* (*Foxc2/Mfh1*) that are expressed anterior to the determination front (Miura et al. 1993; Winnier et al. 1997; Kume et al. 2001).

An important role for *Mesp2* is to position the future somitic boundary (Morimoto et al. 2005). As described earlier, Notch signaling oscillations in the posterior PSM generate waves of NICD production that control *Lfng* expression in the mouse (see Fig. 2A–D). When the NICD/*Lfng* wave reaches the determination front level, *Mesp2* becomes activated in the future segmental domain where it takes over *Lfng* regulation, thus stabilizing its expression in the *Mesp2* expression domain (see Figs. 2D,H and 5). Because Lfng negatively regulates Notch activation, this results in the creation of an interface between a *Mesp2*-positive domain (the future somitic domain) in which Notch activation is suppressed, and an adjacent posterior *Mesp2*-negative domain in which Notch is activated. This interface marks the level of the future somitic boundary (Fig. 5). This model is supported by grafting experiments in the chicken, which suggest that Lfng activity at the posterior border of the interface could instruct PSM fissure formation (Sato et al. 2002). However, mutant mouse embryos that overexpress NICD throughout the PSM still generate stripes of *Mesp2* expression and do form somites, although they exhibit rostrocaudal polarity defects (Feller et al. 2008). Similarly, overexpression of NICD in the zebrafish PSM does not prevent

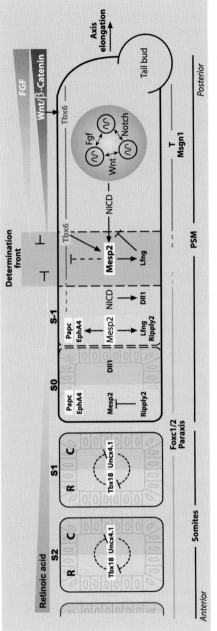

Figure 5. The Notch/*Mesp2* genetic network involved in segment formation and somite rostrocaudal patterning. The system of opposing gradients of Fgf/Wnt and RA plays a key role in positioning the determination front. The posterior PSM expresses a specific set of transcription factors including *Brachyury (T)*, *Tbx6*, and *Msgn1*, and undergoes periodic activation of the Notch, Wnt, and Fgf signaling pathways driven by the segmentation clock. At the determination front level, synergistic action of *Tbx6* and of the pulse of Notch signaling (NICD) downstream of the clock activates *Mesp2* in cells that have reached the determination front during the preceding oscillation cycle, in a striped pattern. Mesp2 activates *Lfng* in the future segmental domain, creating an interface between a domain where Notch is activated (gray) and a domain where Notch is inhibited (green). This interface marks the presumptive somite boundary. Rostrocaudal somite polarity is subsequently established in the newly specified segment by repressing *Mesp2* expression in the future caudal compartment (which reactivates Notch and expresses *Dll1*), while maintaining *Mesp2* in the rostral compartment. Mesp2 then activates downstream targets such as *EphA4* and *Papc. Ripply2* activation by *Mesp2* results in the termination of the segmentation program by a negative feedback loop mechanism. Concomitantly, PSM cells progressively acquire epithelial characteristics after the determination front. The anterior PSM expresses a distinct set of transcription factors including *Paraxis* and *Foxc1* and *-2*. The rostrocaudal polarity of the newly formed somite is maintained by the antagonism between *Tbx18* and *Uncx4.1*. (R) Rostral, (C) caudal. (Somite nomenclature according to Pourquié and Tam 2001).

the formation of somite boundaries (Ozbudak and Lewis 2008), arguing that more than an interface between NICD-positive and -negative domains is necessary for somite boundary specification.

The *Ripply1* and *-2* genes code for transcription factors that interact with the Groucho-related family of corepressors (Kawamura et al. 2005; Chan et al. 2006, 2007; Biris et al. 2007; Morimoto et al. 2007). *Ripply* genes are expressed in the anterior PSM downstream of *Mesp2* and *Tbox* genes (Kawamura et al. 2005; Morimoto et al. 2007; Hitachi et al. 2008). In the mouse, *Ripply2* acts as a negative feedback inhibitor, turning off *Mesp2* expression once the segment is specified (see Fig. 5) (Morimoto et al. 2007). Therefore, *Ripply* genes are an essential component in the termination of the segmentation program, allowing subsequent somite differentiation to proceed (Takahashi et al. 2005, 2007b).

Morphogenesis of the Epithelial Somite

Somite boundary morphogenesis primarily involves a localized medio-lateral fissure that forms across the epithelialized anterior PSM tissue (see Fig. 4B,C). This process has been examined in vivo using time-lapse video microscopy that reveals complex cellular rearrangements (Kulesa and Fraser 2002). It has been proposed that boundary formation involves cell repulsion and differential cell adhesion mechanisms driven by large families of conserved transmembrane molecules, such as Cadherins, Integrins, Eph/ephrins, Immunoglobulin-like Cell Adhesion Molecules (CAMs), and by cytoskeleton remodeling factors such as the Rho small GTPases.

The Cadherins are transmembrane, calcium-dependent, cell-adhesion proteins required for cell sorting, boundary formation, epithelial integrity, and cell movements during embryo morphogenesis (Halbleib and Nelson 2006). Cell–cell contacts and adhesion in epithelial cells are primarily mediated by adherens junction complexes in which classical cadherins are the major cell–cell-connecting elements (Gumbiner 2005; Halbleib and Nelson 2006). At the cell membrane, cadherins are tethered to the actin cytoskeleton by a complex of proteins including α-catenin, β-catenin, and several actin filament-binding proteins (Gumbiner 2005). In amniotes, mesodermal tissues are devoid of E-cadherin and express predominantly N-cadherin (*Cdh2*) (Duband et al. 1987; Horikawa et al. 1999; Cano et al. 2000). N-cadherin is essential for proper somite formation, and the mouse null mutants exhibit significant defects in somite morphogenesis (Duband et al. 1987; Radice et al. 1997; Linask et al. 1998). Furthermore, in the double *N-cadherin* and *Cadherin11* mouse mutant, somites disaggregate into small cell clusters (Horikawa et al. 1999). These results suggest a functional overlap and compensation mechanism among the cadherin family members (Kimura et al. 1995; Horikawa et al. 1999).

The Integrin-extracellular matrix (ECM) interactions play a major role in cell-substrate adhesion. Integrins are transmembrane heterodimers that act as receptors for major components of the ECM, such as fibronectins, laminins, and collagens (Juliano 2002). The interactions between PSM cells and the fibronectin matrix are essential for proper somitogenesis (see Fig. 4C) (Ostrovsky et al. 1983; Lash et al. 1984; Lash and Yamada 1986; Duband et al. 1987; George et al. 1993; Georges-Labouesse et al. 1996). Mice mutant for *integrin α5* and *fibronectin1* exhibit severe defects in mesoderm formation and cell migration, resulting in disruption of somite formation (George et al. 1993; Georges-Labouesse et al. 1996; Goh et al. 1997; Yang et al. 1999). In the chicken embryo, fibronectin is produced mainly by the dorsal ectoderm, a tissue required for proper epithelialization of somites (Palmeirim et al. 1998; Correia and Conlon 2000; Rifes et al. 2007). Other integrin ligands, such as laminin1 and collagen IV, form the main components of somitic basal lamina (see Fig. 4C) (Duband et al. 1987). The anchoring of ECM components to the cell by integrins leads to the formation of focal adhesion sites that act as cell-adhesion and cell-signaling centers (Mitra et al. 2005). Mice mutant for *focal adhesion kinase* exhibit general mesoderm development defects that are similar to those observed in mice mutant for *laminin* (Ilic et al. 1995; Miner and Yurchenco 2004). Thus, multiple interactions between cell-surface integrins and the ECM components are essential for proper PSM and somite morphogenesis.

The small GTPases of the Rho family are major regulators of the actin cytoskeleton and of the endocytic traffic (Ellis and Mellor 2000). Among the Rho GTPases, Ras-related C3 botulinum substrate 1 (Rac1), cell division cycle 42 homolog (cdc42), and RhoA have been shown to control cytoskeletal dynamics, cell polarity, and motility (Malbon 2005). In chicken embryos, the effect of *Rac1* and *cdc42* overexpression suggests that these Rho GTPases control the epithelial versus mesenchymal status of cells in the anterior PSM (Nakaya et al. 2004). Thus, these small GTPases likely integrate Cadherin- and Integrin-mediated signals during somite morphogenesis.

FROM SOMITES TO VERTEBRAE

Resegmentation of Somites Generates the Vertebrae

As originally proposed by Remak (1850), one vertebra does not directly derive from one somite, but rather, from the fusion of the rostral compartment of one sclerotome to the caudal compartment of the following sclerotome through a process called resegmentation (Fig. 6) (Bagnall et al. 1988; Goldstein and Kalcheim 1992; Aoyama and Asamoto 2000; Christ et al. 2007). In contrast, the myotome maintains its original somitic segmentation that results in the intervertebral musculature connecting two successive vertebrae. While the concept of resegmentation has been challenged by some authors (Verbout 1976), experimental evi-

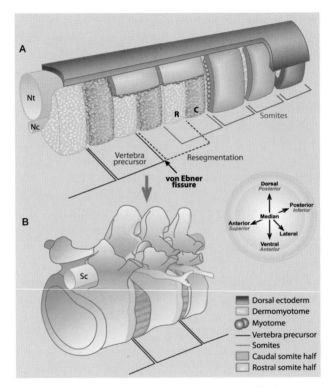

Figure 6. Resegmentation of the sclerotome and its contribution to vertebrae. (*A*) Schematic temporal sequence of sclerotome resegmentation (side view). Sclerotome rostral and caudal compartments are separated by the von Ebner fissure. The rostral compartment of one somite/sclerotome (yellow) fuses to the caudal compartment of the consecutive somite/sclerotome (blue) to form one vertebra. Thus, the somites and the vertebrae are out of register by one half of a segment. The dorsal ectoderm (brown) and dermomyotome that do not resegment (purple) have been removed to visualize the underlying sclerotome. (*B*) Fate of the rostral and caudal sclerotome compartments projected onto adult human vertebrae. Respective contribution of the somite caudal and rostral compartments is shown. The orientation of the embryonic axes is indicated in black bold in the circle and the corresponding medical terminology is shown in gray italics. (R) Rostral somite sclerotome compartments. (C) caudal somite sclerotome compartments, (Nt) neural tube, (Nc) notochord, (Sc) spinal cord.

dence largely supports this hypothesis (Bagnall et al. 1988, 1989; Ewan and Everett 1992). For example, quail somites grafted into a chicken embryo host (Bagnall et al. 1988, 1989; Huang et al. 2000a) or retroviral labeling of single somites (Ewan and Everett 1992) clearly indicate a contribution of each somite to two consecutive vertebrae.

The early rostrocaudal organization of the somite, first established in the anterior PSM, is maintained in the sclerotome in which the rostral and caudal compartments become separated by the von Ebner fissure (Figs. 6 and 7B). The rostral and caudal sclerotome differ notably by their extracellular matrix and cell membrane composition. Thus, the caudal sclerotome has been shown to specifically express molecules such as T-cadherin, versican, collagen type XI, peanut agglutinin binding molecules, and chondroitin-6-sulphate-proteoglycan (Bagnall and Sanders 1989; Oakley and Tosney 1991; Ranscht and Bronner-Fraser 1991; Landolt et al. 1995; Ring et al. 1996; Henderson et al. 1997). The sclerotome compartments also differ in the expression of neural guidance molecules, such as Eph/ephrins, F-sponding, neuropilin2, or Semaphorin3F (Debby-Brafman et al. 1999; Baker and Antin 2003; Kuan et al. 2004; Gammill et al. 2006). As a result, neural crest cells and motor axons are repelled from the caudal sclerotome and their migration is confined to the rostral compartment of the sclerotome (Keynes and Stern 1984; Rickmann et al. 1985). Hence, the rostrocaudal subdivision of the sclerotome controls the segmentation of the peripheral nervous system (Fig. 6) (for review, see Kuan et al. 2004).

Fate mapping studies of the rostral and caudal somite compartments using lineage tracing and quail-chicken chimeras indicate that the rostral somite/sclerotome compartment gives rise to the caudal half of the vertebral body and a small part of the neural arch (Figs. 6 and 7B,D). Conversely, the caudal somite/sclerotome compartment gives rise to the rostral half of the vertebral body, to the pedicles and most of the neural arch (Figs. 6 and 7B,D) (Bagnall et al. 1988, 1989; Ewan and Everett 1992; Goldstein and Kalcheim 1992; Huang et al. 1994, 1996, 2000c; Aoyama and Asamoto 2000). The intervertebral discs are contributed by the somitocoele cells, which integrate the caudal sclerotome compartment (Figs. 6 and 7B,D) (Christ et al. 2007). The ribs also derive from the fusion of two consecutive somites and thus exhibit resegmentation (Chevallier 1975; Evans 2003; Aoyama et al. 2005).

Establishment of Somite Rostrocaudal Polarity

In amniotes, newly formed somites can be subdivided into a rostral and a caudal compartment (Saga and Takeda 2001). The future rostral and caudal compartments of the somite first acquire their identity in the ante-

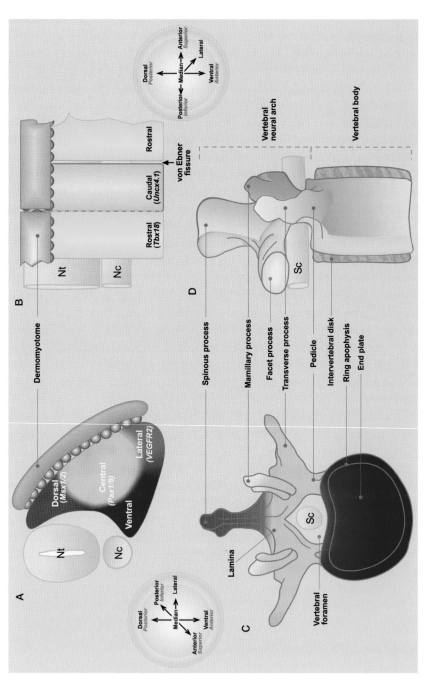

Figure 7. (*See facing page for legend.*)

rior PSM immediately after segment determination and before morphological somite individualization (see Fig. 1C). Rostrocaudal identity of the newly formed segment is materialized by stripes of gene expression restricted either to the future rostral or caudal compartments (see Fig. 5) (Saga and Takeda 2001). In amniotes, *Dll1* is initially expressed in a broad domain of the posterior PSM under the control of Wnt, Notch, and Tbx6 (Galceran et al. 2004; Hofmann et al. 2004; White and Chapman 2005; White et al. 2005). *Dll1* expression becomes down-regulated by Mesp2 in the newly specified segmental domain, and its expression is reactivated in the caudal domain of the forming somite in response to Notch activation (see Fig. 5) (Takahashi et al. 2000). Hence, *Dll1* expression becomes restricted to the caudal compartment of the forming somite, while Mesp2 becomes restricted to the rostral domain (Takahashi et al. 2003). The functional significance of the caudal restriction of *Dll1* expression in somites, however, is unclear since overexpression of *Dll1* throughout the PSM does not alter somite polarity (Serth et al. 2003; Teppner et al. 2007).

Analysis of mice mutant for components of the Notch/*Mesp2* genetic network reveals striking vertebral abnormalities involving specific absence of structures derived from either the rostral or the caudal sclerotome. Such phenotypes can be interpreted as resulting from specific defects in rostrocaudal polarity. Thus, in mice, mutations leading to a loss of Notch activity, such as null mutations for *Notch1*, *Dll1*, *Ofut1*, *PS1*, *RBP-Jk*, and *Ripply2*, exhibit rostralized paraxial mesoderm (Evrard et al. 1998; Kusumi et al. 1998; Zhang and Gridley 1998; del Barco Barrantes et al. 1999; Koizumi et al. 2001; Takahashi et al. 2003; Morimoto et al. 2007; Feller et al. 2008). Conversely, mice harboring mutations that interfere with Mesp1/2 functions (such as null mutants for *Mesp1/2*, *Dll3*, *Lfng* *Hes7*, *Tbx6*, or constitutive expression of NICD) exhibit only caudal sclerotome derivatives (Chapman et al. 1996; Saga et al. 1997; Takahashi et al. 2000, 2003, 2005; Bessho and Kageyama 2003; Bussen et al. 2004;

Figure 7. Sclerotome compartments and their contribution to vertebrae. (*A, B*) Schematic representation of the sclerotome compartments as defined by molecular markers and fate map. (*A*) transversal view. (*B*) lateral view, dermomyotome has been removed to visualize the underlying sclerotome. (*C, D*) Color-coded projection of the sclerotomal compartments (shown in *A, B*, respectively) onto an adult human lumbar vertebra. (*C*) Transversal (superior) view; (*D*) Lateral (right) view. The orientation of the embryonic axes is indicated in black bold in the circle and the corresponding medical terminology is shown in gray italics. (Nt) Neural tube, (Nc) notochord, (Sc) spinal cord.

Cordes et al. 2004; Morimoto et al. 2005, 2006; Yasuhiko et al. 2006; Oginuma et al. 2008). Such genetic evidence argues that *Mesp2* promotes the rostral identity of somites, while Notch activity is required to specify their caudal identity (see Fig. 5). In amniotes, this Notch/*Mesp2* system is downstream of the Foxc1 and -2 transcription factors in the anterior PSM. Mice double mutant for *Foxc1* and -2 do not exhibit any rostro-caudal polarity (Kume et al. 2001). Whereas both the somite boundary formation and the establishment of the rostrocaudal somitic identities are concommitent and use similar regulatory loops and downstream effectors, the two processes can be separated genetically (Nomura-Kitabayashi et al. 2002; Feller et al. 2008). Thus, somite boundaries can form in the absence of defined rostrocaudal polarity.

In humans, *MESP2* mutations have been found to be associated with congenital scoliosis (SCD02 [MIM 608681]) or with Spondylothoracic dysostosis/Jarcho-Levin Syndrome. STD has been characterized as an autosomal recessive disorder in which patients exhibit severe disruption of the spine segmentation. These patients are short in stature due to a short rigid neck and thorax, with a "crab like" chest, resulting from the fusion of the ribs at the costovertebral junctions (see Fig. 3C) (Cornier et al. 2004). This phenotype is strikingly similar to the mouse *Mesp2* null mutation in which the axial skeleton is caudalized (Saga et al. 1997). Sequencing the *MESP2* genes in these patients led to the identification of a recessive null mutation in this gene, suggesting that a similar mechanism controls the establishment of the rostrocaudal identity of somites in mice and humans (Cornier et al. 2008).

The downstream cellular effectors of the Notch/*Mesp2* system are still poorly characterized. It has been shown that in mice, *Mesp2* controls the segmental expression of the ephrin ligand *EphA4* (see Fig. 5) (Nakajima et al. 2006). Eph-ephrins are the largest family of membrane-bound tyrosine kinase receptors (RTKs), with 14 to 16 Eph receptors and eight to nine ephrin ligands identified in mammals (Kullander and Klein 2002). Receptor/ligand associations result in cell–cell attraction, repulsion, or modulation of cell adhesion (Durbin et al. 1998; Xu et al. 1999; Barrios et al. 2003; Cooke et al. 2005). The *ephrin-B2* receptor is expressed in the anterior PSM in a conserved segmental and complementary manner to *EphA4* (Bergemann et al. 1995; Flenniken et al. 1996; Irving et al. 1996; Baker and Antin 2003). While alteration of Eph/ephrin signaling in zebrafish leads to segmentation defects, in mice mutant for *EphA4* or *EphrinB2*, somitogenesis is normal, suggesting the existence of compensatory mechanisms in amniotes (Dottori et al. 1998; Adams et al. 1999; Barrios et al. 2003).

Protocadherins are a large family of cadherin-like molecules that act as putative signaling receptors rather than adhesion molecules (Sano et al. 1993; Redies et al. 2005). *Paraxial protocadherin (Papc)* is implicated in several biological processes, including convergent-extension movements during gastrulation (Kim et al. 1998; Yamamoto et al. 1998; Hukriede et al. 2003; Medina et al. 2004; Unterseher et al. 2004) and paraxial meso-derm patterning in *Xenopus*, zebrafish, and mice (Kim et al. 1998, 2000; Rhee et al. 2003). During somite formation, *Papc* has been shown to be regulated by *Mesp2*, *Ripply*, and paraxial mesoderm-specific *T-box* genes (see Fig. 5) (Kim et al. 2000; Sawada et al. 2000; Nomura-Kitabayashi et al. 2002; Rhee et al. 2003; Kawamura et al. 2005; Oates et al. 2005b; Muyskens and Kimmel 2007). While interfering with PAPC expression results in somitogenesis defects, the mouse *Papc* null mutant, however, is viable without major developmental defects, suggesting a functional redundancy with other protocadherins in amniotes (Yamamoto et al. 1998, 2000; Kim et al. 2000; Rhee et al. 2003). Therefore, Ephrin receptors and ligands, as well as protocadherins, are potentially involved in both the process of boundary establishment and the acquisition of rostrocaudal identities downstream of the *Mesp2* regulatory network.

Maintenance of Somite Rostrocaudal Identity by *Tbx18* and *Uncx4.1*

The two transcription factors *Tbx18* and *Uncx4.1* act downstream of the Notch/*Mesp2* system to maintain the rostral and the caudal identities of somites, respectively (see Figs. 5 and 7B). *Tbx18* expression is initiated at the level of the forming somite and is restricted to the rostral compart-ment (Kraus et al. 2001; Bussen et al. 2004; Haenig and Kispert 2004; Tanaka and Tickle 2004). Mice mutant for *Tbx18* exhibit severe rib fusions and vertebral malformations that result from somite caudaliza-tion (Bussen et al. 2004). Somite polarity is primarily established in the mutant, but is then subsequently lost. In addition, ectopic expression of *Tbx18* is sufficient to confer rostral identity to somite derivatives; thus, *Tbx18* appears necessary and sufficient for somite rostral identity (see Fig. 5) (Bussen et al. 2004). *Uncx4.1,* a paired-related homeobox gene, is expressed in the caudal compartment of formed somites (see Fig. 5) (Rovescalli et al. 1996; Saito et al. 1996; Mansouri et al. 1997). Mice mutant for *Uncx4.1* exhibit skeletal malformations of the spine due to sclerotome condensation defects. They lack the caudal sclerotome com-partment derivatives, including the pedicles and transverse processes, as well as the proximal ribs, while the initial segmental pattern remains unaffected (see Fig. 7B,D) (Leitges et al. 2000; Mansouri et al. 2000; Schrägle et al. 2004). While *Tbx18* and *Uncx4.1* emerge as key factors in somite rostrocaudal identity, their downstream targets remain unknown. The segmental pattern of the axial skeleton also requires maintenance of the sclerotome boundaries, a process involving the TGF-β type II receptor (*Tgfbr2*) (Baffi et al. 2006).

The Ventral Somite Forms the Vertebral Precursors of the Sclerotome

Sclerotome Compartments and Their Derivatives

Newly formed somites are epithelial spheres that rapidly subdivide into the ventral mesenchymal sclerotome and the dorsal epithelial dermomyotome. Cells of the newly formed somites are not yet committed with respect to their future differentiation (Aoyama and Asamoto 1988). The commitment of somite cells to the osteo-chondrogenic lineage is progressively determined in response to local inducers (Aoyama and Asamoto 1988; Dockter and Ordahl 1998; Dockter 2000). Patterning of the epithelial somites is controlled by inductive signals produced by the adjacent structures, namely the notochord, neural tube, dorsal ectoderm, and the lateral plate (Watterson et al. 1954; Strudel 1955; Hall 1977; Brand-Saberi et al. 1993; Pourquié et al. 1993, 1995, 1996; Sosic et al. 1997; Dockter 2000). In response to this combinatorial signal, the sclerotome becomes subdivided into a ventral, a central, a lateral, and a dorsal compartment (see Fig. 7A) (Marcelle et al. 1997; Brand-Saberi and Christ 2000; Brent and Tabin 2002; Kalcheim and Ben-Yair 2005; Christ et al. 2007).

Schematically, the ventral part of the sclerotome expresses predominantly the *Paired box transcription factor* (*Pax1*) and forms the vertebral bodies and intervertebral disks (see Fig. 7A,C). The neural arch essentially derives from the central sclerotome, which also contributes to the proximal ribs. The spinous process and the dorsal neural arch are contributed by the *Msx1* and *-2*-positive dorsal compartment of the sclerotome (see Fig. 7A,C) (Monsoro-Burq 2005). The lateral sclerotome, which strongly expresses *Vascular endothelial growth factor receptor 2* (*VEGFR2*) and *Sim1*, gives rise to the distal ribs, tendons, and endothelial cells (Eichmann et al. 1993; Pourquié et al. 1996). Another small compartment, the syndetome, lies at the interface between the central sclerotome and the dermomyotome. This compartment is marked by *Scleraxis* expression and forms the tendons linking the segmental muscles to the vertebrae (Cserjesi et al. 1995; Schweitzer et al. 2001; Brent et al. 2003). Fate mapping studies have demonstrated that the somitocoele forms a compartment called the arthrotome, which contributes to intervertebral joints (synovial joints) and disks, as well as to the proximal ribs (Huang et al. 1996; Christ et al. 2004; Mittapalli et al. 2005). The sclerotome also contributes to both the endothelium and smooth muscles of the aorta (Wiegreffe et al. 2007; Pouget et al. 2008). Interestingly, while the pelvic girdle is derived entirely from the lateral plate mesoderm, some skeletal elements of the pectoral girdle are contributed by the somites (Chevallier et al. 1977). Thus, in the chicken, the long blade of the scapula is formed by somite-derived precursors that express *Pax1* and exhibit a segmental organization (Huang et al. 2000b; Wang et al. 2005; Christ et al. 2007).

The Sclerotome Is Patterned by Diffusible Signals from Surrounding Structures

The somite ventral compartment differentiates in the sclerotome in response to diffusible signals produced by the notochord and the floor plate (Fig. 8) (Watterson et al. 1954; Strudel 1955; Hall 1977; Brand-Saberi et al. 1993; Pourquié et al. 1993; Dockter 2000). Major components of this ventralizing signal are the proteins Sonic hedgehog (Shh) (Fan and Tessier-Lavigne 1994; Johnson et al. 1994; Fan et al. 1995) and the BMP antagonist, Noggin (Nog) (Hirsinger et al. 1997; Marcelle et al. 1997; McMahon et al. 1998). These signals induce expression of sclerotome-specific transcription factors, such as *Pax1* and -9, which initiate the skeletal differentiation program (Fig. 8). In response to these signals, the ventral somite undergoes an epithelio-mesenchymal transition to generate the mesenchymal sclerotome. This transition involves the down-regulation of N-cadherin and requires the MMP-2 metalloprotease activity (Hatta and Takeichi 1986; Duband et al. 1987; Duong and Erickson 2004). Subsequently, the sclerotome cells form the peri-notochordal tissue that gives rise to the vertebral centra and then to the vertebral bodies and intervertebral discs (see Figs. 7A,C, 8C, and 9C) (Christ et al. 2007). *Pax3*, which is expressed in all cells of the anterior PSM, becomes down-regulated in the sclerotome cells in response to Shh (Dietrich et al. 1993; Fan and Tessier-Lavigne 1994). *Pax3* is maintained in cells of the somite dorsolateral compartment, which retains an epithelial structure and differentiates into the dermomyotome (Fig. 8A) (Schmidt et al. 2004; Christ et al. 2007). The dermomyotome subsequently gives rise to the skeletal musculature of the trunk and the dorsal dermis (Buckingham et al. 2006).

Mice mutant for *Shh* lack the entire axial skeleton, but initiate *Pax1* expression in the ventral somite (Chiang et al. 1996). However, this weak induction of *Pax1* in the ventral somite appears to result from Indian

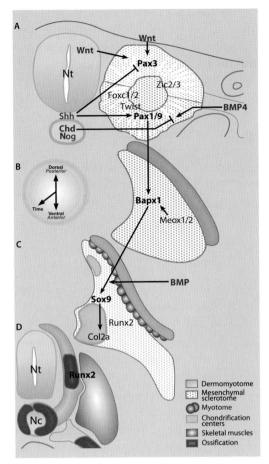

Figure 8. Signals patterning the sclerotome and transcriptional program underlying sclerotome differentiation. Temporal sequence from the differentiation of the epithelial somite to the ossification of the vertebra (signaling factors are shown in red). (*A*) Epithelial somites are patterned in two compartments by diffusible signals produced by adjacent structures. The sclerotome is primarily specified by "ventralizing signals," such as Shh produced by the notochord and the floor plate, and BMP antagonists Chordin (Chd) and Noggin (Nog) produced by the notochord. While Chd and Nog antagonize BMP4 (from the lateral plate mesoderm), Shh signaling activates the expression of the transcription factors *Pax1* and *-9*. The dermomyotome is specificied by "dorsalizing" signals, namely Wnts, produced by the dorsal ectoderm and neural tube, which activate *Pax3* and antagonize the action of Shh. (*B*) Pax1 and -9 control a cascade of transcription factors that are responsible for sclerotomal differentiation. Thus, Pax1 and -9 activate the key chondrogenesis factor Bapx1 (Nkx3.2) in synergy with Meox1 and -2. (*C*) Bapx1 allows the sclerotome to respond to BMP signals by triggering the expression of Sox9. This initiates the chondrogenesis program leading to the expression of *Col2a* and *Runx2*. (*D*) The final stage of sclerotomal differentiation consists of the ossification of the cartilaginous vertebral model, controlled by the Runx2 genetic network. The orientation of the embryonic axis and time axis is indicated in black bold and the corresponding medical terminology is indicated in gray italics.

hedgehog (Ihh) activity produced by the dorsal endoderm (Zhang et al. 2001). In mice mutant for the Shh coreceptor *Smoothened* (*Smo*) or for both *Shh* and *Ihh*, induction of sclerotome markers, such as *Pax1* or *Nkx3.2* (*Bapx1*), is not observed, suggesting that Hedgehog signaling is an absolute requirement for sclerotome induction (Zhang et al. 2001). In the ventral somite, Shh activates the expression of its receptor *Patched*, the *Gli2* and *-3* genes, and the Wnt antagonist *Sfrp2* (Goodrich et al. 1996; Hahn et al. 1996; Borycki et al. 1998; Motoyama et al. 1998; Buttitta et al. 2003). The Gli proteins are zinc-finger transcription factors acting downstream of Hedgehog signaling in vertebrates. *Gli1*, *-2*, and *-3* are expressed in somites, and mice mutant for *Gli* genes exhibit skeletal defects (Hui et al. 1994; Mo et al. 1997; Platt et al. 1997; Borycki et al. 1998). Shh signaling in the ventral somite antagonizes the dorsalizing influence from the neural tube and the dorsal ectoderm mediated by Wnt signals (see Fig. 8A) (Fan and Tessier-Lavigne 1994; Fan et al. 1995). In addition to its patterning function, Shh is also required as a trophic factor for sclerotomal cells to promote their survival and proliferation (Borycki et al. 1998; Teillet et al. 1998; Marcelle et al. 1999).

The secreted BMP4 growth factor is expressed by the lateral plate at the time of somite formation (Pourquié et al. 1996). The right level of BMP signaling along the medio-lateral axis is critical to maintain paraxial mesoderm and sclerotome identity (Hirsinger et al. 1997; Tonegawa et al. 1997). At this stage, BMP4 induces lateral markers, such as *Sim1*, in the somite and it antagonizes *Pax1* expression, contributing to its restriction to the medial somite domain (Balling et al. 1996; Pourquié et al. 1996; Tonegawa et al. 1997). The BMP signal is counteracted by the axial structures that produce the noggin (Nog) antagonist (Fig. 8A) (Hirsinger et al. 1997; Marcelle et al. 1997; Reshef et al. 1998; Tonegawa and Takahashi 1998). In addition, other BMP antagonists (such as Chordin [Chd], Cerberus, and Follistatin) may counteract the BMP4 effects (Patel et al. 1996; Belo et al. 1997; Biben et al. 1998; Shawlot et al. 1998). Therefore, sclerotome specification involves antagonism between the Shh and the BMP4/Wnt signaling pathways.

Cross talk between the myotome and sclerotome compartments involving platelet-derived growth factor (PDGF) signaling is also required for proper sclerotomal development. While the receptor *PDGFRα* is expressed mostly in the sclerotome, PDGFA and C ligands are produced by the myotome (Orr-Urtreger et al. 1992; Orr-Urtreger and Lonai 1992; Ding et al. 2000; Aase et al. 2002). Mice mutant for *PDGFRα* exhibit spina bifida and fusion of skeletal elements along the axis (Soriano 1997; Klinghoffer et al. 2002). PDGF signaling is required for regulating chondrogenesis and sclerotomal cell migration, leading to formation of the neural arches (Hoch and Soriano 2003; Pickett et al. 2008).

A Transcriptional Network Underlying Sclerotome Development

A network of transcription factors that acts downstream of Shh and BMP signaling, controls sclerotomal fate, and cartilage and bone differentiation. Shh induces *Pax1* and *Pax9*, which are essential for sclerotomal cell proliferation and differentiation in the ventral somite (see Fig. 8A) (Balling et al. 1996; Müller et al. 1996; McMahon et al. 1998; Peters et al. 1999). *Pax1* expression precedes the epithelio-mesenchymal transition that characterizes the sclerotome formation, while *Pax9* is expressed later during sclerotomal differentiation (Goulding et al. 1993; Love and Tuan 1993). While mice mutant for *Pax9* are essentially normal, *Pax1* mutants (*Undulated-short-tail* [*Uns*]) lack vertebral bodies and intervertebral disks (Wallin et al. 1994). Mice double mutant for *Pax1* and *Pax9* form only neural arches, demonstrating the key role played by these factors in vertebral body formation (see Figs. 7A,C, and 8) (Peters et al. 1999). *Pax1* and *Pax9* (and also *Meox1* and *-2*; discussed later in this chapter) directly activate expression of the homeobox-containing transcription factor *Nkx3.2* (*Bapx1*) in the sclerotome (see Fig. 8B) (Lettice et al. 1999; Tribioli and Lufkin 1999; Akazawa et al. 2000; Herbrand et al. 2002; Rodrigo et al. 2003, 2004). As in the *Pax1* and *-9* double mutant, mice mutant for *Nkx3.2* lack vertebral bodies and intervertebral disks (Lettice et al. 1999; Tribioli and Lufkin 1999; Akazawa et al. 2000). The induction of *Nkx3.2*, downstream of Shh and *Pax1* and *-9*, is a key step in the sclerotome determination process as it endows sclerotomal cells with the competence to respond to the BMP signal by activating *Sox9* (see Fig. 8C) (Murtaugh et al. 2001). *Nkx3.2* forms with *Sox9*, a BMP-dependent, autoregulatory loop that promotes sclerotomal chondrogenesis in amniotes (Murtaugh et al. 2001; Zeng et al. 2002). Subsequently, *Sox9* activates the chondrogenic program that leads to the formation of the vertebral cartilaginous template (see Fig. 8C,D) (Healy et al. 1996, 1999; Bell et al. 1997; Bi et al. 1999, 2001). During later stages, the Runx2 transcriptional network (which controls bone formation) is activated during vertebral ossification (see Fig. 8D). The role of Sox9 or Runx2 in the control of cartilage and bone development is discussed extensively in other chapters of this book and is not detailed here.

Several other transcription factors have been implicated in sclerotome differentiation. *Nkx3.1* is a gene closely related to *Nkx3.2*, which is also strongly expressed in the sclerotome (Herbrand et al. 2002). Whereas the *Nkx3.1* null mutant does not show an axial skeleton phenotype, the double *Nkx3.1* and *3.2* null mutant exhibits more severe skeletal defects compared to the single *Nkx3.2* mutant, suggesting that the two genes play partly redundant roles (Herbrand et al. 2002). The *Meox 1* and *2* homeobox transcription factors are expressed in somites of vertebrate embryos (see Fig. 8B) (Candia et al. 1992; Candia and Wright 1996; Mankoo et al. 2003; Reijntjes et al. 2007). *Meox1* and *-2* have partially redundant functions in mice. While *Meox1* mutants present defects in sclerotomal derivatives (such as fusion of vertebrae) and *Meox2* mutants do not exhibit an axial skeleton phenotype, the double *Meox1/2* null mutants present a severe disruption of somite organization and differentiation (Mankoo et al. 2003). *Foxc2* is expressed in the sclerotome under the control of Shh and is required to promote sclerotomal cell proliferation (see Fig. 8A) (Winnier et al. 1997; Furumoto et al. 1999). In addition, it has also been demonstrated that *Twist,* coding for a bHLH transcription factor activated by Shh and Fgf signaling, controls sclerotomal cell viability and proliferation, while inhibiting bone formation (see Fig. 8A) (Chen and Behringer 1995; Bialek et al. 2004; Hornik et al. 2004). *Zic1-3* (zinc-finger transcription factors) are expressed in differentiating somites, and mice mutant for these transcription factors exhibit skeletal defects (Aruga et al. 1999; Carrel et al. 2000; Klootwijk et al. 2000; Nagai et al. 2000; Purandare et al. 2002; Merzdorf 2007). *Zic2* and *-3* are specifically expressed in the PSM and newly formed somites, and have been shown to act cooperatively during paraxial mesoderm segmentation (see Fig. 8A) (Inoue et al. 2007). Since the somite boundaries and rostrocaudal polarity are normal in the compound mutant mice, *Zic2* and *-3* are more likely to be implicated in the maintenance and differentiation of the sclerotome (Inoue et al. 2007). The transcription factor *Jun/AP-1* has been implicated in the differentiation of intervertebral disks by controlling degeneration of the notochordal cells (Behrens et al. 2003).

Differentiation of the Vertebrae from the Sclerotome

The sclerotome subsequently differentiates into a cartilaginous model that becomes ossified and then forms the definitive vertebra (Fig. 9C–E) (Bono et al. 2006; Christ et al. 2007). In humans, cartilage-producing centers (identifiable by the deposition of Collagen type 2) form at approximately the 6th week in the embryo (see Figs. 8C and 9C) (Bareggi et al. 1993; Nolting et al. 1998). Primary chondrification centers appear in the neural arches, and they subsequently fuse over the midline dorsally and with the centrum ventrally (see Figs. 8D and 9D). Failure of the neural arch to fuse dorsally is associated with spina bifida. Two primary chondrification cen-

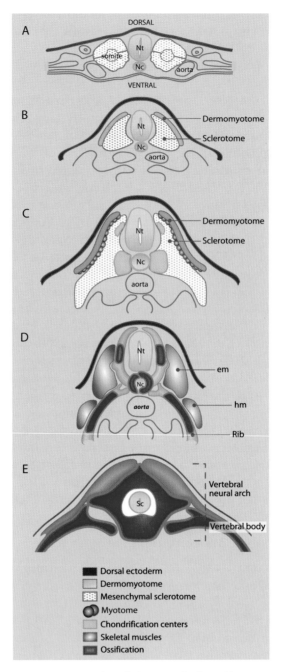

Figure 9. (*See facing page for legend.*)

ters also form on the left and right sides of the regressing notochord and eventually fuse to form the cartilaginous centrum (Fig. 9C). Subsequently, cartilaginous transverse and spinous processes form from the neural arch. In between forming centra, regions of loosely packed cells surrounding the remnant of the notochord will form the cartilaginous intervertebral disks. The disk center, consisting of the *nucleus pulposus*, is produced by a remnant of the notochordal cells, while it is surrounded by the *annulus fibrosus*, formed by sclerotomal cells (Christ et al. 2007).

The cartilaginous template then becomes progressively ossified (endochondral ossification) (Fig. 9D,E) (Bareggi et al. 1993; Nolting et al. 1998). In the human embryo, this process begins around the 8th week when blood vessels invade the cartilaginous centrum (Bono et al. 2006). Three primary ossification centers are typically recognized: one in the centrum, and one on each side of the vertebral arch (Fig. 9D). In mice and humans, ossification of vertebrae follows a specific spatio-temporal sequence along the anteroposterior axis, beginning in the lower thoracic and upper lumbar regions (Bareggi et al. 1993). Five secondary ossification centers develop following the onset of puberty: (1) one at the tip of the spinous process, (2) one at the tip of each transverse process, and (3) one ring apophysis in the superior and inferior endplates (attachment surface disk-vertebrae) of the vertebral bodies (see Fig. 7C) (Bono et al. 2006).

Regionalization of the Axial Skeleton and the Role of the *Hox* Genes

Each vertebra exhibits a distinct morphology depending on its position along the anteroposterior axis. Based on shared morphological features,

---

Figure 9. Sclerotome differentiation and vertebra formation. Schematic temporal sequence of sclerotome differentiation, morphogenesis, chondrogenesis, and osteogenesis leading to the formation of the vertebra (thoracic level; transversal section). (*A*) Two compartments are specified from the epithelial somites: the dermomyotome dorsally and the sclerotome ventrally. (*B*) The epithelial sclerotome undergoes an epithelio-mesenchymal transition, while the dermomyotome maintains its epithelial structure. (*C*) The sclerotome initiates chondrogenesis in centers initially localized around the notochord ventrally and on both neural arches dorsally. Meanwhile, the dermomyotome produces the skeletal muscles and dermis. (*D*) Skeletogenesis further proceeds by the progressive ossification of the cartilaginous model of the vertebrae, in which vertebrae centra, neural arches, and costal elements (thoracic level) are identifiable. (*E*) Ossification centers eventually fuse to produce the final vertebra structure. (Nc) Notochord, (Nt) neural tube, (Sc) spinal cord, (Em) epaxial muscle mass, (Hm) hypaxial muscle mass.

five main regions of the amniote vertebral column can be distinguished: the cervical, thoracic, lumbar, sacral, and caudal (Fig. 10). In humans, the five sacral vertebrae fuse to form a triangular-shaped sacrum. The number of each type of vertebra defines the vertebral formula of a particular species and varies dramatically among vertebrate species (Burke et al. 1995; Burke 2000; Gomez et al. 2008). The acquisition of the distinct vertebral identities is controlled by a class of transcription factors called Hox proteins (for review, see Krumlauf 1994; Duboule 2007; Wellik 2007). In

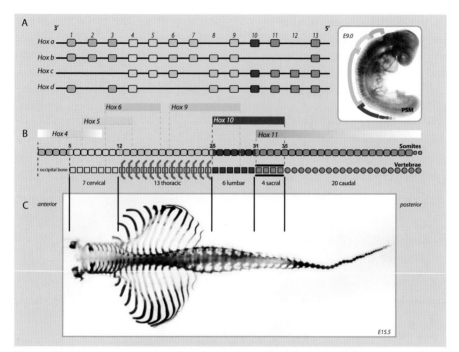

Figure 10. *Hox* gene expression domains and vertebral formula of the mouse embryo. (*A*) Representation of the genomic organization of the four mouse paralog *Hox* clusters. *Hox* gene combinatorial expression controls the vertebral identity along the anteroposterior axis. Genes are color-coded based on their expression and phenotypic domains (the skeletal domain that is affected when a particular *Hox* gene is mutated) along the anteroposterior axis. The inset panel (*upper right*) illustrates a mouse embryo (embryonic day 9.0) with approximate somite identity along the anteroposterior axis (lateral view). (*B*) Schematic representation of the somitic regions compared to the vertebral formula of the mouse. The somites and derived vertebrae are aligned and subdivided by color-coded domains corresponding to distinct vertebrae types and which correlate with specific *Hox* paralog expression (*A*). (*C*) The different regions of the axial skeleton are illustrated in a mouse embryo (embryonic day 15.5) stained with Alcian blue.

mice and humans, there are 39 *Hox* genes that are organized into four paralogous clusters (*Hox1–13*) (Fig. 10A) (Duboule 2007). The position of the genes along a cluster correlates with the order in which each gene will be expressed in time and space along the anteroposterior axis of the embryo, a property called collinearity (Kmita and Duboule 2003). *Hox* genes control the timing of ingression of paraxial mesoderm precursors into the primitive streak (Iimura and Pourquié 2006). Thus, precursors expressing more "posterior" genes (i.e., genes located 5′ in the *Hox* cluster) will ingress later than those expressing more "anterior" (genes located 3′) and, hence, will be positioned more posteriorly along the anteroposterior axis. This mechanism results in a collinear or nested arrangement of the *Hox* gene expression domains along the anteroposterior axis. This nested arrangement of the *Hox* expression domains defines a specific combination of genes expressed for each somite (Kessel and Gruss 1991). The combination of *Hox* genes that are expressed in each somite is involved in the control of the specification of vertebral identities (Fig. 10) (Wellik 2007). However, although the *Hox* gene expression domains often extend from the posterior end of the embryo to their anterior expression limit in the somites, their action is essentially restricted to their anteriormost expression domain. Thus, the identity of a segment is controlled by the posteriormost *Hox* genes expressed in this segment, a property called posterior prevalence in vertebrates (Duboule and Morata 1994; Burke et al. 1995). Grafting experiments in chicken embryos support the idea that vertebral axial identity is already determined at the level of the PSM, before somite individualization (Kieny et al. 1972; Nowicki and Burke 2000). *Hox* loss of function or ectopic expression in the paraxial mesoderm can change vertebral identities, leading to homeotic transformations in the axial skeleton (see Wellik 2007 and references therein). The function and targets of *Hox* genes in the developing spine are virtually unknown.

Several factors involved in the regulation of *Hox* genes in the patterning of the axial skeleton have been identified. In the mouse, the TGF-β family secreted factor, GDF11 (which is expressed in the tail bud), acts upstream of *Hox* genes. Mice mutant for *Gdf11* show homeotic transformation of the vertebral column and tail truncation, correlating with disruption of *Hox* expression domains (McPherron et al. 1999; Nakashima et al. 1999). These patterning defects are partially phenocopied in mice mutant for the GDF11 proprotein convertase *Pcsk5*, specifically expressed in the PSM, as well as TGF-β type I receptor (*Alk5*) and type II receptors (*ActRIIA/B* [*Acvr2/b*]) (Oh and Li 1997; Rancourt and Rancourt 1997; Oh et al. 2002; Andersson et al. 2006; Essalmani et al. 2008; Szumska et al. 2008).

The caudal transcription factors (*Cdx*) play a conserved role in *Hox* regulation during anteroposterior axis formation (van den Akker et al. 2002; Houle et al. 2003a; Chawengsaksophak et al. 2004; Copf et al. 2004; Pilon et al. 2007). In mice, *Cdx* gene expression (*Cdx1, -2,* and *-4*) is initiated at the primitive streak stage and is later restricted to the posterior part of the embryo. *Cdx* gene expression is regulated by Wnt, Fgf, and RA signalings (Isaacs et al. 1994; Pownall et al. 1996; Houle et al. 2000, 2003b; Ikeya and Takada 2001; Prinos et al. 2001; Bel-Vialar et al. 2002; Lohnes 2003; Deschamps and van Nes 2005; Pilon et al. 2006, 2007). Cdx factors act by regulating a subset of *Hox* genes in a dose-dependent manner (van den Akker et al. 2002). Mice mutant for *Cdx1, -2,* and *-4* exhibit severe axis truncation (van den Akker et al. 2002; Chawengsaksophak et al. 2004; van Nes et al. 2006). Other transcription factors, such as *Plzf* (*Zfp145*) and Hox nuclear cofactors such as *Pbx1*, also regulate *Hox* gene-dependent patterning along the anteroposterior axis (Barna et al. 2000; Selleri et al. 2001; Capellini et al. 2008). Importantly, *Hox* gene expression in the axial skeleton also depends on the conserved families of chromatin remodeling proteins of the Polycomb and Trithorax groups, which negatively and positively regulate *Hox* expression, respectively. Mutation of several genes of these families results in homeotic transformations of the axial skeleton (Kim et al. 2006).

The anterior boundary of any given *Hox* gene expression domain in the precursors of the vertebral column is always positioned at the same somitic level, indicating that *Hox* gene expression is tightly coupled to somitogenesis (see Fig. 10B). Several *Hox* genes have been shown to be expressed in a dynamic fashion (similar to *Mesp2*) in the anterior PSM, suggesting that their regulation might be taken over by the segmentation machinery (Zákány et al. 2001). In the chicken embryo, changing the somite size by treating embryos with Fgf results in a corresponding change in the position of Hox expression boundaries (Dubrulle et al. 2001). Interestingly, in the mouse, altering the signals that control the positioning of the determination front—namely Fgf (*FgfR1*), Wnt (*Wnt3a*), and RA (RA, *RAR* receptors, *Cyp26*)—also results in homeotic transformations (Kessel 1992; Lohnes et al. 1994; Partanen et al. 1998; Abu-Abed et al. 2001; Ikeya and Takada 2001; Sakai et al. 2001; Mallo et al. 2008).

ACKNOWLEDGMENTS

We thank members of the Pourquié lab for helpful discussions and comments on the manuscript and for providing illustration material, S. Esteban for artwork, and J. Chatfield for editorial assistance. We thank P.

Turnpenny for providing the radiograph depicted in Figure 3 (panel B) and A. Cornier for the radiograph depicted in Figure 3 (panel C). This work is supported by Stowers Institute for Medical Research. O.P. is a Howard Hughes Medical Institute Investigator.

REFERENCES

Aase, K., Abramsson, A., Karlsson, L., Betsholtz, C., and Eriksson, U. 2002. Expression analysis of PDGF-C in adult and developing mouse tissues. *Mech. Dev.* **110:** 187–191.

Abu-Abed, S., Dollé, P., Metzger, D., Beckett, B., Chambon, P., and Petkovich, M. 2001. The retinoic acid-metabolizing enzyme, CYP26A1, is essential for normal hindbrain patterning, vertebral identity, and development of posterior structures. *Genes Dev.* **15:** 226–240.

Adams, R.H., Wilkinson, G.A., Weiss, C., Diella, F., Gale, N.W., Deutsch, U., Risau, W., and Klein, R. 1999. Roles of ephrinB ligands and EphB receptors in cardiovascular development: Demarcation of arterial/venous domains, vascular morphogenesis, and sprouting angiogenesis. *Genes Dev.* **13:** 295–306.

Afzal, A.R., Rajab, A., Fenske, C.D., Oldridge, M., Elanko, N., Ternes-Pereira, E., Tüysüz, B., Murday, V.A., Patton, M.A., Wilkie, A.O., et al. 2000. Recessive Robinow syndrome, allelic to dominant brachydactyly type B, is caused by mutation of *ROR2*. *Nat. Genet.* **25:** 419–422.

Ahn, U.M., Ahn, N.U., Nallamshetty, L., Buchowski, J.M., Rose, P.S., Miller, N.H., Kostuik, J.P., and Sponseller, P.D. 2002. The etiology of adolescent idiopathic scoliosis. *Am. J. Orthop.* **31:** 387–395.

Akazawa, H., Komuro, I., Sugitani, Y., Yazaki, Y., Nagai, R., and Noda, T. 2000. Targeted disruption of the homeobox transcription factor *Bapx1* results in lethal skeletal dysplasia with asplenia and gastroduodenal malformation. *Genes Cells* **5:** 499–513.

Al-Shawi, R., Ashton, S.V., Underwood, C., and Simons, J.P. 2001. Expression of the *Ror1* and *Ror2* receptor tyrosine kinase genes during mouse development. *Dev. Genes Evol.* **211:** 161–171.

Andersson, O., Reissmann, E., and Ibanez, C.F. 2006. Growth differentiation factor 11 signals through the transforming growth factor-β receptor ALK5 to regionalize the anterior-posterior axis. *EMBO Rep.* **7:** 831–837.

Aoyama, H. and Asamoto, K. 1988. Determination of somite cells: Independence of cell differentiation and morphogenesis. *Development* **104:** 15–28.

Aoyama, H. and Asamoto, K. 2000. The developmental fate of the rostral/caudal half of a somite for vertebra and rib formation: Experimental confirmation of the resegmentation theory using chick-quail chimeras. *Mech. Dev.* **99:** 71–82.

Aoyama, H., Mizutani-Koseki, S., and Koseki, H. 2005. Three developmental compartments involved in rib formation. *Int. J. Dev. Biol.* **49:** 325–333.

Arnold, S.J., Stappert, J., Bauer, A., Kispert, A., Herrmann, B.G., and Kemler, R. 2000. *Brachyury* is a target gene of the Wnt/β-catenin signaling pathway. *Mech. Dev.* **91:** 249–258.

Artavanis-Tsakonas, S., Rand, M.D., and Lake, R.J. 1999. Notch signaling: Cell fate control and signal integration in development. *Science* **284:** 770–776.

Aruga, J., Mizugishi, K., Koseki, H., Imai, K., Balling, R., Noda, T., and Mikoshiba, K. 1999. *Zic1* regulates the patterning of vertebral arches in cooperation with *Gli3*. *Mech. Dev.* **89:** 141–150.

Aulehla, A. and Herrmann, B.G. 2004. Segmentation in vertebrates: Clock and gradient finally joined. *Genes Dev.* **18:** 2060–2067.

Aulehla, A. and Johnson, R.L. 1999. Dynamic expression of *lunatic fringe* suggests a link between *notch* signaling and an autonomous cellular oscillator driving somite segmentation. *Dev. Biol.* **207:** 49–61.

Aulehla, A., Wehrle, C., Brand-Saberi, B., Kemler, R., Gossler, A., Kanzler, B., and Herrmann, B.G. 2003. *Wnt3a* plays a major role in the segmentation clock controlling somitogenesis. *Dev. Cell* **4:** 395–406.

Aulehla, A., Wiegraebe, W., Baubet, V., Wahl, M.B., Deng, C., Taketo, M., Lewandoski, M., and Pourquié, O. 2008. A β-catenin gradient links the clock and wavefront systems in mouse embryo segmentation. *Nat. Cell Biol.* **10:** 186–193.

Baffi, M.O., Moran, M.A., and Serra, R. 2006. *Tgfbr2* regulates the maintenance of boundaries in the axial skeleton. *Dev. Biol.* **296:** 363–374.

Bagnall, K.M. and Sanders, E.J. 1989. The binding pattern of peanut lectin associated with sclerotome migration and the formation of the vertebral axis in the chick embryo. *Anat. Embryol.* **180:** 505–513.

Bagnall, K.M., Higgins, S.J., and Sanders, E.J. 1988. The contribution made by a single somite to the vertebral column: Experimental evidence in support of resegmentation using the chick-quail chimaera model. *Development* **103:** 69–85.

Bagnall, K.M., Higgins, S.J., and Sanders, E.J. 1989. The contribution made by cells from a single somite to tissues within a body segment and assessment of their integration with similar cells from adjacent segments. *Development* **107:** 931–943.

Baker, R.K. and Antin, P.B. 2003. Ephs and ephrins during early stages of chick embryogenesis. *Dev. Dyn.* **228:** 128–142.

Balling, R., Helwig, U., Nadeau, J., Neubüser, A., Schmahl, W., and Imai, K. 1996. Pax genes and skeletal development. *Ann. N.Y. Acad. Sci.* **785:** 27–33.

Baranski, M., Berdougo, E., Sandler, J.S., Darnell, D.K., and Burrus, L.W. 2000. The dynamic expression pattern of *frzb-1* suggests multiple roles in chick development. *Dev. Biol.* **217:** 25–41.

Bareggi, R., Grill, V., Sandrucci, M.A., Baldini, G., De Pol, A., Forabosco, A., and Narducci, P. 1993. Developmental pathways of vertebral centra and neural arches in human embryos and fetuses. *Anat. Embryol.* **187:** 139–144.

Barna, M., Hawe, N., Niswander, L., and Pandolfi, P.P. 2000. Plzf regulates limb and axial skeletal patterning. *Nat. Genet.* **25:** 166–172.

Barnes, G.L., Alexander, P.G., Hsu, C.W., Mariani, B.D., and Tuan, R.S. 1997. Cloning and characterization of chicken *Paraxis:* A regulator of paraxial mesoderm development and somite formation. *Dev. Biol.* **189:** 95–111.

Baron, M. 2003. An overview of the Notch signalling pathway. *Semin. Cell Dev. Biol.* **14:** 113–119.

Barrantes, I.B., Elia, A.J., Wunsch, K., Hrabě de Angelis, M.H., Mak, T.W., Rossant, J., Conlon, R.A., Gossler, A., and de la Pompa, J.L. 1999. Interaction between Notch signalling and Lunatic fringe during somite boundary formation in the mouse. *Curr. Biol.* **9:** 470–480.

Barrios, A., Poole, R.J., Durbin, L., Brennan, C., Holder, N., and Wilson, S.W. 2003. Eph/Ephrin signaling regulates the mesenchymal-to-epithelial transition of the paraxial mesoderm during somite morphogenesis. *Curr. Biol.* **13:** 1571–1582.

Barsi, J.C., Rajendra, R., Wu, J.I., and Artzt, K. 2005. Mind bomb1 is a ubiquitin ligase essential for mouse embryonic development and Notch signaling. *Mech. Dev.* **122:** 1106–1117.

Begemann, G. and Meyer, A. 2001. Hindbrain patterning revisited: Timing and effects of retinoic acid signalling. *Bioessays* **23:** 981–986.

Begemann, G., Schilling, T.F., Rauch, G.J., Geisler, R., and Ingham, P.W. 2001. The zebrafish *neckless* mutation reveals a requirement for *raldh2* in mesodermal signals that pattern the hindbrain. *Development* **128:** 3081–3094.

Behrens, A., Haigh, J., Mechta-Grigoriou, F., Nagy, A., Yaniv, M., and Wagner, E.F. 2003. Impaired intervertebral disc formation in the absence of *Jun. Development* **130:** 103–109.

Bell, D.M., Leung, K.K., Wheatley, S.C., Ng, L.J., Zhou, S., Ling, K.W., Sham, M.H., Koopman, P., Tam, P.P., and Cheah, K.S. 1997. SOX9 directly regulates the type-II collagen gene. *Nat. Genet.* **16:** 174–178.

Bellairs, R. 1986. The primitive streak. *Anat. Embryol.* **174:** 1–14.

Belo, J.A., Bouwmeester, T., Leyns, L., Kertesz, N., Gallo, M., Follettie, M., and De Robertis, E.M. 1997. Cerberus-like is a secreted factor with neutralizing activity expressed in the anterior primitive endoderm of the mouse gastrula. *Mech. Dev.* **68:** 45–57.

Bel-Vialar, S., Itasaki, N., and Krumlauf, R. 2002. Initiating Hox gene expression: In the early chick neural tube differential sensitivity to FGF and RA signaling subdivides the *HoxB* genes in two distinct groups. *Development* **129:** 5103–5115.

Bergemann, A.D., Cheng, H.J., Brambilla, R., Klein, R., and Flanagan, J.G. 1995. ELF-2, a new member of the Eph ligand family, is segmentally expressed in mouse embryos in the region of the hindbrain and newly forming somites. *Mol. Cell. Biol.* **15:** 4921–4929.

Bessho, Y. and Kageyama, R. 2003. Oscillations, clocks and segmentation. *Curr. Opin. Genet. Dev.* **13:** 379–384.

Bessho, Y., Sakata, R., Komatsu, S., Shiota, K., Yamada, S., and Kageyama, R. 2001. Dynamic expression and essential functions of *Hes7* in somite segmentation. *Genes Dev.* **15:** 2642–2647.

Bessho, Y., Hirata, H., Masamizu, Y., and Kageyama, R. 2003. Periodic repression by the bHLH factor Hes7 is an essential mechanism for the somite segmentation clock. *Genes Dev.* **17:** 1451–1456.

Bettenhausen, B., Hrabě de Angelis, M., Simon, D., Guénet, J.L., and Gossler, A. 1995. Transient and restricted expression during mouse embryogenesis of *Dll1*, a murine gene closely related to *Drosophila Delta. Development* **121:** 2407–2418.

Bi, W., Deng, J.M., Zhang, Z., Behringer, R.R., and de Crombrugghe, B. 1999. *Sox9* is required for cartilage formation. *Nat. Genet.* **22:** 85–89.

Bi, W., Huang, W., Whitworth, D.J., Deng, J.M., Zhang, Z., Behringer, R.R., and de Crombrugghe, B. 2001. Haploinsufficiency of *Sox9* results in defective cartilage primordia and premature skeletal mineralization. *Proc. Natl. Acad. Sci.* **98:** 6698–6703.

Bialek, P., Kern, B., Yang, X., Schrock, M., Sosic, D., Hong, N., Wu, H., Yu, K., Ornitz, D.M., Olson, E.N., et al. 2004. A twist code determines the onset of osteoblast differentiation. *Dev. Cell* **6:** 423–435.

Biben, C., Stanley, E., Fabri, L., Kotecha, S., Rhinn, M., Drinkwater, C., Lah, M., Wang, C.C., Nash, A., Hilton, D., et al1998. Murine cerberus homologue *mCer-1*: A candidate anterior patterning molecule. *Dev. Biol.* **194:** 135–151.

Biris, K.K., Dunty Jr., W.C., and Yamaguchi, T.P. 2007. Mouse *Ripply2* is downstream of Wnt3a and is dynamically expressed during somitogenesis. *Dev. Dyn.* **236:** 3167–3172.

Blanar, M.A., Crossley, P.H., Peters, K.G., Steingrimsson, E., Copeland, N.G., Jenkins, N.A., Martin, G.R., and Rutter, W.J. 1995. Meso1, a basic-helix-loop-helix protein involved in mammalian presomitic mesoderm development. *Proc. Natl. Acad. Sci.* **92:** 5870–5874.

Bono, C.M., Parke, W.W., and Garfin, S.R. 2006. Development of the spine. In *Rothman-Simeone the spine* (ed. H.N. Herkowitz et al.), pp. 3–15. Elsevier, Philadelphia.

Borycki, A.G., Mendham, L., and Emerson, C. 1998. Control of somite patterning by Sonic hedgehog and its downstream signal response genes. *Development* **125:** 777–790.

Bottcher, R.T. and Niehrs, C. 2005. Fibroblast growth factor signaling during early vertebrate development. *Endocr. Rev.* **26:** 63–77.

Brand-Saberi, B. and Christ, B. 2000. Evolution and development of distinct cell lineages derived from somites. *Curr. Top. Dev. Biol.* **48:** 1–42.

Brand-Saberi, B., Ebensperger, C., Wilting, J., Balling, R., and Christ, B. 1993. The ventralizing effect of the notochord on somite differentiation in chick embryos. *Anat. Embryol.* **188:** 239–245.

Bray, S.J. 2006. Notch signalling: A simple pathway becomes complex. *Nat. Rev. Mol. Cell Biol.* **7:** 678–689.

Brent, A.E. and Tabin, C.J. 2002. Developmental regulation of somite derivatives: Muscle, cartilage and tendon. *Curr. Opin. Genet. Dev.* **12:** 548–557.

Brent, A.E., Schweitzer, R., and Tabin, C.J. 2003. A somitic compartment of tendon progenitors. *Cell* **113:** 235–248.

Buchberger, A., Seidl, K., Klein, C., Eberhardt, H., and Arnold, H.H. 1998. cMeso-1, a novel bHLH transcription factor, is involved in somite formation in chicken embryos. *Dev. Biol.* **199:** 201–215.

Buchberger, A., Bonneick, S., and Arnold, H. 2000. Expression of the novel basic-helix-loop-helix transcription factor cMespo in presomitic mesoderm of chicken embryos. *Mech. Dev.* **97:** 223–226.

Buchberger, A., Bonneick, S., Klein, C., and Arnold, H.H. 2002. Dynamic expression of chicken cMeso2 in segmental plate and somites. *Dev. Dyn.* **223:** 108–118.

Buckingham, M., Bajard, L., Daubas, P., Esner, M., Lagha, M., Relaix, F., and Rocancourt, D. 2006. Myogenic progenitor cells in the mouse embryo are marked by the expression of Pax3/7 genes that regulate their survival and myogenic potential. *Anat. Embryol.* (suppl. 1) **211:** 51–56.

Bulman, M.P., Kusumi, K., Frayling, T.M., McKeown, C., Garrett, C., Lander, E.S., Krumlauf, R., Hattersley, A.T., Ellard, S., and Turnpenny, P.D. 2000. Mutations in the human *Delta* homologue, *DLL3*, cause axial skeletal defects in spondylocostal dysostosis. *Nat. Genet.* **24:** 438–441.

Burgess, R., Cserjesi, P., Ligon, K.L., and Olson, E.N. 1995. Paraxis: a basic helix-loop-helix protein expressed in paraxial mesoderm and developing somites. *Dev. Biol.* **168:** 296–306.

Burgess, R., Rawls, A., Brown, D., Bradley, A., and Olson, E.N. 1996. Requirement of the paraxis gene for somite formation and musculoskeletal patterning. *Nature* **384:** 570–573.

Burke, A.C. 2000. *Hox* genes and the global patterning of the somitic mesoderm. *Curr. Top. Dev. Biol.* **47:** 155–181.

Burke, A.C., Nelson, C.E., Morgan, B.A., and Tabin, C. 1995. *Hox* genes and the evolution of vertebrate axial morphology. *Development* **121:** 333–346.

Bussen, M., Petry, M., Schuster-Gossler, K., Leitges, M., Gossler, A., and Kispert, A. 2004. The T-box transcription factor Tbx18 maintains the separation of anterior and posterior somite compartments. *Genes Dev.* **18:** 1209–1221.

Buttitta, L., Mo, R., Hui, C.C., and Fan, C.M. 2003. Interplays of Gli2 and Gli3 and their requirement in mediating Shh-dependent sclerotome induction. *Development* **130:** 6233–6243.

Cambray, N. and Wilson, V. 2002. Axial progenitors with extensive potency are localised to the mouse chordoneural hinge. *Development* **129:** 4855–4866.

Cambray, N. and Wilson, V. 2007. Two distinct sources for a population of maturing axial progenitors. *Development* **134:** 2829–2840.

Candia, A.F. and Wright, C.V. 1996. Differential localization of Mox-1 and Mox-2 proteins indicates distinct roles during development. *Int. J. Dev. Biol.* **40:** 1179–1184.

Candia, A.F., Hu, J., Crosby, J., Lalley, P.A., Noden, D., Nadeau, J.H., and Wright, C.V. 1992. *Mox-1* and *Mox-2* define a novel homeobox gene subfamily and are differentially expressed during early mesodermal patterning in mouse embryos. *Development* **116:** 1123–1136.

Cano, A., Perez-Moreno, M.A., Rodrigo, I., Locascio, A., Blanco, M.J., del Barrio, M.G., Portillo, F., and Nieto, M.A. 2000. The transcription factor snail controls epithelial-mesenchymal transitions by repressing E-cadherin expression. *Nat. Cell Biol.* **2:** 76–83.

Capellini, T.D., Zewdu, R., Di Giacomo, G., Asciutti, S., Kugler, J.E., Di Gregorio, A., and Selleri, L. 2008. Pbx1/Pbx2 govern axial skeletal development by controlling Polycomb and Hox in mesoderm and Pax1/Pax9 in sclerotome. *Dev. Biol.* **321:** 500–514.

Carpio, R., Honoré, S.M., Araya, C., and Mayor, R. 2004. *Xenopus paraxis* homologue shows novel domains of expression. *Dev. Dyn.* **231:** 609–613.

Carrel, T., Purandare, S.M., Harrison, W., Elder, F., Fox, T., Casey, B., and Herman, G.E. 2000. The X-linked mouse mutation Bent tail is associated with a deletion of the *Zic3* locus. *Hum. Mol. Genet.* **9:** 1937–1942.

Catala, M. 2002. Genetic control of caudal development. *Clin. Genet.* **61:** 89–96.

Catala, M., Teillet, M.A., and Le Douarin, N.M. 1995. Organization and development of the tail bud analyzed with the quail-chick chimaera system. *Mech. Dev.* **51:** 51–65.

Cauthen, C.A., Berdougo, E., Sandler, J., and Burrus, L.W. 2001. Comparative analysis of the expression patterns of *Wnts* and *Frizzleds* during early myogenesis in chick embryos. *Mech. Dev.* **104:** 133–138.

Chambers, D., Wilson, L., Maden, M., and Lumsden, A. 2007. RALDH-independent generation of retinoic acid during vertebrate embryogenesis by CYP1B1. *Development* **134:** 1369–1383.

Chan, T., Satow, R., Kitagawa, H., Kato, S., and Asashima, M. 2006. *Ledgerline*, a novel *Xenopus laevis* gene, regulates differentiation of presomitic mesoderm during somitogenesis. *Zoolog. Sci.* **23:** 689–697.

Chan, T., Kondow, A., Hosoya, A., Hitachi, K., Yukita, A., Okabayashi, K., Nakamura, H., Ozawa, H., Kiyonari, H., Michiue, T., et al. 2007. Ripply2 is essential for precise somite formation during mouse early development. *FEBS Lett.* **581:** 2691–2696.

Chapman, D.L. and Papaioannou, V.E. 1998. Three neural tubes in mouse embryos with mutations in the T-box gene *Tbx6*. *Nature* **391:** 695–697.

Chapman, D.L., Agulnik, I., Hancock, S., Silver, L.M., and Papaioannou, V.E. 1996. *Tbx6*, a mouse T-Box gene implicated in paraxial mesoderm formation at gastrulation. *Dev. Biol.* **180:** 534–542.

Charrier, J.B., Teillet, M.A., Lapointe, F., and Le Douarin, N.M. 1999. Defining subregions of Hensen's node essential for caudalward movement, midline development and cell survival. *Development* **126:** 4771–4783.

Chawengsaksophak, K., de Graaff, W., Rossant, J., Deschamps, J., and Beck, F. 2004. Cdx2 is essential for axial elongation in mouse development. *Proc. Natl. Acad. Sci.* **101:** 7641–7645.

Chen, Z.F. and Behringer, R.R. 1995. twist is required in head mesenchyme for cranial neural tube morphogenesis. *Genes Dev.* **9**: 686–699.

Chen, W. and Casey Corliss, D. 2004. Three modules of zebrafish Mind bomb work cooperatively to promote Delta ubiquitination and endocytosis. *Dev. Biol.* **267**: 361–373.

Chen, Y., Bellamy, W.P., Seabra, M.C., Field, M.C., and Ali, B.R. 2005. ER-associated protein degradation is a common mechanism underpinning numerous monogenic diseases including Robinow syndrome. *Hum. Mol. Genet.* **14**: 2559–2569.

Chevallier, A. 1975. Role of the somitic mesoderm in the development of the rib cage of bird embryos. I. Origin of the sternal component and conditions for the development of the ribs (author's transl.). *J. Embryol. Exp. Morphol.* **33**: 291–311.

Chevallier, A., Kieny, M., and Mauger, A. 1977. Limb-somite relationship: Origin of the limb musculature. *J. Embryol. Exp. Morphol.* **41**: 245–258.

Chiang, C., Litingtung, Y., Lee, E., Young, K.E., Corden, J.L., Westphal, H., and Beachy, P.A. 1996. Cyclopia and defective axial patterning in mice lacking Sonic hedgehog gene function. *Nature* **383**: 407–413.

Christ, B., Jacob, H.J., and Jacob, M. 1974. Somitogenesis in the chick embryo. Determination of the segmentation direction. *Verh. Anat. Ges.* **68**: 573–579.

Christ, B., Huang, R., and Scaal, M. 2004. Formation and differentiation of the avian sclerotome. *Anat. Embryol.* **208**: 333–350.

Christ, B., Huang, R., and Scaal, M. 2007. Amniote somite derivatives. *Dev. Dyn.* **236**: 2382–2396.

Ciruna, B. and Rossant, J. 2001. FGF signaling regulates mesoderm cell fate specification and morphogenetic movement at the primitive streak. *Dev. Cell* **1**: 37–49.

Ciruna, B.G., Schwartz, L., Harpal, K., Yamaguchi, T.P., and Rossant, J. 1997. Chimeric analysis of *fibroblast growth factor receptor-1* (*Fgfr1*) function: A role for FGFR1 in morphogenetic movement through the primitive streak. *Development* **124**: 2829–2841.

Cole, S.E., Levorse, J.M., Tilghman, S.M., and Vogt, T.F. 2002. Clock regulatory elements control cyclic expression of *Lunatic fringe* during somitogenesis. *Dev. Cell* **3**: 75–84.

Conlon, R.A., Reaume, A.G., and Rossant, J. 1995. Notch1 is required for the coordinate segmentation of somites. *Development* **121**: 1533–1545.

Cooke, J. and Zeeman, E.C. 1976. A clock and wavefront model for control of the number of repeated structures during animal morphogenesis. *J. Theor. Biol.* **58**: 455–476.

Cooke, J.E., Kemp, H.A., and Moens, C.B. 2005. EphA4 is required for cell adhesion and rhombomere-boundary formation in the zebrafish. *Curr. Biol.* **15**: 536–542.

Copf, T., Schroder, R., and Averof, M. 2004. Ancestral role of caudal genes in axis elongation and segmentation. *Proc. Natl. Acad. Sci.* **101**: 17711–17715.

Cordes, R., Schuster-Gossler, K., Serth, K., and Gossler, A. 2004. Specification of vertebral identity is coupled to Notch signalling and the segmentation clock. *Development* **131**: 1221–1233.

Cornier, A.S., Ramirez, N., Arroyo, S., Acevedo, J., Garcia, L., Carlo, S., and Korf, B. 2004. Phenotype characterization and natural history of spondylothoracic dysplasia syndrome: A series of 27 new cases. *Am. J. Med. Genet.* **128**: 120–126.

Cornier, A.S., Staehling-Hampton, K., Delventhal, K.M., Saga, Y., Caubet, J.F., Sasaki, N., Ellard, S., Young, E., Ramirez, N., Carlo, S.E., et al. 2008. Mutations in the MESP2 gene cause spondylothoracic dysostosis/Jarcho-Levin syndrome. *Am. J. Hum. Genet.* **82**: 1334–1341.

Correia, K.M. and Conlon, R.A. 2000. Surface ectoderm is necessary for the morphogenesis of somites. *Mech. Dev.* **91**: 19–30.

Crossley, P.H. and Martin, G.R. 1995. The mouse *Fgf8* gene encodes a family of polypep-

tides and is expressed in regions that direct outgrowth and patterning in the developing embryo. *Development* **121**: 439–451.

Cserjesi, P., Brown, D., Ligon, K.L., Lyons, G.E., Copeland, N.G., Gilbert, D.J., Jenkins, N.A.,
and Olson, E.N. 1995. Scleraxis: A basic helix-loop-helix protein that prefigures skeletal formation during mouse embryogenesis. *Development* **121**: 1099–1110.

Dale, J.K. and Pourquié, O. 2000. A clock-work somite. *Bioessays* **22**: 72–83.

Dale, J.K., Maroto, M., Dequeant, M.L., Malapert, P., McGrew, M., and Pourquié, O. 2003.
Periodic notch inhibition by lunatic fringe underlies the chick segmentation clock.
Nature **421**: 275–278.

Dale, J.K., Malapert, P., Chal, J., Vilhais-Neto, G., Maroto, M., Johnson, T., Jayasinghe, S.,
Trainor, P., Herrmann, B., and Pourquié, O. 2006. Oscillations of the snail genes in the
presomitic mesoderm coordinate segmental patterning and morphogenesis in vertebrate somitogenesis. *Dev. Cell* **10**: 355–366.

Davis, R.L. and Kirschner, M.W. 2000. The fate of cells in the tailbud of *Xenopus laevis*.
Development **127**: 255–267.

Debby-Brafman, A., Burstyn-Cohen, T., Klar, A., and Kalcheim, C. 1999. F-Spondin,
expressed in somite regions avoided by neural crest cells, mediates inhibition of distinct somite domains to neural crest migration. *Neuron* **22**: 475–488.

DeChiara, T.M., Kimble, R.B., Poueymirou, W.T., Rojas, J., Masiakowski, P., Valenzuela,
D.M., and Yancopoulos, G.D. 2000. Ror2, encoding a receptor-like tyrosine kinase, is
required for cartilage and growth plate development. *Nat. Genet.* **24**: 271–274.

del Barco Barrantes, I., Elia, A., Wünsch, K., Hrabě de Angelis, M., Mak, T., Rossant, J.,
Conlon, R., Gossler, A., and Luis de la Pompa, J. 1999. Interaction between Notch signalling and Lunatic Fringe during somite boundary formation in the mouse. *Curr. Biol.*
9: 470–480.

Delfini, M.C., Dubrulle, J., Malapert, P., Chal, J., and Pourquié, O. 2005. Control of the
segmentation process by graded MAPK/ERK activation in the chick embryo. *Proc. Natl.
Acad. Sci.* **102**: 11343–11348.

Deng, C.X., Wynshaw-Boris, A., Shen, M.M., Daugherty, C., Ornitz, D.M., and Leder, P.
1994. Murine FGFR-1 is required for early postimplantation growth and axial organization. *Genes Dev.* **8**: 3045–3057.

Dequeant, M.L. and Pourquié, O. 2008. Segmental patterning of the vertebrate embryonic
axis. *Nat. Rev. Genet.* **9**: 370–382.

Dequeant, M.L., Glynn, E., Gaudenz, K., Wahl, M., Chen, J., Mushegian, A., and Pourquié,
O. 2006. A complex oscillating network of signaling genes underlies the mouse segmentation clock. *Science* **314**: 1595–1598.

Deschamps, J. and van Nes, J. 2005. Developmental regulation of the Hox genes during
axial morphogenesis in the mouse. *Development* **132**: 2931–2942.

Dietrich, S., Schubert, F.R., and Gruss, P. 1993. Altered Pax gene expression in murine notochord mutants: The notochord is required to initiate and maintain ventral identity in
the somite. *Mech. Dev.* **44**: 189–207.

Diez del Corral, R., Olivera-Martinez, I., Goriely, A., Gale, E., Maden, M., and Storey, K.
2003. Opposing FGF and retinoid pathways control ventral neural pattern, neuronal
differentiation, and segmentation during body axis extension. *Neuron* **40**: 65–79.

Ding, H., Wu, X., Kim, I., Tam, P.P., Koh, G.Y., and Nagy, A. 2000. The mouse Pdgfc gene:
Dynamic expression in embryonic tissues during organogenesis. *Mech. Dev.* **96**: 209–213.

Dockter, J.L. 2000. Sclerotome induction and differentiation. *Curr. Top. Dev. Biol.* **48**: 77–127.

Dockter, J.L. and Ordahl, C.P. 1998. Determination of sclerotome to the cartilage fate.

Development **125:** 2113–2124.

Donoviel, D.B., Hadjantonakis, A.K., Ikeda, M., Zheng, H., Hyslop, P.S., and Bernstein, A. 1999. Mice lacking both presenilin genes exhibit early embryonic patterning defects. *Genes Dev.* **13:** 2801–2810.

Dottori, M., Hartley, L., Galea, M., Paxinos, G., Polizzotto, M., Kilpatrick, T., Bartlett, P.F., Murphy, M., Kontgen, F., and Boyd, A.W. 1998. EphA4 (Sek1) receptor tyrosine kinase is required for the development of the corticospinal tract. *Proc. Natl. Acad. Sci.* **95:** 13248–13253.

Duband, J.L., Dufour, S., Hatta, K., Takeichi, M., Edelman, G.M., and Thiery, J.P. 1987. Adhesion molecules during somitogenesis in the avian embryo. *J. Cell Biol.* **104:** 1361–1374.

Duboule, D. 2007. The rise and fall of Hox gene clusters. *Development* **134:** 2549–2560.

Duboule, D. and Morata, G. 1994. Colinearity and functional hierarchy among genes of the homeotic complexes. *Trends Genet.* **10:** 358–364.

Dubrulle, J. and Pourquié, O. 2004a. Coupling segmentation to axis formation. *Development* **131:** 5783–5793.

Dubrulle, J. and Pourquié, O. 2004b. *fgf8* mRNA decay establishes a gradient that couples axial elongation to patterning in the vertebrate embryo. *Nature* **427:** 419–422.

Dubrulle, J., McGrew, M.J., and Pourquié, O. 2001. FGF signaling controls somite boundary position and regulates segmentation clock control of spatiotemporal *Hox* gene activation. *Cell* **106:** 219–232.

Dunty Jr., W.C., Biris, K.K., Chalamalasetty, R.B., Taketo, M.M., Lewandoski, M., and Yamaguchi, T.P. 2008. Wnt3a/β-catenin signaling controls posterior body development by coordinating mesoderm formation and segmentation. *Development* **135:** 85–94.

Dunwoodie, S.L., Henrique, D., Harrison, S.M., and Beddington, R.S. 1997. Mouse *Dll3*: A novel divergent *Delta* gene which may complement the function of other *Delta* homologues during early pattern formation in the mouse embryo. *Development* **124:** 3065–3076.

Dunwoodie, S.L., Clements, M., Sparrow, D.B., Sa, X., Conlon, R.A., and Beddington, R.S. 2002. Axial skeletal defects caused by mutation in the spondylocostal dysplasia/pudgy gene *Dll3* are associated with disruption of the segmentation clock within the presomitic mesoderm. *Development* **129:** 1795–1806.

Duong, T.D. and Erickson, C.A. 2004. MMP-2 plays an essential role in producing epithelial-mesenchymal transformations in the avian embryo. *Dev. Dyn.* **229:** 42–53.

Durbin, L., Brennan, C., Shiomi, K., Cooke, J., Barrios, A., Shanmugalingam, S., Guthrie, B., Lindberg, R., and Holder, N. 1998. Eph signaling is required for segmentation and differentiation of the somites. *Genes Dev.* **12:** 3096–3109.

Eichmann, A., Marcelle, C., Bréant, C., and Le Douarin, N.M. 1993. Two molecules related to the VEGF receptor are expressed in early endothelial cells during avian embryonic development. *Mech. Dev.* **42:** 33–48.

Ellis, S. and Mellor, H. 2000. Regulation of endocytic traffic by rho family GTPases. *Trends Cell Biol.* **10:** 85–88.

Eloy-Trinquet, S. and Nicolas, J.F. 2002. Cell coherence during production of the presomitic mesoderm and somitogenesis in the mouse embryo. *Development* **129:** 3609–3619.

Essalmani, R., Zaid, A., Marcinkiewicz, J., Chamberland, A., Pasquato, A., Seidah, N.G., and Prat, A. 2008. In vivo functions of the proprotein convertase PC5/6 during mouse development: Gdf11 is a likely substrate. *Proc. Natl. Acad. Sci.* **105:** 5750–5755.

Evans, D.J. 2003. Contribution of somitic cells to the avian ribs. *Dev. Biol.* **256:** 114–126.

Evrard, Y.A., Lun, Y., Aulehla, A., Gan, L., and Johnson, R.L. 1998. *lunatic fringe* is an essential mediator of somite segmentation and patterning. *Nature* **394:** 377–381.

Ewan, K.B. and Everett, A.W. 1992. Evidence for resegmentation in the formation of the vertebral column using the novel approach of retroviral-mediated gene transfer. *Exp. Cell Res.* **198:** 315–320.

Fan, C.M. and Tessier-Lavigne, M. 1994. Patterning of mammalian somites by surface ectoderm and notochord: Evidence for sclerotome induction by a hedgehog homolog. *Cell* **79:** 1175–1186.

Fan, C.M., Porter, J.A., Chiang, C., Chang, D.T., Beachy, P.A., and Tessier-Lavigne, M. 1995. Long-range sclerotome induction by sonic hedgehog: Direct role of the amino-terminal cleavage product and modulation by the cyclic AMP signaling pathway. *Cell* **81:** 457–465.

Feller, J., Schneider, A., Schuster-Gossler, K., and Gossler, A. 2008. Noncyclic Notch activity in the presomitic mesoderm demonstrates uncoupling of somite compartmentalization and boundary formation. *Genes Dev.* **22:** 2166–2171.

Filipe, M., Goncalves, L., Bento, M., Silva, A.C., and Belo, J.A. 2006. Comparative expression of mouse and chicken Shisa homologues during early development. *Dev. Dyn.* **235:** 2567–2573.

Flenniken, A.M., Gale, N.W., Yancopoulos, G.D., and Wilkinson, D.G. 1996. Distinct and overlapping expression patterns of ligands for Eph-related receptor tyrosine kinases during mouse embryogenesis. *Dev. Biol.* **179:** 382–401.

Forsberg, H., Crozet, F., and Brown, N.A. 1998. Waves of mouse *Lunatic fringe* expression, in four-hour cycles at two-hour intervals, precede somite boundary formation. *Curr. Biol.* **8:** 1027–1030.

Freitas, C., Rodrigues, S., Charrier, J.B., Teillet, M.A., and Palmeirim, I. 2001. Evidence for medial/lateral specification and positional information within the presomitic mesoderm. *Development* **128:** 5139–5147.

Fryer, C.J., White, J.B., and Jones, K.A. 2004. Mastermind recruits CycC:CDK8 to phosphorylate the Notch ICD and coordinate activation with turnover. *Mol. Cell* **16:** 509–520.

Fuerer, C., Nusse, R., and Ten Berge, D. 2008. Wnt signalling in development and disease. Max Delbrück Center for Molecular Medicine meeting on Wnt Signaling in Development and Disease. *EMBO Rep.* **9:** 134–138.

Furumoto, T.A., Miura, N., Akasaka, T., Mizutani-Koseki, Y., Sudo, H., Fukuda, K., Maekawa, M., Yuasa, S., Fu, Y., Moriya, H., et al. 1999. Notochord-dependent expression of MFH1 and PAX1 cooperates to maintain the proliferation of sclerotome cells during the vertebral column development. *Dev. Biol* **210:** 15–29.

Furushima, K., Yamamoto, A., Nagano, T., Shibata, M., Miyachi, H., Abe, T., Ohshima, N., Kiyonari, H., and Aizawa, S. 2007. Mouse homologues of Shisa antagonistic to Wnt and Fgf signalings. *Dev. Biol.* **306:** 480–492.

Gajewski, M., Sieger, D., Alt, B., Leve, C., Hans, S., Wolff, C., Rohr, K.B., and Tautz, D. 2003. Anterior and posterior waves of cyclic her1 gene expression are differentially regulated in the presomitic mesoderm of zebrafish. *Development* **130:** 4269–4278.

Galceran, J., Fariñas, I., Depew, M.J., Clevers, H., and Grosschedl, R. 1999. Wnt3a$^{-/-}$-like phenotype and limb deficiency in Lef1$^{-/-}$Tcf1$^{-/-}$ mice. *Genes Dev.* **13:** 709–717.

Galceran, J., Sustmann, C., Hsu, S.C., Folberth, S., and Grosschedl, R. 2004. LEF1-mediated regulation of *Delta-like1* links Wnt and Notch signaling in somitogenesis. *Genes Dev.* **18:** 2718–2723.

Gammill, L.S., Gonzalez, C., Gu, C., and Bronner-Fraser, M. 2006. Guidance of trunk neural crest migration requires neuropilin 2/semaphorin 3F signaling. *Development* **133:** 99–106.

Geffers, I., Serth, K., Chapman, G., Jaekel, R., Schuster-Gossler, K., Cordes, R., Sparrow, D.B., Kremmer, E., Dunwoodie, S.L., Klein, T., et al. 2007. Divergent functions and distinct localization of the Notch ligands DLL1 and DLL3 in vivo. *J. Cell Biol.* **178:** 465–476.

Geling, A., Steiner, H., Willem, M., Bally-Cuif, L., and Haass, C. 2002. A γ-secretase inhibitor blocks Notch signaling in vivo and causes a severe neurogenic phenotype in zebrafish. *EMBO Rep.* **3:** 688–694.

George, E.L., Georges-Labouesse, E.N., Patel-King, R.S., Rayburn, H., and Hynes, R.O. 1993. Defects in mesoderm, neural tube and vascular development in mouse embryos lacking fibronectin. *Development* **119:** 1079–1091.

Georges-Labouesse, E.N., George, E.L., Rayburn, H., and Hynes, R.O. 1996. Mesodermal development in mouse embryos mutant for fibronectin. *Dev. Dyn.* **207:** 145–156.

Giudicelli, F., Ozbudak, E.M., Wright, G.J., and Lewis, J. 2007. Setting the tempo in development: An investigation of the zebrafish somite clock mechanism. *PLoS Biol.* **5:** e150.

Goh, K.L., Yang, J.T., and Hynes, R.O. 1997. Mesodermal defects and cranial neural crest apoptosis in α5 integrin-null embryos. *Development* **124:** 4309–4319.

Goldbeter, A. and Pourquié, O. 2008. Modeling the segmentation clock as a network of coupled oscillations in the Notch, Wnt and FGF signaling pathways. *J. Theor. Biol.* **252:** 574–585.

Goldbeter, A., Gonze, D., and Pourquié, O. 2007. Sharp developmental thresholds defined through bistability by antagonistic gradients of retinoic acid and FGF signaling. *Dev. Dyn.* **236:** 1495–1508.

Goldstein, R.S. and Kalcheim, C. 1992. Determination of epithelial half-somites in skeletal morphogenesis. *Development* **116:** 441–445.

Gomez, C., Ozbudak, E.M., Wunderlich, J., Baumann, D., Lewis, J., and Pourquié, O. 2008. Control of segment number in vertebrate embryos. *Nature* **254:** 335–339.

Gont, L.K., Steinbeisser, H., Blumberg, B., and de Robertis, E.M. 1993. Tail formation as a continuation of gastrulation: The multiple cell populations of the *Xenopus* tailbud derive from the late blastopore lip. *Development* **119:** 991–1004.

Goodrich, L.V., Johnson, R.L., Milenkovic, L., McMahon, J.A., and Scott, M.P. 1996. Conservation of the hedgehog/patched signaling pathway from flies to mice: Induction of a mouse patched gene by Hedgehog. *Genes Dev.* **10:** 301–312.

Goulding, M.D., Lumsden, A., and Gruss, P. 1993. Signals from the notochord and floor plate regulate the region-specific expression of two Pax genes in the developing spinal cord. *Development* **117:** 1001–1016.

Greco, T.L., Takada, S., Newhouse, M.M., McMahon, J.A., McMahon, A.P., and Camper, S.A. 1996. Analysis of the vestigial tail mutation demonstrates that Wnt-3a gene dosage regulates mouse axial development. *Genes Dev.* **10:** 313–324.

Gridley, T. 2006. The long and short of it: Somite formation in mice. *Dev. Dyn.* **235:** 2330–2336.

Gumbiner, B.M. 2005. Regulation of cadherin-mediated adhesion in morphogenesis. *Nat. Rev. Mol. Cell Biol.* **6:** 622–634.

Haenig, B. and Kispert, A. 2004. Analysis of TBX18 expression in chick embryos. *Dev. Genes Evol.* **214:** 407–411.

Hahn, H., Christiansen, J., Wicking, C., Zaphiropoulos, P.G., Chidambaram, A., Gerrard, B., Vorechovsky, I., Bale, A.E., Toftgard, R., Dean, M., et al. 1996. A mammalian patched homolog is expressed in target tissues of sonic hedgehog and maps to a region associated with developmental abnormalities. *J. Biol. Chem.* **271:** 12125–12128.

Haines, N. and Irvine, K.D. 2003. Glycosylation regulates Notch signalling. *Nat. Rev. Mol.*

Cell Biol. **4:** 786–797.

Halbleib, J.M. and Nelson, W.J. 2006. Cadherins in development: Cell adhesion, sorting, and tissue morphogenesis. *Genes Dev.* **20:** 3199–3214.

Hall, B.K. 1977. Chondrogenesis of the somitic mesoderm. *Adv. Anat. Embryol. Cell Biol.* **53:** 3–47.

Hamada, Y., Kadokawa, Y., Okabe, M., Ikawa, M., Coleman, J.R., and Tsujimoto, Y. 1999. Mutation in ankyrin repeats of the mouse Notch2 gene induces early embryonic lethality. *Development* **126:** 3415–3424.

Hamblet, N.S., Lijam, N., Ruiz-Lozano, P., Wang, J., Yang, Y., Luo, Z., Mei, L., Chien, K.R., Sussman, D.J., and Wynshaw-Boris, A. 2002. Dishevelled 2 is essential for cardiac outflow tract development, somite segmentation and neural tube closure. *Development* **129:** 5827–5838.

Hamburger, V. and Hamilton, H.L. 1992. A series of normal stages in the development of the chick embryo. 1951. *Dev. Dyn.* **195:** 231–272.

Hatada, Y. and Stern, C.D. 1994. A fate map of the epiblast of the early chick embryo. *Development* **120:** 2879–2889.

Hatta, K. and Takeichi, M. 1986. Expression of N-cadherin adhesion molecules associated with early morphogenetic events in chick development. *Nature* **320:** 447–449.

He, X. 2005. Antagonizing Wnt and FGF receptors: An enemy from within (the ER). *Cell* **120:** 156–158.

Healy, C., Uwanogho, D., and Sharpe, P.T. 1996. Expression of the chicken Sox9 gene marks the onset of cartilage differentiation. *Ann. N.Y. Acad. Sci.* **785:** 261–262.

Healy, C., Uwanogho, D., and Sharpe, P.T. 1999. Regulation and role of Sox9 in cartilage formation. *Dev. Dyn.* **215:** 69–78.

Henderson, D.J., Ybot-Gonzalez, P., and Copp, A.J. 1997. Over-expression of the chondroitin sulphate proteoglycan versican is associated with defective neural crest migration in the Pax3 mutant mouse (splotch). *Mech. Dev.* **69:** 39–51.

Henrique, D., Adam, J., Myat, A., Chitnis, A., Lewis, J., and Ish-Horowicz, D. 1995. Expression of a *Delta* homologue in prospective neurons in the chick. *Nature* **375:** 787–790.

Henry, C.A., Urban, M.K., Dill, K.K., Merlie, J.P., Page, M.F., Kimmel, C.B., and Amacher, S.L. 2002. Two linked *hairy/Enhancer of split*-related zebrafish genes, *her1* and *her7*, function together to refine alternating somite boundaries. *Development* **129:** 3693–3704.

Herbrand, H., Pabst, O., Hill, R., and Arnold, H.H. 2002. Transcription factors *Nkx3.1* and *Nkx3.2* (*Bapx1*) play an overlapping role in sclerotomal development of the mouse. *Mech. Dev.* **117:** 217–224.

Herreman, A., Serneels, L., Annaert, W., Collen, D., Schoonjans, L., and De Strooper B. 2000. Total inactivation of γ-secretase activity in presenilin-deficient embryonic stem cells. *Nat. Cell Biol.* **2:** 461–462.

Herrmann, B.G., Labeit, S., Poustka, A., King, T.R., and Lehrach, H. 1990. Cloning of the *T* gene required in mesoderm formation in the mouse. *Nature* **343:** 617–622.

Hirata, H., Bessho, Y., Kokubu, H., Masamizu, Y., Yamada, S., Lewis, J., and Kageyama, R. 2004. Instability of Hes7 protein is crucial for the somite segmentation clock. *Nat. Genet.* **36:** 750–754.

Hirsinger, E., Duprez, D., Jouve, C., Malapert, P., Cooke, J., and Pourquié, O. 1997. Noggin acts downstream of Wnt and Sonic Hedgehog to antagonize BMP4 in avian somite patterning. *Development* **124:** 4605–4614.

Hitachi, K., Kondow, A., Danno, H., Inui, M., Uchiyama, H., and Asashima, M. 2008. Tbx6, Thylacine1, and E47 synergistically activate bowline expression in *Xenopus* somitogenesis. *Dev. Biol.* **313**: 816–828.

Hoang, B.H., Thomas, J.T., Abdul-Karim, F.W., Correia, K.M., Conlon, R.A., Luyten, F.P., and Ballock, R.T. 1998. Expression pattern of two *Frizzled*-related genes, *Frzb-1* and *Sfrp-1*, during mouse embryogenesis suggests a role for modulating action of *Wnt* family members. *Dev. Dyn.* **212**: 364–372.

Hoch, R.V. and Soriano, P. 2003. Roles of PDGF in animal development. *Development* **130**: 4769–4784.

Hoch, R.V. and Soriano, P. 2006. Context-specific requirements for Fgfr1 signaling through Frs2 and Frs3 during mouse development. *Development* **133**: 663–673.

Hofmann, M., Schuster-Gossler, K., Watabe-Rudolph, M., Aulehla, A., Herrmann, B.G., and Gossler, A. 2004. WNT signaling, in synergy with T/TBX6, controls Notch signaling by regulating *Dll1* expression in the presomitic mesoderm of mouse embryos. *Genes Dev.* **18**: 2712–2717.

Holley, S.A. 2007. The genetics and embryology of zebrafish metamerism. *Dev. Dyn.* **236**: 1422–1449.

Holley, S.A., Geisler, R., and Nüsslein-Volhard, C. 2000. Control of *her1* expression during zebrafish somitogenesis by a *Delta*-dependent oscillator and an independent wave-front activity. *Genes Dev.* **14**: 1678–1690.

Holley, S.A., Jülich, D., Rauch, G.J., Geisler, R., and Nüsslein-Volhard, C. 2002. *her1* and the *notch* pathway function within the oscillator mechanism that regulates zebrafish somitogenesis. *Development* **129**: 1175–1183.

Hoppler, S. and Kavanagh, C.L. 2007. Wnt signalling: Variety at the core. *J. Cell Sci.* **120**: 385–393.

Horikawa, K., Radice, G., Takeichi, M., and Chisaka, O. 1999. Adhesive subdivisions intrinsic to the epithelial somites. *Dev. Biol.* **215**: 182–189.

Horikawa, K., Ishimatsu, K., Yoshimoto, E., Kondo, S., and Takeda, H. 2006. Noise-resistant and synchronized oscillation of the segmentation clock. *Nature* **441**: 719–723.

Hornik, C., Brand-Saberi, B., Rudloff, S., Christ, B., and Fuchtbauer, E.M. 2004. Twist is an integrator of SHH, FGF, and BMP signaling. *Anat. Embryol.* **209**: 31–39.

Houle, M., Prinos, P., Iulianella, A., Bouchard, N., and Lohnes, D. 2000. Retinoic acid regulation of Cdx1: An indirect mechanism for retinoids and vertebral specification. *Mol. Cell. Biol.* **20**: 6579–6586.

Houle, M., Allan, D., Lohnes, D., and Lufkin, T. 2003a. Cdx homeodomain proteins in vertebral patterning. *Adv. Dev. Biol. Biochem.* **13**: 69–105.

Houle, M., Sylvestre, J.R., and Lohnes, D. 2003b. Retinoic acid regulates a subset of Cdx1 function in vivo. *Development* **130**: 6555–6567.

Hrabě de Angelis, M., McIntyre II, J., and Gossler, A. 1997. Maintenance of somite borders in mice requires the *Delta* homologue *Dll1*. *Nature* **386**: 717–721.

Hsieh, J.C., Kodjabachian, L., Rebbert, M.L., Rattner, A., Smallwood, P.M., Samos, C.H., Nusse, R., Dawid, I.B., and Nathans, J. 1999. A new secreted protein that binds to Wnt proteins and inhibits their activities. *Nature* **398**: 431–436.

Huang, R., Zhi, Q., Wilting, J., and Christ, B. 1994. The fate of somitocoele cells in avian embryos. *Anat. Embryol.* **190**: 243–250.

Huang, R., Zhi, Q., Neubüser, A., Müller, T.S., Brand-Saberi, B., Christ, B., and Wilting, J. 1996. Function of somite and somitocoele cells in the formation of the vertebral motion segment in avian embryos. *Acta Anat.* **155**: 231–241.

Huang, R., Zhi, Q., Ordahl, C.P., and Christ, B. 1997. The fate of the first avian somite. *Anat. Embryol.* **195**: 435–449.

Huang, R., Zhi, Q., Brand-Saberi, B., and Christ, B. 2000a. New experimental evidence for somite resegmentation. *Anat. Embryol.* **202:** 195–200.

Huang, R., Zhi, Q., Patel, K., Wilting, J., and Christ, B. 2000b. Dual origin and segmental organisation of the avian scapula. *Development* **127:** 3789–3794.

Huang, R., Zhi, Q., Schmidt, C., Wilting, J., Brand-Saberi, B., and Christ, B. 2000c. Sclerotomal origin of the ribs. *Development* **127:** 527–532.

Hui, C.C., Slusarski, D., Platt, K.A., Holmgren, R., and Joyner, A.L. 1994. Expression of three mouse homologs of the *Drosophila* segment polarity gene *cubitus interruptus, Gli, Gli-2, and Gli-3*, in ectoderm- and mesoderm-derived tissues suggests multiple roles during postimplantation development. *Dev. Biol.* **162:** 402–413.

Hukriede, N.A., Tsang, T.E., Habas, R., Khoo, P.L., Steiner, K., Weeks, D.L., Tam, P.P., and Dawid, I.B. 2003. Conserved requirement of Lim1 function for cell movements during gastrulation. *Dev. Cell* **4:** 83–94.

Huppert, S.S., Ilagan, M.X., De Strooper, B., and Kopan, R. 2005. Analysis of Notch function in presomitic mesoderm suggests a γ-secretase-independent role for presenilins in somite differentiation. *Dev. Cell* **8:** 677–688.

Iimura, T. and Pourquié, O. 2006. Collinear activation of *Hoxb* genes during gastrulation is linked to mesoderm cell ingression. *Nature* **442:** 568–571.

Iimura, T., Yang, X., Weijer, C.J., and Pourquié, O. 2007. Dual mode of paraxial mesoderm formation during chick gastrulation. *Proc. Natl. Acad. Sci.* **104:** 2744–2749.

Ikeya, M. and Takada, S. 2001. Wnt-3a is required for somite specification along the antero-posterior axis of the mouse embryo and for regulation of *cdx-1* expression. *Mech. Dev.* **103:** 27–33.

Ilic, D., Furuta, Y., Kanazawa, S., Takeda, N., Sobue, K., Nakatsuji, N., Nomura, S., Fujimoto, J., Okada, M., and Yamamoto, T. 1995. Reduced cell motility and enhanced focal adhesion contact formation in cells from FAK-deficient mice. *Nature* **377:** 539–544.

Inoue, T., Ota, M., Mikoshiba, K., and Aruga, J. 2007. *Zic2* and *Zic3* synergistically control neurulation and segmentation of paraxial mesoderm in mouse embryo. *Dev. Biol.* **306:** 669–684.

Irving, C., Nieto, M.A., DasGupta, R., Charnay, P., and Wilkinson, D.G. 1996. Progressive spatial restriction of *Sek-1* and *Krox-20* gene expression during hindbrain segmentation. *Dev. Biol.* **173:** 26–38.

Isaacs, H.V., Pownall, M.E., and Slack, J.M. 1994. eFGF regulates Xbra expression during *Xenopus* gastrulation. *EMBO J.* **13:** 4469–4481.

Ishikawa, A., Kitajima, S., Takahashi, Y., Kokubo, H., Kanno, J., Inoue, T., and Saga, Y. 2004. Mouse Nkd1, a Wnt antagonist, exhibits oscillatory gene expression in the PSM under the control of Notch signaling. *Mech. Dev.* **121:** 1443–1453.

Itoh, M., Kim, C.H., Palardy, G., Oda, T., Jiang, Y.J., Maust, D., Yeo, S.Y., Lorick, K., Wright, G.J., Ariza-McNaughton, L., et al. 2003. Mind bomb is a ubiquitin ligase that is essential for efficient activation of Notch signaling by Delta. *Dev. Cell* **4:** 67–82.

Jho, E.H., Zhang, T., Domon, C., Joo, C.K., Freund, J.N., and Costantini, F. 2002. Wnt/β-catenin/Tcf signaling induces the transcription of Axin2, a negative regulator of the signaling pathway. *Mol. Cell. Biol.* **22:** 1172–1183.

Jiang, Y.J., Aerne, B.L., Smithers, L., Haddon, C., Ish-Horowicz, D., and Lewis, J. 2000. Notch signalling and the synchronization of the somite segmentation clock. *Nature* **408:** 475–479.

Johnson, R.L., Laufer, E., Riddle, R.D., and Tabin, C. 1994. Ectopic expression of Sonic hedgehog alters dorsal-ventral patterning of somites. *Cell* **79:** 1165–1173.

Joutel, A. and Tournier-Lasserve, E. 1998. Notch signalling pathway and human diseases. *Semin. Cell Dev. Biol.* **9:** 619–625.

Jouve, C., Palmeirim, I., Henrique, D., Beckers, J., Gossler, A., Ish-Horowicz, D., and Pourquié, O. 2000. Notch signalling is required for cyclic expression of the hairy-like gene *HES1* in the presomitic mesoderm. *Development* **127:** 1421–1429.

Juliano, R.L. 2002. Signal transduction by cell adhesion receptors and the cytoskeleton: Functions of integrins, cadherins, selectins, and immunoglobulin-superfamily members. *Annu. Rev. Pharmacol. Toxicol.* **42:** 283–323.

Jülich, D., Hwee Lim, C., Round, J., Nicolaije, C., Schroeder, J., Davies, A., Geisler, R., Lewis, J., Jiang, Y.J., and Holley, S.A. 2005. *beamter/deltaC* and the role of Notch ligands in the zebrafish somite segmentation, hindbrain neurogenesis and hypochord differentiation. *Dev. Biol.* **286:** 391–404.

Kageyama, R., Ohtsuka, T., and Kobayashi, T. 2007. The Hes gene family: Repressors and oscillators that orchestrate embryogenesis. *Development* **134:** 1243–1251.

Kalcheim, C. and Ben-Yair, R. 2005. Cell rearrangements during development of the somite and its derivatives. *Curr. Opin. Genet. Dev.* **15:** 371–380.

Kanki, J.P. and Ho, R.K. 1997. The development of the posterior body in zebrafish. *Development* **124:** 881–893.

Kawakami, Y., Raya, A., Raya, R.M., Rodriguez-Esteban, C., and Belmonte, J.C. 2005. Retinoic acid signalling links left-right asymmetric patterning and bilaterally symmetric somitogenesis in the zebrafish embryo. *Nature* **435:** 165–171.

Kawamura, A., Koshida, S., Hijikata, H., Ohbayashi, A., Kondoh, H., and Takada, S. 2005. Groucho-associated transcriptional repressor ripply1 is required for proper transition from the presomitic mesoderm to somites. *Dev. Cell* **9:** 735–744.

Kawano, Y. and Kypta, R. 2003. Secreted antagonists of the Wnt signalling pathway. *J. Cell Sci.* **116:** 2627–2634.

Kessel, M. 1992. Respecification of vertebral identities by retinoic acid. *Development* **115:** 487–501.

Kessel, M. and Gruss, P. 1991. Homeotic transformations of murine vertebrae and concomitant alteration of Hox codes induced by retinoic acid. *Cell* **67:** 89–104.

Keynes, R.J. and Stern, C.D. 1984. Segmentation in the vertebrate nervous system. *Nature* **310:** 786–789.

Kieny, M., Mauger, A., and Sengel, P. 1972. Early regionalization of somitic mesoderm as studied by the development of axial skeleton of the chick embryo. *Dev. Biol.* **28:** 142–161.

Kim, S.H., Yamamoto, A., Bouwmeester, T., Agius, E., and Robertis, E.M. 1998. The role of paraxial protocadherin in selective adhesion and cell movements of the mesoderm during *Xenopus* gastrulation. *Development* **125:** 4681–4690.

Kim, S.H., Jen, W.C., De Robertis, E.M., and Kintner, C. 2000. The protocadherin PAPC establishes segmental boundaries during somitogenesis in *Xenopus* embryos. *Curr. Biol.* **10:** 821–830.

Kim, S.Y., Paylor, S.W., Magnuson, T., and Schumacher, A. 2006. Juxtaposed Polycomb complexes co-regulate vertebral identity. *Development* **133:** 4957–4968.

Kimura, Y., Matsunami, H., Inoue, T., Shimamura, K., Uchida, N., Ueno, T., Miyazaki, T., and Takeichi, M. 1995. Cadherin-11 expressed in association with mesenchymal morphogenesis in the head, somite, and limb bud of early mouse embryos. *Dev. Biol.* **169:** 347–358.

Klaus, A. and Birchmeier, W. 2008. Wnt signalling and its impact on development and cancer. *Nat. Rev.* **8:** 387–398.

Klinghoffer, R.A., Hamilton, T.G., Hoch, R., and Soriano, P. 2002. An allelic series at the PDGFαR locus indicates unequal contributions of distinct signaling pathways during development. *Dev. Cell* **2:** 103–113.

Klootwijk, R., Franke, B., van der Zee, C.E., de Boer, R.T., Wilms, W., Hol, F.A., and Mariman, E.C. 2000. A deletion encompassing Zic3 in bent tail, a mouse model for X-linked neural tube defects. *Hum. Mol. Genet.* **9:** 1615–1622.

Kmita, M. and Duboule, D. 2003. Organizing axes in time and space; 25 years of colinear tinkering. *Science* **301:** 331–333.

Knezevic, V., De Santo, R., and Mackem, S. 1998. Continuing organizer function during chick tail development. *Development* **125:** 1791–1801.

Koizumi, K., Nakajima, M., Yuasa, S., Saga, Y., Sakai, T., Kuriyama, T., Shirasawa, T., and Koseki, H. 2001. The role of presenilin 1 during somite segmentation. *Development* **128:** 1391–1402.

Kokubu, C., Heinzmann, U., Kokubu, T., Sakai, N., Kubota, T., Kawai, M., Wahl, M.B., Galceran, J., Grosschedl, R., Ozono, K., et al. 2004. Skeletal defects in *ringelschwanz* mutant mice reveal that Lrp6 is required for proper somitogenesis and osteogenesis. *Development* **131:** 5469–5480.

Koo, B.K., Lim, H.S., Song, R., Yoon, M.J., Yoon, K.J., Moon, J.S., Kim, Y.W., Kwon, M.C., Yoo, K.W., Kong, M.P., et al. 2005. Mind bomb 1 is essential for generating functional Notch ligands to activate Notch. *Development* **132:** 3459–3470.

Kopan, R. and Ilagan, M.X. 2004. γ-Secretase: Proteasome of the membrane? *Nat. Rev. Mol. Cell Biol.* **5:** 499–504.

Kraus, F., Haenig, B., and Kispert, A. 2001. Cloning and expression analysis of the mouse T-box gene Tbx18. *Mech. Dev.* **100:** 83–86.

Krebs, L.T., Deftos, M.L., Bevan, M.J., and Gridley, T. 2001. The *Nrarp* gene encodes an ankyrin-repeat protein that is transcriptionally regulated by the notch signaling pathway. *Dev. Biol.* **238:** 110–119.

Krumlauf, R. 1994. Hox genes in vertebrate development. *Cell* **78:** 191–201.

Kuan, C.Y., Tannahill, D., Cook, G.M., and Keynes, R.J. 2004. Somite polarity and segmental patterning of the peripheral nervous system. *Mech. Dev.* **121:** 1055–1068.

Kulesa, P.M. and Fraser, S.E. 2002. Cell dynamics during somite boundary formation revealed by time-lapse analysis. *Science* **298:** 991–995.

Kulesa, P.M., Schnell, S., Rudloff, S., Baker, R.E., and Maini, P.K. 2007. From segment to somite: Segmentation to epithelialization analyzed within quantitative frameworks. *Dev. Dyn.* **236:** 1392–1402.

Kullander, K. and Klein, R. 2002. Mechanisms and functions of Eph and ephrin signalling. *Nat. Rev. Mol. Cell Biol.* **3:** 475–486.

Kume, T., Jiang, H., Topczewska, J.M., and Hogan, B.L. 2001. The murine winged helix transcription factors, Foxc1 and Foxc2, are both required for cardiovascular development and somitogenesis. *Genes Dev.* **15:** 2470–2482.

Kusumi, K., Sun, E.S., Kerrebrock, A.W., Bronson, R.T., Chi, D.C., Bulotsky, M.S., Spencer, J.B., Birren, B.W., Frankel, W.N., and Lander, E.S. 1998. The mouse pudgy mutation disrupts *Delta* homologue *Dll3* and initiation of early somite boundaries. *Nat. Genet* **19:** 274–278.

Ladher, R.K., Church, V.L., Allen, S., Robson, L., Abdelfattah, A., Brown, N.A., Hattersley, G., Rosen, V., Luyten, F.P., Dale, L., et al. 2000. Cloning and expression of the Wnt antagonists *Sfrp-2* and *Frzb* during chick development. *Dev. Biol.* **218:** 183–198.

Ladi, E., Nichols, J.T., Ge, W., Miyamoto, A., Yao, C., Yang, L.T., Boulter, J., Sun, Y.E.,

Kintner, C., and Weinmaster, G. 2005. The divergent DSL ligand Dll3 does not activate Notch signaling but cell autonomously attenuates signaling induced by other DSL ligands. *J. Cell Biol.* **170:** 983–992.

Lamar, E., Deblandre, G., Wettstein, D., Gawantka, V., Pollet, N., Niehrs, C., and Kintner, C. 2001. Nrarp is a novel intracellular component of the Notch signaling pathway. *Genes Dev.* **15:** 1885–1899.

Landolt, R.M., Vaughan, L., Winterhalter, K.H., and Zimmermann, D.R. 1995. Versican is selectively expressed in embryonic tissues that act as barriers to neural crest cell migration and axon outgrowth. *Development* **121:** 2303–2312.

Lash, J.W. and Yamada, K. 1986. The adhesion signal of fibronectin: A possible trigger mechanism for compaction during somitogenesis. In *Somites in developing embryos* (ed. R. Bellairs et al.), pp. 201–208. Plenum, New York.

Lash, J.W., Seitz, A.W., Cheney, C.M., and Ostrovsky, D. 1984. On the role of fibronectin during the compaction stage of somitogenesis in the chick embryo. *J. Exp. Zool.* **232:** 197–206.

Le Borgne, R. 2006. Regulation of Notch signalling by endocytosis and endosomal sorting. *Curr. Opin. Cell Biol.* **18:** 213–222.

Leimeister, C., Dale, K., Fischer, A., Klamt, B., Hrabě de Angelis, M., Radtke, F., McGrew, M.J., Pourquié, O., and Gessler, M. 2000. Oscillating expression of *c-hey2* in the presomitic mesoderm suggests that the segmentation clock may use combinatorial signaling through multiple interacting bHLH factors. *Dev. Biol.* **227:** 91–103.

Leitges, M., Neidhardt, L., Haenig, B., Herrmann, B.G., and Kispert, A. 2000. The paired homeobox gene *Uncx4.1* specifies pedicles, transverse processes and proximal ribs of the vertebral column. *Development* **127:** 2259–2267.

Lettice, L.A., Purdie, L.A., Carlson, G.J., Kilanowski, F., Dorin, J., and Hill, R.E. 1999. The mouse bagpipe gene controls development of axial skeleton, skull, and spleen. *Proc. Natl. Acad. Sci.* **96:** 9695–9700.

Leung, J.Y., Kolligs, F.T., Wu, R., Zhai, Y., Kuick, R., Hanash, S., Cho, K.R., and Fearon, E.R. 2002. Activation of AXIN2 expression by β-catenin-T cell factor. A feedback repressor pathway regulating Wnt signaling. *J. Biol. Chem.* **277:** 21657–21665.

Leve, C., Gajewski, M., Rohr, K.B., and Tautz, D. 2001. Homologues of *c-hairy1* (*her9*) and *lunatic fringe* in zebrafish are expressed in the developing central nervous system, but not in the presomitic mesoderm. *Dev. Genes Evol.* **211:** 493–500.

Lewis, J. 2003. Autoinhibition with transcriptional delay: A simple mechanism for the zebrafish somitogenesis oscillator. *Curr. Biol.* **13:** 1398–1408.

Lewis, J. and Ozbudak, E.M. 2007. Deciphering the somite segmentation clock: Beyond mutants and morphants. *Dev. Dyn.* **236:** 1410–1415.

Li, L., Krantz, I.D., Deng, Y., Genin, A., Banta, A.B., Collins, C.C., Qi, M., Trask, B.J., Kuo, W.L., Cochran, J., et al. 1997. Alagille syndrome is caused by mutations in human *Jagged1*, which encodes a ligand for Notch1. *Nat. Genet.* **16:** 243–251.

Li, T., Ma, G., Cai, H., Price, D.L., and Wong, P.C. 2003a. Nicastrin is required for assembly of presenilin/γ-secretase complexes to mediate Notch signaling and for processing and trafficking of β-amyloid precursor protein in mammals. *J. Neurosci* **23:** 3272–3277.

Li, Y., Fenger, U., Niehrs, C., and Pollet, N. 2003b. Cyclic expression of *esr9* gene in *Xenopus* presomitic mesoderm. *Differentiation* **71:** 83–89.

Linask, K.K., Ludwig, C., Han, M.D., Liu, X., Radice, G.L., and Knudsen, K.A. 1998. N-cadherin/catenin-mediated morphoregulation of somite formation. *Dev. Biol.* **202:** 85–102.

Lindsell, C.E., Shawber, C.J., Boulter, J., and Weinmaster, G. 1995. Jagged: A mammalian ligand that activates Notch1. *Cell* **80:** 909–917.

Linker, C., Lesbros, C., Gros, J., Burrus, L.W., Rawls, A., and Marcelle, C. 2005. β-Catenin-dependent Wnt signalling controls the epithelial organisation of somites through the activation of paraxis. *Development* **132**: 3895–3905.

Lohnes, D. 2003. The Cdx1 homeodomain protein: An integrator of posterior signaling in the mouse. *Bioessays* **25**: 971–980.

Lohnes, D., Mark, M., Mendelsohn, C., Dollé, P., Dierich, A., Gorry, P., Gansmuller, A., and Chambon, P. 1994. Function of the retinoic acid receptors (RARs) during development. I. Craniofacial and skeletal abnormalities in RAR double mutants. *Development* **120**: 2723–2748.

Love, J.M. and Tuan, R.S. 1993. Pair-rule gene expression in the somitic stage chick embryo: Association with somite segmentation and border formation. *Differentiation* **54**: 73–83.

Lunn, J.S., Fishwick, K.J., Halley, P.A., and Storey, K.G. 2007. A spatial and temporal map of FGF/Erk1/2 activity and response repertoires in the early chick embryo. *Dev. Biol.* **302**: 536–552.

Ma, G., Li, T., Price, D.L., and Wong, P.C. 2005. APH-1a is the principal mammalian APH-1 isoform present in γ-secretase complexes during embryonic development. *J. Neurosci.* **25**: 192–198.

MacDonald, B.T., Adamska, M., and Meisler, M.H. 2004. Hypomorphic expression of Dkk1 in the doubleridge mouse: Dose dependence and compensatory interactions with Lrp6. *Development* **131**: 2543–2552.

Malbon, C.C. 2005. G proteins in development. *Nat. Rev. Mol. Cell Biol.* **6**: 689–701.

Mallo, M., Vinagre, T., and Carapuco, M. 2008. The road to the vertebral formula. *Int. J. Dev. Biol.* **52**: (in press).

Mankoo, B.S., Skuntz, S., Harrigan, I., Grigorieva, E., Candia, A., Wright, C.V., Arnheiter, H., and Pachnis, V. 2003. The concerted action of Meox homeobox genes is required upstream of genetic pathways essential for the formation, patterning and differentiation of somites. *Development* **130**: 4655–4664.

Mansouri, A., Yokota, Y., Wehr, R., Copeland, N.G., Jenkins, N.A., and Gruss, P. 1997. Paired-related murine homeobox gene expressed in the developing sclerotome, kidney, and nervous system. *Dev. Dyn.* **210**: 53–65.

Mansouri, A., Voss, A.K., Thomas, T., Yokota, Y., and Gruss, P. 2000. *Uncx4.1* is required for the formation of the pedicles and proximal ribs and acts upstream of *Pax9*. *Development* **127**: 2251–2258.

Mara, A., Schroeder, J., Chalouni, C., and Holley, S.A. 2007. Priming, initiation and synchronization of the segmentation clock by *deltaD* and *deltaC*. *Nat. Cell Biol.* **9**: 523–530.

Marcelle, C., Stark, M.R., and Bronner-Fraser, M. 1997. Coordinate actions of BMPs, Wnts, Shh and Noggin mediate patterning of the dorsal somite. *Development* **124**: 3955–3963.

Marcelle, C., Ahlgren, S., and Bronner-Fraser, M. 1999. In vivo regulation of somite differentiation and proliferation by Sonic Hedgehog. *Dev. Biol.* **214**: 277–287.

Maruhashi, M., Van De Putte, T., Huylebroeck, D., Kondoh, H., and Higashi, Y. 2005. Involvement of SIP1 in positioning of somite boundaries in the mouse embryo. *Dev. Dyn.* **234**: 332–338.

Masamizu, Y., Ohtsuka, T., Takashima, Y., Nagahara, H., Takenaka, Y., Yoshikawa, K., Okamura, H., and Kageyama, R. 2006. Real-time imaging of the somite segmentation clock: Revelation of unstable oscillators in the individual presomitic mesoderm cells. *Proc. Natl. Acad. Sci.* **103**: 1313–1318.

Matsuda, T., Nomi, M., Ikeya, M., Kani, S., Oishi, I., Terashima, T., Takada, S., and Minami, Y. 2001. Expression of the receptor tyrosine kinase genes, *Ror1* and *Ror2*, during mouse development. *Mech. Dev.* **105:** 153–156.

McGrew, M.J., Dale, J.K., Fraboulet, S., and Pourquié, O. 1998. The *lunatic Fringe* gene is a target of the molecular clock linked to somite segmentation in avian embryos. *Curr. Biol.* **8:** 979–982.

McGrew, M.J., Sherman, A., Lillico, S.G., Ellard, F.M., Radcliffe, P.A., Gilhooley, H.J., Mitrophanous, K.A., Cambray, N., Wilson, V., and Sang, H. 2008. Localised axial progenitor cell populations in the avian tail bud are not committed to a posterior Hox identity. *Development* **135:** 2289–2299.

McMahon, J.A., Takada, S., Zimmerman, L.B., Fan, C.M., Harland, R.M., and McMahon, A.P. 1998. Noggin-mediated antagonism of BMP signaling is required for growth and patterning of the neural tube and somite. *Genes Dev.* **12:** 1438–1452.

McPherron, A.C., Lawler, A.M., and Lee, S.J. 1999. Regulation of anterior/posterior patterning of the axial skeleton by growth/differentiation factor 11. *Nat. Genet.* **22:** 260–264.

Medina, A., Swain, R.K., Kuerner, K.M., and Steinbeisser, H. 2004. *Xenopus* paraxial protocadherin has signaling functions and is involved in tissue separation. *EMBO J.* **23:** 3249–3258.

Menkes, B. and Sandor, S. 1969. Researches on the development of axial organs. V. *Rev. Roum. Embryol. Cytol.* **6:** 65–72.

Merzdorf, C.S. 2007. Emerging roles for zic genes in early development. *Dev. Dyn.* **236:** 922–940.

Miner, J.H. and Yurchenco, P.D. 2004. Laminin functions in tissue morphogenesis. *Annu. Rev. Cell Dev. Biol.* **20:** 255–284.

Mitra, S.K., Hanson, D.A., and Schlaepfer, D.D. 2005. Focal adhesion kinase: In command and control of cell motility. *Nat. Rev. Mol. Cell Biol.* **6:** 56–68.

Mittapalli, V.R., Huang, R., Patel, K., Christ, B., and Scaal, M. 2005. Arthrotome: A specific joint forming compartment in the avian somite. *Dev. Dyn.* **234:** 48–53.

Miura, N., Wanaka, A., Tohyama, M., and Tanaka, K. 1993. MFH-1, a new member of the fork head domain family, is expressed in developing mesenchyme. *FEBS Lett.* **326:** 171–176.

Miura, S., Davis, S., Klingensmith, J., and Mishina, Y. 2006. BMP signaling in the epiblast is required for proper recruitment of the prospective paraxial mesoderm and development of the somites. *Development* **133:** 3767–3775.

Mo, R., Freer, A.M., Zinyk, D.L., Crackower, M.A., Michaud, J., Heng, H.H., Chik, K.W., Shi, X.M., Tsui, L.C., et al. 1997. Specific and redundant functions of *Gli2* and *Gli3* zinc finger genes in skeletal patterning and development. *Development* **124:** 113–123.

Monsoro-Burq, A.H. 2005. Sclerotome development and morphogenesis: When experimental embryology meets genetics. *Int. J. Dev. Biol.* **49:** 301–308.

Morales, A.V., Yasuda, Y., and Ish-Horowicz, D. 2002. Periodic *Lunatic fringe* expression is controlled during segmentation by a cyclic transcriptional enhancer responsive to notch signaling. *Dev. Cell* **3:** 63–74.

Moreno, T.A. and Kintner, C. 2004. Regulation of segmental patterning by retinoic acid signaling during *Xenopus* somitogenesis. *Dev. Cell* **6:** 205–218.

Morimoto, M., Takahashi, Y., Endo, M., and Saga, Y. 2005. The Mesp2 transcription factor establishes segmental borders by suppressing Notch activity. *Nature* **435:** 354–359.

Morimoto, M., Kiso, M., Sasaki, N., and Saga, Y. 2006. Cooperative Mesp activity is required for normal somitogenesis along the anterior-posterior axis. *Dev. Biol.* **300:** 687–698.

Morimoto, M., Sasaki, N., Oginuma, M., Kiso, M., Igarashi, K., Aizaki, K., Kanno, J., and Saga, Y. 2007. The negative regulation of Mesp2 by mouse Ripply2 is required to establish the rostro-caudal patterning within a somite. *Development* **134**: 1561–1569.

Motoyama, J., Takabatake, T., Takeshima, K., and Hui, C. 1998. Ptch2, a second mouse Patched gene is co-expressed with Sonic hedgehog (letter). *Nat. Genet.* **18**: 104–106.

Mukhopadhyay, M., Shtrom, S., Rodriguez-Esteban, C., Chen, L., Tsukui, T., Gomer, L., Dorward, D.W., Glinka, A., Grinberg, A., Huang, S.P., et al. 2001. *Dickkopf1* is required for embryonic head induction and limb morphogenesis in the mouse. *Dev. Cell* **1**: 423–434.

Müller, T.S., Ebensperger, C., Neubüser, A., Koseki, H., Balling, R., Christ, B., and Wilting, J. 1996. Expression of avian *Pax1* and *Pax9* is intrinsically regulated in the pharyngeal endoderm, but depends on environmental influences in the paraxial mesoderm. *Dev. Biol.* **178**: 403–417.

Murtaugh, L.C., Zeng, L., Chyung, J.H., and Lassar, A.B. 2001. The chick transcriptional repressor *Nkx3.2* acts downstream of Shh to promote BMP-dependent axial chondrogenesis. *Dev. Cell* **1**: 411–422.

Muyskens, J.B. and Kimmel, C.B. 2007. Tbx16 cooperates with Wnt11 in assembling the zebrafish organizer. *Mech. Dev.* **124**: 35–42.

Myat, A., Henrique, D., Ish-Horowicz, D., and Lewis, J. 1996. A chick homologue of *Serrate* and its relationship with *Notch* and *Delta* homologues during central neurogenesis. *Dev. Biol.* **174**: 233–247.

Nagai, T., Aruga, J., Minowa, O., Sugimoto, T., Ohno, Y., Noda, T., and Mikoshiba, K. 2000. Zic2 regulates the kinetics of neurulation. *Proc. Natl. Acad. Sci.* **97**: 1618–1623.

Nagano, T., Takehara, S., Takahashi, M., Aizawa, S., and Yamamoto, A. 2006. Shisa2 promotes the maturation of somitic precursors and transition to the segmental fate in *Xenopus* embryos. *Development* **133**: 4643–4654.

Nakajima, Y., Morimoto, M., Takahashi, Y., Koseki, H., and Saga, Y. 2006. Identification of Epha4 enhancer required for segmental expression and the regulation by Mesp2. *Development* **133**: 2517–2525.

Nakashima, M., Toyono, T., Akamine, A., and Joyner, A. 1999. Expression of growth/differentiation factor 11, a new member of the BMP/TGFβ superfamily during mouse embryogenesis. *Mech. Dev.* **80**: 185–189.

Nakaya, Y., Kuroda, S., Katagiri, Y.T., Kaibuchi, K., and Takahashi, Y. 2004. Mesenchymal-epithelial transition during somitic segmentation is regulated by differential roles of Cdc42 and Rac1. *Dev. Cell* **7**: 425–438.

Nicolas, J.F., Mathis, L., Bonnerot, C., and Saurin, W. 1996. Evidence in the mouse for self-renewing stem cells in the formation of a segmented longitudinal structure, the myotome. *Development* **122**: 2933–2946.

Nicolet, G. 1970. Analyse autoradiographique de la localisation des differentes ebauches presomptives dans la ligne primitive de l'embryon de poulet. *J. Embryol. Exp. Morphol.* **23**: 79–108.

Nicolet, G. 1971. Avian gastrulation. *Adv. Morphog.* **9**: 231–262.

Niederreither, K. and Dollé, P. 2008. Retinoic acid in development: Towards an integrated view. *Nat. Rev. Genet.* **9**: 541–553.

Niederreither, K., Subbarayan, V., Dollé, P., and Chambon, P. 1999. Embryonic retinoic acid synthesis is essential for early mouse post-implantation development. *Nat. Genet.* **21**: 444–448.

Niederreither, K., Abu-Abed, S., Schuhbaur, B., Petkovich, M., Chambon, P., and Dollé, P. 2002a. Genetic evidence that oxidative derivatives of retinoic acid are not involved in retinoid signaling during mouse development. *Nat. Genet.* **31**: 84–88.

Niederreither, K., Fraulob, V., Garnier, J.M., Chambon, P., and Dollé, P. 2002b. Differential expression of retinoic acid-synthesizing (RALDH) enzymes during fetal development and organ differentiation in the mouse. *Mech. Dev.* **110**: 165–171.

Niwa, Y., Masamizu, Y., Liu, T., Nakayama, R., Deng, C.X., and Kageyama, R. 2007. The initiation and propagation of Hes7 oscillation are cooperatively regulated by Fgf and notch signaling in the somite segmentation clock. *Dev. Cell* **13**: 298–304.

Nolting, D., Hansen, B.F., Keeling, J., and Kjaer, I. 1998. Prenatal development of the normal human vertebral corpora in different segments of the spine. *Spine* **23**: 2265–2271.

Nomi, M., Oishi, I., Kani, S., Suzuki, H., Matsuda, T., Yoda, A., Kitamura, M., Itoh, K., Takeuchi, S., Takeda, K., et al. 2001. Loss of *mRor1* enhances the heart and skeletal abnormalities in *mRor2*-deficient mice: Redundant and pleiotropic functions of mRor1 and mRor2 receptor tyrosine kinases. *Mol. Cell. Biol.* **21**: 8329–8335.

Nomura-Kitabayashi, A., Takahashi, Y., Kitajima, S., Inoue, T., Takeda, H., and Saga, Y. 2002. Hypomorphic Mesp allele distinguishes establishment of rostrocaudal polarity and segment border formation in somitogenesis. *Development* **129**: 2473–2481.

Nowicki, J.L. and Burke, A.C. 2000. *Hox* genes and morphological identity: Axial versus lateral patterning in the vertebrate mesoderm. *Development* **127**: 4265–4275.

Nusse, R. 2005. Wnt signaling in disease and in development. *Cell Res.* **15**: 28–32.

Oakley, R.A. and Tosney, K.W. 1991. Peanut agglutinin and chondroitin-6-sulfate are molecular markers for tissues that act as barriers to axon advance in the avian embryo. *Dev. Biol.* **147**: 187–206.

Oates, A.C. and Ho, R.K. 2002. *Hairy/E(spl)*-related (*Her*) genes are central components of the segmentation oscillator and display redundancy with the Delta/Notch signaling pathway in the formation of anterior segmental boundaries in the zebrafish. *Development* **129**: 2929–2946.

Oates, A.C., Mueller, C., and Ho, R.K. 2005a. Cooperative function of *deltaC* and *her7* in anterior segment formation. *Dev. Biol.* **280**: 133–149.

Oates, A.C., Rohde, L.A., and Ho, R.K. 2005b. Generation of segment polarity in the paraxial mesoderm of the zebrafish through a T-box-dependent inductive event. *Dev. Biol.* **283**: 204–214.

Oda, T., Elkahloun, A.G., Pike, B.L., Okajima, K., Krantz, I.D., Genin, A., Piccoli, D.A., Meltzer, P.S., Spinner, N.B., Collins, F.S., et al. 1997. Mutations in the human *Jagged1* gene are responsible for Alagille syndrome. *Nat. Genet.* **16**: 235–242.

Oginuma, M., Hirata, T., and Saga, Y. 2008. Identification of presomitic mesoderm (PSM)-specific Mesp1 enhancer and generation of a PSM-specific Mesp1/Mesp2-null mouse using BAC-based rescue technology. *Mech. Dev.* **125**: 432–440.

Oh, S.P. and Li, E. 1997. The signaling pathway mediated by the type IIB activin receptor controls axial patterning and lateral asymmetry in the mouse. *Genes Dev* **11**: 1812–1826.

Oh, S.P., Yeo, C.Y., Lee, Y., Schrewe, H., Whitman, M., and Li, E. 2002. Activin type IIA and IIB receptors mediate Gdf11 signaling in axial vertebral patterning. *Genes Dev.* **16**: 2749–2754.

Oishi, I., Suzuki, H., Onishi, N., Takada, R., Kani, S., Ohkawara, B., Koshida, I., Suzuki, K., Yamada, G., Schwabe, G.C., et al. 2003. The receptor tyrosine kinase Ror2 is involved in non-canonical Wnt5a/JNK signalling pathway. *Genes Cells* **8**: 645–654.

Oka, C., Nakano, T., Wakeham, A., de la Pompa, J.L., Mori, C., Sakai, T., Okazaki, S., Kawaichi, M., Shiota, K., Mak, T.W., et al. 1995. Disruption of the mouse *RBP-Jκ* gene results in early embryonic death. *Development* **121**: 3291–3301.

Olivera-Martinez, I., Coltey, M., Dhouailly, D., and Pourquié, O. 2000. Mediolateral somitic origin of ribs and dermis determined by quail-chick chimeras. *Development* **127:** 4611–4617.

O'Rahilly, R. and Müller, F. 2003. Somites, spinal ganglia, and centra. Enumeration and interrelationships in staged human embryos, and implications for neural tube defects. *Cells Tissues Organs* **173:** 75–92.

Ordahl, C.P. and Le Douarin, N.M. 1992. Two myogenic lineages within the developing somite. *Development* **114:** 339–353.

Orr-Urtreger, A. and Lonai, P. 1992. Platelet-derived growth factor-A and its receptor are expressed in separate, but adjacent cell layers of the mouse embryo. *Development* **115:** 1045–1058.

Orr-Urtreger, A., Bedford, M.T., Do, M.S., Eisenbach, L., and Lonai, P. 1992. Developmental expression of the α receptor for platelet-derived growth factor, which is deleted in the embryonic lethal *Patch* mutation. *Development* **115:** 289–303.

Ostrovsky, D., Cheney, C.M., Seitz, A.W., and Lash, J.W. 1983. Fibronectin distribution during somitogenesis in the chick embryo. *Cell Differ.* **13:** 217–223.

Ozbudak, E.M. and Lewis, J. 2008. Notch signalling synchronizes the zebrafish segmentation clock but is not needed to create somite boundaries. *PLoS Genet.* **4:** e15.

Ozbudak, E.M. and Pourquié, O. 2008. The vertebrate segmentation clock: The tip of the iceberg. *Curr. Opin. Genet. Dev.* **18:** 317–323.

Packard, D.S.J. 1978. Chick somite determination: The role of factors in young somites and the segmental plate. *J. Exp. Zool.* **203:** 295–306.

Palmeirim, I., Henrique, D., Ish-Horowicz, D., and Pourquié, O. 1997. Avian hairy gene expression identifies a molecular clock linked to vertebrate segmentation and somitogenesis. *Cell* **91:** 639–648.

Palmeirim, I., Dubrulle, J., Henrique, D., Ish-Horowicz, D., and Pourquié, O. 1998. Uncoupling segmentation and somitogenesis in the chick presomitic mesoderm. *Dev. Genet.* **23:** 77–85.

Partanen, J., Schwartz, L., and Rossant, J. 1998. Opposite phenotypes of hypomorphic and Y766 phosphorylation site mutations reveal a function for *Fgfr1* in anteroposterior patterning of mouse embryos. *Genes Dev.* **12:** 2332–2344.

Pasteels, J. 1937. Etudes sur la gastrulation des Vertébrés méroblastiques. III. Oiseaux. IV. Conclusions générales. *Arch. Biol.* **48:** 381–488.

Patel, K., Connolly, D.J., Amthor, H., Nose, K., and Cooke, J. 1996. Cloning and early dorsal axial expression of Flik, a chick follistatin-related gene: Evidence for involvement in dorsalization/neural induction. *Dev. Biol.* **178:** 327–342.

Patton, M.A. and Afzal, A.R. 2002. Robinow syndrome. *J. Med. Genet* **39:** 305–310.

Perantoni, A.O., Timofeeva, O., Naillat, F., Richman, C., Pajni-Underwood, S., Wilson, C., Vainio, S., Dove, L.F., and Lewandoski, M. 2005. Inactivation of FGF8 in early mesoderm reveals an essential role in kidney development. *Development* **132:** 3859–3871.

Peters, H., Wilm, B., Sakai, N., Imai, K., Maas, R., and Balling, R. 1999. Pax1 and Pax9 synergistically regulate vertebral column development. *Development* **126:** 5399–5408.

Pickett, E.A., Olsen, G.S., and Tallquist, M.D. 2008. Disruption of PDGFRα-initiated PI3K activation and migration of somite derivatives leads to spina bifida. *Development* **135:** 589–598.

Pilon, N., Oh, K., Sylvestre, J.R., Bouchard, N., Savory, J., and Lohnes, D. 2006. *Cdx4* is a direct target of the canonical Wnt pathway. *Dev Biol* **289:** 55–63.

Pilon, N., Oh, K., Sylvestre, J.R., Savory, J.G., and Lohnes, D. 2007. Wnt signaling is a key mediator of *Cdx1* expression in vivo. *Development* **134:** 2315–2323.

Pirot, P., van Grunsven, L.A., Marine, J.C., Huylebroeck, D., and Bellefroid, E.J. 2004. Direct regulation of the *Nrarp* gene promoter by the Notch signaling pathway. *Biochem. Biophys. Res. Commun.* **322:** 526–534.

Platt, K.A., Michaud, J., and Joyner, A.L. 1997. Expression of the mouse *Gli* and *Ptc* genes is adjacent to embryonic sources of hedgehog signals suggesting a conservation of pathways between flies and mice. *Mech. Dev.* **62:** 121–135.

Pouget, C., Pottin, K., and Jaffredo, T. 2008. Sclerotomal origin of vascular smooth muscle cells and pericytes in the embryo. *Dev. Biol.* **315:** 437–447.

Pourquié, O. 2001. Vertebrate somitogenesis. *Annu. Rev. Cell Dev. Biol.* **17:** 311–350.

Pourquié, O. 2004. The chick embryo: A leading model in somitogenesis studies. *Mech. Dev.* **121:** 1069–1079.

Pourquié, O. and Tam, P.P. 2001. A nomenclature for prospective somites and phases of cyclic gene expression in the presomitic mesoderm. *Dev. Cell* **1:** 619–620.

Pourquié, O., Coltey, M., Teillet, M.A., Ordahl, C., and Le Douarin, N.M. 1993. Control of dorsoventral patterning of somitic derivatives by notochord and floor plate. *Proc. Natl. Acad. Sci.* **90:** 5242–5246.

Pourquié, O., Coltey, M., Breant, C., and Le Douarin, N.M. 1995. Control of somite patterning by signals from the lateral plate. *Proc. Natl. Acad. Sci.* **92:** 3219–3223.

Pourquié, O., Fan, C.M., Coltey, M., Hirsinger, E., Watanabe, Y., Breant, C., Francis-West, P., Brickell, P., Tessier-Lavigne, M., and Le Douarin, N.M. 1996. Lateral and axial signals involved in avian somite patterning: A role for BMP4. *Cell* **84:** 461–471.

Pownall, M.E., Tucker, A.S., Slack, J.M., and Isaacs, H.V. 1996. eFGF, Xcad3 and Hox genes form a molecular pathway that establishes the anteroposterior axis in *Xenopus*. *Development* **122:** 3881–3892.

Prinos, P., Joseph, S., Oh, K., Meyer, B.I., Gruss, P., and Lohnes, D. 2001. Multiple pathways governing *Cdx1* expression during murine development. *Dev. Biol.* **239:** 257–269.

Psychoyos, D. and Stern, C.D. 1996. Fates and migratory routes of primitive streak cells in the chick embryo. *Development* **122:** 1523–1534.

Purandare, S.M., Ware, S.M., Kwan, K.M., Gebbia, M., Bassi, M.T., Deng, J.M., Vogel, H., Behringer, R.R., Belmont, J.W., and Casey, B. 2002. A complex syndrome of left-right axis, central nervous system and axial skeleton defects in *Zic3* mutant mice. *Development* **129:** 2293–2302.

Qiu, X., Xu, H., Haddon, C., Lewis, J., and Jiang, Y.J. 2004. Sequence and embryonic expression of three zebrafish *fringe* genes: *lunatic fringe, radical fringe,* and *manic fringe*. *Dev. Dyn.* **231:** 621–630.

Quertermous, E.E., Hidai, H., Blanar, M.A., and Quertermous, T. 1994. Cloning and characterization of a basic helix-loop-helix protein expressed in early mesoderm and the developing somites. *Proc. Natl. Acad. Sci.* **91:** 7066–7070.

Radice, G.L., Rayburn, H., Matsunami, H., Knudsen, K.A., Takeichi, M., and Hynes, R.O. 1997. Developmental defects in mouse embryos lacking N-cadherin. *Dev. Biol.* **181:** 64–78.

Rancourt, S.L. and Rancourt, D.E. 1997. Murine subtilisin-like proteinase SPC6 is expressed during embryonic implantation, somitogenesis, and skeletal formation. *Dev. Genet.* **21:** 75–81.

Ranscht, B. and Bronner-Fraser, M. 1991. T-cadherin expression alternates with migrating neural crest cells in the trunk of the avian embryo. *Development* **111:** 15–22.

Raz, R., Stricker, S., Gazzerro, E., Clor, J.L., Witte, F., Nistala, H., Zabski, S., Pereira, R.C., Stadmeyer, L., Wang, X., et al. 2008. The mutation $ROR2^{W749X}$, linked to human BDB, is a recessive mutation in the mouse, causing brachydactyly, mediating patterning of joints and modeling recessive Robinow syndrome. *Development* **135**: 1713–1723.

Redies, C., Vanhalst, K., and Roy, F. 2005. δ-Protocadherins: Unique structures and functions. *Cell. Mol. Life Sci.* **62**: 2840–2852.

Reijntjes, S., Stricker, S., and Mankoo, B.S. 2007. A comparative analysis of *Meox1* and *Meox2* in the developing somites and limbs of the chick embryo. *Int. J. Dev. Biol.* **51**: 753–759.

Remak, R. 1850. *Untersuchungen über die entwicklung der Wirbeltiere*. Reimer, Berlin.

Reshef, R., Maroto, M., and Lassar, A.B. 1998. Regulation of dorsal somitic cell fates: BMPs and Noggin control the timing and pattern of myogenic regulator expression. *Genes Dev.* **12**: 290–303.

Rhee, J., Takahashi, Y., Saga, Y., Wilson-Rawls, J., and Rawls, A. 2003. The protocadherin papc is involved in the organization of the epithelium along the segmental border during mouse somitogenesis. *Dev. Biol.* **254**: 248–261.

Rickmann, M., Fawcett, J.W., and Keynes, R.J. 1985. The migration of neural crest cells and the growth of motor axons through the rostral half of the chick somite. *J. Embryol. Exp. Morphol.* **90**: 437–455.

Rida, P.C., Le Minh, N., and Jiang, Y.J. 2004. A Notch feeling of somite segmentation and beyond. *Dev. Biol.* **265**: 2–22.

Riedel-Kruse, I.H., Muller, C., and Oates, A.C. 2007. Synchrony dynamics during initiation, failure, and rescue of the segmentation clock. *Science* **317**: 1911–1915.

Rifes, P., Carvalho, L., Lopes, C., Andrade, R.P., Rodrigues, G., Palmeirim, I., and Thorsteinsdottir, S. 2007. Redefining the role of ectoderm in somitogenesis: A player in the formation of the fibronectin matrix of presomitic mesoderm. *Development* **134**: 3155–3165.

Ring, C., Hassell, J., and Halfter, W. 1996. Expression pattern of collagen IX and potential role in the segmentation of the peripheral nervous system. *Dev. Biol.* **180**: 41–53.

Rodrigo, I., Hill, R.E., Balling, R., Munsterberg, A., and Imai, K. 2003. Pax1 and Pax9 activate *Bapx1* to induce chondrogenic differentiation in the sclerotome. *Development* **130**: 473–482.

Rodrigo, I., Bovolenta, P., Mankoo, B.S., and Imai, K. 2004. Meox homeodomain proteins are required for *Bapx1* expression in the sclerotome and activate its transcription by direct binding to its promoter. *Mol. Cell. Biol.* **24**: 2757–2766.

Romanoff, A. 1960. *The avian embryo: Structural and functional development*. Macmillan, New York.

Rosenquist, G.C. 1966. A radioautographic study of labeled grafts in the chick blastoderm. Development from primitive-streak stages to stage 12. *Contrib. Embryol. Carnegie Inst. Wash.* **38**: 71–110.

Rossant, J., Zirngibl, R., Cado, D., Shago, M., and Giguère, V. 1991. Expression of a retinoic acid response element-hsplacZ transgene defines specific domains of transcriptional activity during mouse embryogenesis. *Genes Dev.* **5**: 1333–1344.

Rovescalli, A.C., Asoh, S., and Nirenberg, M. 1996. Cloning and characterization of four murine homeobox genes. *Proc. Natl. Acad. Sci.* **93**: 10691–10696.

Saga, Y. 2007. Segmental border is defined by the key transcription factor Mesp2, by means of the suppression of Notch activity. *Dev. Dyn.* **236**: 1450–1455.

Saga, Y. and Takeda, H. 2001. The making of the somite: Molecular events in vertebrate segmentation. *Nat. Rev. Genet.* **2**: 835–845.

Saga, Y., Hata, N., Kobayashi, S., Magnuson, T., Seldin, M.F., and Taketo, M.M. 1996.

MesP1: A novel basic helix-loop-helix protein expressed in the nascent mesodermal cells during mouse gastrulation. *Development* **122:** 2769–2778.

Saga, Y., Hata, N., Koseki, H., and Taketo, M.M. 1997. Mesp2: A novel mouse gene expressed in the presegmented mesoderm and essential for segmentation initiation. *Genes Dev.* **11:** 1827–1839.

Saito, T., Lo, L., Anderson, D.J., and Mikoshiba, K. 1996. Identification of novel paired homeodomain protein related to *C. elegans* unc-4 as a potential downstream target of MASH1. *Dev. Biol.* **180:** 143–155.

Sakai, Y., Meno, C., Fujii, H., Nishino, J., Shiratori, H., Saijoh, Y., Rossant, J., and Hamada, H. 2001. The retinoic acid-inactivating enzyme CYP26 is essential for establishing an uneven distribution of retinoic acid along the anterio-posterior axis within the mouse embryo. *Genes Dev.* **15:** 213–225.

Sandell, L.L., Sanderson, B.W., Moiseyev, G., Johnson, T., Mushegian, A., Young, K., Rey, J.P., Ma, J.X., Staehling-Hampton, K., and Trainor, P.A. 2007. RDH10 is essential for synthesis of embryonic retinoic acid and is required for limb, craniofacial, and organ development. *Genes Dev.* **21:** 1113–1124.

Sano, K., Tanihara, H., Heimark, R.L., Obata, S., Davidson, M., St John, T., Taketani, S., and Suzuki, S. 1993. Protocadherins: A large family of cadherin-related molecules in central nervous system. *EMBO J.* **12:** 2249–2256.

Sato, Y., Yasuda, K., and Takahashi, Y. 2002. Morphogical boundary forms by a novel inductive enevt mediated by Lunatic fringe and Notch during somitic segmentation. *Development* **129:** 3633–3644.

Satoh, W., Gotoh, T., Tsunematsu, Y., Aizawa, S., and Shimono, A. 2006. *Sfrp1* and *Sfrp2* regulate anteroposterior axis elongation and somite segmentation during mouse embryogenesis. *Development* **133:** 989–999.

Satoh, W., Matsuyama, M., Takemura, H., Aizawa, S., and Shimono, A. 2008. Sfrp1, Sfrp2, and Sfrp5 regulate the Wnt/β-catenin and the planar cell polarity pathways during early trunk formation in mouse. *Genesis* **46:** 92–103.

Sawada, A., Fritz, A., Jiang, Y., Yamamoto, A., Yamasu, K., Kuroiwa, A., Saga, Y., and Takeda, H. 2000. Zebrafish *Mesp* family genes, *mesp-a* and *mesp-b* are segmentally expressed in the presomitic mesoderm, and Mesp-b confers the anterior identity to the developing somites. *Development* **127:** 1691–1702.

Sawada, A., Shinya, M., Jiang, Y.J., Kawakami, A., Kuroiwa, A., and Takeda, H. 2001. Fgf/MAPK signalling is a crucial positional cue in somite boundary formation. *Development* **128:** 4873–4880.

Schmidt, C., Stoeckelhuber, M., McKinnell, I., Putz, R., Christ, B., and Patel, K. 2004. Wnt 6 regulates the epithelialisation process of the segmental plate mesoderm leading to somite formation. *Dev. Biol.* **271:** 198–209.

Schoenwolf, G.C., Garcia-Martinez, V., and Dias, M.S. 1992. Mesoderm movement and fate during avian gastrulation and neurulation. *Dev. Dyn.* **193:** 235–248.

Schrägle, J., Huang, R., Christ, B., and Pröls, F. 2004. Control of the temporal and spatial *Uncx4.1* expression in the paraxial mesoderm of avian embryos. *Anat. Embryol.* **208:** 323–332.

Schubert, F.R., Tremblay, P., Mansouri, A., Faisst, A.M., Kammandel, B., Lumsden, A., Gruss, P., and Dietrich, S. 2001. Early mesodermal phenotypes in *splotch* suggest a role for *Pax3* in the formation of epithelial somites. *Dev. Dyn.* **222:** 506–521.

Schulte-Merker, S., van Eeden, F.J., Halpern, M.E., Kimmel, C.B., and Nüsslein-Volhard, C. 1994. *no tail* (*ntl*) is the zebrafish homologue of the mouse T (*Brachyury*) gene.

Development **120:** 1009–1015.

Schwabe, G.C., Trepczik, B., Suring, K., Brieske, N., Tucker, A.S., Sharpe, P.T., Minami, Y., and Mundlos, S. 2004. *Ror2* knockout mouse as a model for the developmental pathology of autosomal recessive Robinow syndrome. *Dev. Dyn.* **229:** 400–410.

Schweitzer, R., Chyung, J.H., Murtaugh, L.C., Brent, A.E., Rosen, V., Olson, E.N., Lassar, A., and Tabin, C.J. 2001. Analysis of the tendon cell fate using Scleraxis, a specific marker for tendons and ligaments. *Development* **128:** 3855–3866.

Selleck, M.A. and Stern, C.D. 1991. Fate mapping and cell lineage analysis of Hensen's node in the chick embryo. *Development* **112:** 615–626.

Selleri, L., Depew, M.J., Jacobs, Y., Chanda, S.K., Tsang, K.Y., Cheah, K.S., Rubenstein, J.L., O'Gorman, S., and Cleary, M.L. 2001. Requirement for *Pbx1* in skeletal patterning and programming chondrocyte proliferation and differentiation. *Development* **128:** 3543–3557.

Sëmenov, M.V., Tamai, K., Brott, B.K., Kühl, M., Sokol, S., and He, X. 2001. Head inducer Dickkopf-1 is a ligand for Wnt coreceptor LRP6. *Curr. Biol.* **11:** 951–961.

Serth, K., Schuster-Gossler, K., Cordes, R., and Gossler, A. 2003. Transcriptional oscillation of Lunatic fringe is essential for somitogenesis. *Genes Dev.* **17:** 912–925.

Shanmugalingam, S. and Wilson, S.W. 1998. Isolation, expression and regulation of a zebrafish paraxis homologue. *Mech. Dev.* **78:** 85–89.

Shawlot, W., Deng, J.M., and Behringer, R.R. 1998. Expression of the mouse *cerberus*-related gene, *Cerr1*, suggests a role in anterior neural induction and somitogenesis. *Proc. Natl. Acad. Sci.* **95:** 6198–6203.

Shen, J., Bronson, R.T., Chen, D.F., Xia, W., Selkoe, D.J., and Tonegawa, S. 1997. Skeletal and CNS defects in Presenilin-1-deficient mice. *Cell* **89:** 629–639.

Shen, M.M. 2007. Nodal signaling: Developmental roles and regulation. *Development* **134:** 1023–1034.

Shi, S. and Stanley, P. 2003. Protein O-fucosyltransferase 1 is an essential component of Notch signaling pathways. *Proc. Natl. Acad. Sci.* **100:** 5234–5239.

Shifley, E.T., Vanhorn, K.M., Perez-Balaguer, A., Franklin, J.D., Weinstein, M., and Cole, S.E. 2008. Oscillatory lunatic fringe activity is crucial for segmentation of the anterior but not posterior skeleton. *Development* **135:** 899–908.

Shiratori, H. and Hamada, H. 2006. The left-right axis in the mouse: From origin to morphology. *Development* **133:** 2095–2104.

Sirbu, I.O. and Duester, G. 2006. Retinoic-acid signalling in node ectoderm and posterior neural plate directs left-right patterning of somitic mesoderm. *Nat. Cell Biol.* **8:** 271–277.

Soriano, P. 1997. The PDGF α receptor is required for neural crest cell development and for normal patterning of the somites. *Development* **124:** 2691–2700.

Sosic, D., Brand-Saberi, B., Schmidt, C., Christ, B., and Olson, E.N. 1997. Regulation of paraxis expression and somite formation by ectoderm- and neural tube-derived signals. *Dev. Biol.* **185:** 229–243.

Sparrow, D.B., Chapman, G., Wouters, M.A., Whittock, N.V., Ellard, S., Fatkin, D., Turnpenny, P.D., Kusumi, K., Sillence, D., and Dunwoodie, S.L. 2006. Mutation of the *LUNATIC FRINGE* gene in humans causes spondylocostal dysostosis with a severe vertebral phenotype. *Am. J. Hum. Genet.* **78:** 28–37.

Sparrow, D.B., Guillén-Navarro, E., Fatkin, D., and Dunwoodie, S.L. 2008. Mutation of *HAIRY-AND-ENHANCER-OF-SPLIT-7* in humans causes spondylocostal dysostosis. *Hum. Mol. Genet.* **17:** 3761–3766.

Speder, P., Petzoldt, A., Suzanne, M., and Noselli, S. 2007. Strategies to establish left/right asymmetry in vertebrates and invertebrates. *Curr. Opin. Genet. Dev.* **17**: 351–358.

Spörle, R. and Schughart, K. 1997. System to identify individual somites and their derivatives in the developing mouse embryo. *Dev. Dyn.* **210**: 216–226.

Stern, C.D. 2006. Evolution of the mechanisms that establish the embryonic axes. *Curr. Opin. Genet. Dev.* **16**: 413–418.

Strudel, G. 1955. L'action morphogène du tube nerveux et de la corde sur la différenciation des Vertébrés et des muscles vertébraux chez l'embryon de Poulet. *Arch. Anat. Microsc. Morphol. Exp.* **44**: 209–235.

Sun, X., Meyers, E.N., Lewandoski, M., and Martin, G.R. 1999. Targeted disruption of *Fgf8* causes failure of cell migration in the gastrulating mouse embryo. *Genes Dev.* **13**: 1834–1846.

Suriben, R., Fisher, D.A., and Cheyette, B.N. 2006. *Dact1* presomitic mesoderm expression oscillates in phase with *Axin2* in the somitogenesis clock of mice. *Dev. Dyn.* **235**: 3177–3183.

Szumska, D., Pieles, G., Essalmani, R., Bilski, M., Mesnard, D., Kaur, K., Franklyn, A., El Omari, K., Jefferis, J., Bentham, J., et al. 2008. VACTERL/caudal regression/Currarino syndrome-like malformations in mice with mutation in the proprotein convertase *Pcsk5*. *Genes Dev.* **22**: 1465–1477.

Takada, S., Stark, K.L., Shea, M.J., Vassileva, G., McMahon, J.A., and McMahon, A.P. 1994. Wnt-3a regulates somite and tailbud formation in the mouse embryo. *Genes Dev.* **8**: 174–189.

Takahashi, Y., Koizumi, K., Takagi, A., Kitajima, S., Inoue, T., Koseki, H., and Saga, Y. 2000. Mesp2 initiates somite segmentation through the Notch signalling pathway. *Nat. Genet.* **25**: 390–396.

Takahashi, Y., Inoue, T., Gossler, A., and Saga, Y. 2003. Feedback loops comprising Dll1, Dll3 and Mesp2, and differential involvement of Psen1 are essential for rostrocaudal patterning of somites. *Development* **130**: 4259–4268.

Takahashi, Y., Kitajima, S., Inoue, T., Kanno, J., and Saga, Y. 2005. Differential contributions of Mesp1 and Mesp2 to the epithelialization and rostro-caudal patterning of somites. *Development* **132**: 787–796.

Takahashi, Y., Takagi, A., Hiraoka, S., Koseki, H., Kanno, J., Rawls, A., and Saga, Y. 2007a. Transcription factors Mesp2 and Paraxis have critical roles in axial musculoskeletal formation. *Dev. Dyn.* **236**: 1484–1494.

Takahashi, Y., Yasuhiko, Y., Kitajima, S., Kanno, J., and Saga, Y. 2007b. Appropriate suppression of Notch signaling by Mesp factors is essential for stripe pattern formation leading to segment boundary formation. *Dev. Biol.* **304**: 593–603.

Takeuchi, S., Takeda, K., Oishi, I., Nomi, M., Ikeya, M., Itoh, K., Tamura, S., Ueda, T., Hatta, T., Otani, H., et al. 2000. Mouse Ror2 receptor tyrosine kinase is required for the heart development and limb formation. *Genes Cells* **5**: 71–78.

Takke, C. and Campos-Ortega, J.A. 1999. *her1*, a zebrafish pair-rule like gene, acts downstream of notch signalling to control somite development. *Development* **126**: 3005–3014.

Tam, P.P. 1981. The control of somitogenesis in mouse embryos. *J. Embryol. Exp. Morphol.* (suppl.) **65**: 103–128.

Tam, P.P. and Beddington, R.S. 1987. The formation of mesodermal tissues in the mouse embryo during gastrulation and early organogenesis. *Development* **99**: 109–126.

Tam, P.P. and Trainor, P.A. 1994. Specification and segmentation of the paraxial mesoderm. *Anat. Embryol.* **189**: 275–305.

Tanaka, M. and Tickle, C. 2004. *Tbx18* and boundary formation in chick somite and wing development. *Dev. Biol.* **268**: 470–480.

Teillet, M., Watanabe, Y., Jeffs, P., Duprez, D., Lapointe, F., and Le Douarin, N.M. 1998. Sonic hedgehog is required for survival of both myogenic and chondrogenic somitic lineages. *Development* **125**: 2019–2030.

Teppner, I., Becker, S., de Angelis, M.H., Gossler, A., and Beckers, J. 2007. Compartmentalised expression of *Delta-like 1* in epithelial somites is required for the formation of intervertebral joints. *BMC Dev. Biol.* **7**: 68.

Terry, K., Magan, H., Baranski, M., and Burrus, L.W. 2000. *Sfrp-1* and *sfrp-2* are expressed in overlapping and distinct domains during chick development. *Mech. Dev.* **97**: 177–182.

Tonegawa, A. and Takahashi, Y. 1998. Somitogenesis controlled by Noggin. *Dev. Biol.* **202**: 172–182.

Tonegawa, A., Funayama, N., Ueno, N., and Takahashi, Y. 1997. Mesodermal subdivision along the mediolateral axis in chicken controlled by different concentrations of BMP-4. *Development* **124**: 1975–1984.

Tribioli, C. and Lufkin, T. 1999. The murine *Bapx1* homeobox gene plays a critical role in embryonic development of the axial skeleton and spleen. *Development* **126**: 5699–5711.

Tseng, H.T. and Jamrich, M. 2004. Identification and developmental expression of *Xenopus* paraxis. *Int. J. Dev. Biol.* **48**: 1155–1158.

Turnpenny, P.D., Alman, B., Cornier, A.S., Giampietro, P.F., Offiah, A., Tassy, O., Pourquié, O., Kusumi, K., and Dunwoodie, S. 2007. Abnormal vertebral segmentation and the notch signaling pathway in man. *Dev. Dyn.* **236**: 1456–1474.

Unterseher, F., Hefele, J.A., Giehl, K., De Robertis, E.M., Wedlich, D., and Schambony, A. 2004. Paraxial protocadherin coordinates cell polarity during convergent extension via Rho A and JNK. *EMBO J.* **23**: 3259–3269.

van Bokhoven, H., Celli, J., Kayserili, H., van Beusekom, E., Balci, S., Brussel, W., Skovby, F., Kerr, B., Percin, E.F., Akarsu, N., et al. 2000. Mutation of the gene encoding the ROR2 tyrosine kinase causes autosomal recessive Robinow syndrome. *Nat. Genet.* **25**: 423–426.

van den Akker, E., Forlani, S., Chawengsaksophak, K., de Graaff, W., Beck, F., Meyer, B.I., and Deschamps, J. 2002. *Cdx1* and *Cdx2* have overlapping functions in anteroposterior patterning and posterior axis elongation. *Development* **129**: 2181–2193.

van Eeden, F.J., Holley, S.A., Haffter, P., and Nüsslein-Volhard, C. 1998. Zebrafish segmentation and pair-rule patterning. *Dev. Genet.* **23**: 65–76.

van Nes, J., de Graaff, W., Lebrin, F., Gerhard, M., Beck, F., and Deschamps, J. 2006. The *Cdx4* mutation affects axial development and reveals an essential role of Cdx genes in the ontogenesis of the placental labyrinth in mice. *Development* **133**: 419–428.

Verbout, A.J. 1976. A critical review of the 'neugliederung' concept in relation to the development of the vertebral column. *Acta Biotheor.* **25**: 219–258.

Vermot, J. and Pourquié, O. 2005. Retinoic acid coordinates somitogenesis and left-right patterning in vertebrate embryos. *Nature* **435**: 215–220.

Vermot, J., Llamas, J.G., Fraulob, V., Niederreither, K., Chambon, P., and Dollé, P. 2005. Retinoic acid controls the bilateral symmetry of somite formation in the mouse embryo. *Science* **308**: 563–566.

Vilhais-Neto, G.C. and Pourquié, O. 2008. Retinoic acid. *Curr. Biol.* **18**: R191–R192.

Waddington, C.H. 1952. *The epigenetics of birds.* Cambridge University Press, Cambridge.

Wahl, M.B., Deng, C., Lewandoski, M., and Pourquié, O. 2007. FGF signaling acts upstream of the NOTCH and WNT signaling pathways to control segmentation clock oscillations in mouse somitogenesis. *Development* **134:** 4033–4041.

Wallin, J., Wilting, J., Koseki, H., Fritsch, R., Christ, B., and Balling, R. 1994. The role of *Pax-1* in axial skeleton development. *Development* **120:** 1109–1121.

Wang, B., He, L., Ehehalt, F., Geetha-Loganathan, P., Nimmagadda, S., Christ, B., Scaal, M., and Huang, R. 2005. The formation of the avian scapula blade takes place in the hypaxial domain of the somites and requires somatopleure-derived BMP signals. *Dev. Biol.* **287:** 11–18.

Wang, J., Hamblet, N.S., Mark, S., Dickinson, M.E., Brinkman, B.C., Segil, N., Fraser, S.E., Chen, P., Wallingford, J.B., and Wynshaw-Boris, A. 2006. Dishevelled genes mediate a conserved mammalian PCP pathway to regulate convergent extension during neurulation. *Development* **133:** 1767–1778.

Wardle, F.C. and Papaioannou, V.E. 2008. Teasing out T-box targets in early mesoderm. *Curr. Opin. Genet. Dev.* **8:** 418–425.

Watterson, R., Fowler, I., and Fowler, B.J. 1954. The role of the neural tube and notochord in development of the axial skeleton of the chick. *Am. J. Anat.* **95:** 337–399.

Weinmaster, G. and Kintner, C. 2003. Modulation of notch signaling during somitogenesis. *Annu. Rev. Cell Dev. Biol.* **19:** 367–395.

Wellik, D.M. 2007. *Hox* patterning of the vertebrate axial skeleton. *Dev. Dyn.* **236:** 2454–2463.

White, P.H. and Chapman, D.L. 2005. *Dll1* is a downstream target of Tbx6 in the paraxial mesoderm. *Genesis* **42:** 193–202.

White, P.H., Farkas, D.R., McFadden, E.E., and Chapman, D.L. 2003. Defective somite patterning in mouse embryos with reduced levels of *Tbx6*. *Development* **130:** 1681–1690.

White, P.H., Farkas, D.R., and Chapman, D.L. 2005. Regulation of *Tbx6* expression by Notch signaling. *Genesis* **42:** 61–70.

Whittock, N.V., Sparrow, D.B., Wouters, M.A., Sillence, D., Ellard, S., Dunwoodie, S.L., and Turnpenny, P.D. 2004. Mutated *MESP2* causes spondylocostal dysostosis in humans. *Am. J. Hum. Genet.* **74:** 1249–1254.

Wiegreffe, C., Christ, B., Huang, R., and Scaal, M. 2007. Sclerotomal origin of smooth muscle cells in the wall of the avian dorsal aorta. *Dev. Dyn.* **236:** 2578–2585.

William, D.A., Saitta, B., Gibson, J.D., Traas, J., Markov, V., Gonzalez, D.M., Sewell, W., Anderson, D.M., Pratt, S.C., Rappaport, E.F., et al. 2007. Identification of oscillatory genes in somitogenesis from functional genomic analysis of a human mesenchymal stem cell model. *Dev. Biol.* **305:** 172–186.

Wilson, V., Manson, L., Skarnes, W.C., and Beddington, R.S. 1995. The *T* gene is necessary for normal mesodermal morphogenetic cell movements during gastrulation. *Development* **121:** 877–886.

Wilson-Rawls, J., Rhee, J.M., and Rawls, A. 2004. Paraxis is a basic helix-loop-helix protein that positively regulates transcription through binding to specific E-box elements. *J. Biol. Chem.* **279:** 37685–37692.

Winnier, G.E., Hargett, L., and Hogan, B.L. 1997. The winged helix transcription factor MFH1 is required for proliferation and patterning of paraxial mesoderm in the mouse embryo. *Genes Dev.* **11:** 926–940.

Wittler, L., Shin, E.H., Grote, P., Kispert, A., Beckers, A., Gossler, A., Werber, M., and Herrmann, B.G. 2007. Expression of *Msgn1* in the presomitic mesoderm is controlled by synergism of WNT signalling and *Tbx6*. *EMBO Rep.* **8:** 784–789.

Wong, P.C., Zheng, H., Chen, H., Becher, M.W., Sirinathsinghji, D.J., Trumbauer, M.E., Chen, H.Y., Price, D.L., Van der Ploeg, L.H., and Sisodia, S.S. 1997. Presenilin 1 is required for *Notch1* and *DlI1* expression in the paraxial mesoderm. *Nature* **387:** 288–292.

Xu, Q., Mellitzer, G., Robinson, V., and Wilkinson, D.G. 1999. In vivo cell sorting in complementary segmental domains mediated by Eph receptors and ephrins. *Nature* **399:** 267–271.

Yamaguchi, T.P., Conlon, R.A., and Rossant, J. 1992. Expression of the fibroblast growth factor receptor FGFR-1/flg during gastrulation and segmentation in the mouse embryo. *Dev. Biol.* **152:** 75–88.

Yamaguchi, T.P., Harpal, K., Henkemeyer, M., and Rossant, J. 1994. fgfr-1 is required for embryonic growth and mesodermal patterning during mouse gastrulation. *Genes Dev.* **8:** 3032–3044.

Yamaguchi, T.P., Bradley, A., McMahon, A.P., and Jones, S. 1999a. A *Wnt5a* pathway underlies outgrowth of multiple structures in the vertebrate embryo. *Development* **126:** 1211–1223.

Yamaguchi, T.P., Takada, S., Yoshikawa, Y., Wu, N., and McMahon, A.P. 1999b. *T* (*Brachyury*) is a direct target of Wnt3a during paraxial mesoderm specification. *Genes Dev.* **13:** 3185–3190.

Yamamoto, A., Amacher, S.L., Kim, S.H., Geissert, D., Kimmel, C.B., and De Robertis, E.M. 1998. Zebrafish paraxial protocadherin is a downstream target of spadetail involved in morphogenesis of gastrula mesoderm. *Development* **125:** 3389–3397.

Yamamoto, A., Kemp, C., Bachiller, D., Geissert, D., and De Robertis, E.M. 2000. Mouse paraxial protocadherin is expressed in trunk mesoderm and is not essential for mouse development. *Genesis* **27:** 49–57.

Yamamoto, A., Nagano, T., Takehara, S., Hibi, M., and Aizawa, S. 2005. Shisa promotes head formation through the inhibition of receptor protein maturation for the caudalizing factors, Wnt and FGF. *Cell* **120:** 223–235.

Yang, J.T., Bader, B.L., Kreidberg, J.A., Ullman-Cullere, M., Trevithick, J.E., and Hynes, R.O. 1999. Overlapping and independent functions of fibronectin receptor integrins in early mesodermal development. *Dev. Biol.* **215:** 264–277.

Yasuhiko, Y., Haraguchi, S., Kitajima, S., Takahashi, Y., Kanno, J., and Saga, Y. 2006. Tbx6-mediated Notch signaling controls somite-specific *Mesp2* expression. *Proc. Natl. Acad. Sci.* **103:** 3651–3656.

Yokouchi, Y., Vogan, K.J., Pearse II, R.V., and Tabin, C.J. 1999. Antagonistic signaling by *Caronte*, a novel *Cerberus*-related gene, establishes left-right asymmetric gene expression. *Cell* **98:** 573–583.

Yoon, J.K., Moon, R.T., and Wold, B. 2000. The bHLH class protein pMesogenin1 can specify paraxial mesoderm phenotypes. *Dev. Biol.* **222:** 376–391.

Yoshikawa, Y., Fujimori, T., McMahon, A.P., and Takada, S. 1997. Evidence that absence of *Wnt-3a* signaling promotes neuralization instead of paraxial mesoderm development in the mouse. *Dev. Biol.* **183:** 234–242.

Yu, H.M., Jerchow, B., Sheu, T.J., Liu, B., Costantini, F., Puzas, J.E., Birchmeier, W., and Hsu, W. 2005. The role of Axin2 in calvarial morphogenesis and craniosynostosis. *Development* **132:** 1995–2005.

Zákány, J., Kmita, M., Alarcon, P., de la Pompa, J.L., and Duboule, D. 2001. Localized and transient transcription of *Hox* genes suggests a link between patterning and the segmentation clock. *Cell* **106:** 207–217.

Zeng, L., Fagotto, F., Zhang, T., Hsu, W., Vasicek, T.J., Perry III, W.L., Lee, J.J., Tilghman, S.M., Gumbiner, B.M., and Costantini, F. 1997. The mouse *Fused* locus encodes Axin, an inhibitor of the Wnt signaling pathway that regulates embryonic axis formation. *Cell* **90:** 181–192.

Zeng, L., Kempf, H., Murtaugh, L.C., Sato, M.E., and Lassar, A.B. 2002. Shh establishes an Nkx3.2/Sox9 autoregulatory loop that is maintained by BMP signals to induce somitic chondrogenesis. *Genes Dev.* **16:** 1990–2005.

Zhang, N. and Gridley, T. 1998. Defects in somite formation in *lunatic fringe*-deficient mice. *Nature* **394:** 374–377.

Zhang, N., Norton, C.R., and Gridley, T. 2002. Segmentation defects of Notch pathway mutants and absence of a synergistic phenotype in lunatic fringe/radical fringe double mutant mice. *Genesis* **33:** 21–28.

Zhang, X.M., Ramalho-Santos, M., and McMahon, A.P. 2001. Smoothened mutants reveal redundant roles for Shh and Ihh signaling including regulation of L/R asymmetry by the mouse node. *Cell* **105:** 781–792.

Zhang, C., Li, Q., and Jiang, Y.J. 2007. Zebrafish Mib and Mib2 are mutual E3 ubiquitin ligases with common and specific Delta substrates. *J. Mol. Biol.* **366:** 1115–1128.

4

Craniofacial Patterning

Nicole M. Le Douarin[1,2] and Sophie E. Creuzet[2]

[1]Collège de France,
 3, rue d'Ulm, 75005 Paris, France

[2]Institut de Neurobiologie Alfred Fessard
 Laboratoire Développement, Evolution et Plasticité du Système Nerveux,
 CNRS-UPR2197, 91198 Gif-sur-Yvette, France

HISTORICAL BACKGROUND

One of the most striking characteristics about the craniofacial bones is that, contrary to the rest of the vertebrate skeleton, they are not entirely of mesodermal origin. Embryological studies, which started at the end of the 19th century with the observations of Kastschenko (1888, for selacians) and Goronovitch (1892, 1893, for teleosts and birds), have established that mesenchymal cells can arise, not only from the mesodermal, but also from the ectodermal germ layer. During this period, Julia Platt was the first to propose in 1893 that ectoderm contributed not only to the mesenchyme, but also to the cartilage of the visceral arches and to the dentine of the teeth in the mud puppy, *Necturus*. This derivation of mesenchyme, bones and cartilages from the ectoderm, was shown to occur via a transient structure, the Neural Crest (NC), which was first described in the chick embryo by the German Histologist Wilhem His in 1868.

These observations contradicted the germ layer theory first put forward by Christian Heinrich Pander (1817), who described the formation of three layers of cells from the chick blastoderm. Later, Karl von Baer (1828) extended Pander's findings to all vertebrate embryos. In 1849, Thomas Huxley generalized the presence of germ layers to invertebrates and the terms *ectoderm, mesoderm,* and *endoderm* were first used to designate the vertebrate germ layers by Ernst Haeckel in 1874, in the context of the *Gastrea* concept.

The observation that formation of germ layers precedes organ morphogenesis and cellular differentiation was followed by the rather dogmatic view that each of these groups of cells was exclusively devoted to yield a definite set of tissue derivatives. Ectoderm formed nerves and epidermis; mesoderm formed muscles, mesenchyme, cartilages, bones, connective tissues, kidneys, and cardiovascular structures; and endoderm formed the alimentary canal and associated glands. Therefore, the contention put forward by Julia Platt that facial bones and cartilages could receive a contribution from cells ectodermal in origin seemed heretical for most biologists at the time. She coined the term "mesectoderm" to designate mesenchymal cells of NC origin and "mesentoderm" for the mesenchyme derived from the mesoderm. Her observations, although supported later by several workers in the late 19th century, gained acceptance very slowly.

Because an NC origin of the facial skeleton was so contentious, a long gap separates Platt's papers from a series of seminal experiments by Hörstadius and Sellman, carried out on *Amblystoma jeffersonianum* (1941, 1946), which definitively demonstrated the skeletogenic capacities of the cephali NC cells.

The well acknowledged monograph of Sven Hörstadius (1950), entitled *The Neural Crest: Its properties and derivatives in the light of experimental research*, represents a critical step on the road to understanding the role of the NC in vertebrate development and evolution.

THE MODERN ERA

The next decades were marked by the use of cell markers to study the migration of neural crest cells (NCC) in higher forms of vertebrates. One of the techniques consisted of labeling dividing cells of an embryo with tritiated thymidine (^{3}HTdR) and substituting definite regions of an unlabelled host by their equivalent, coming from a labeled donor. This technique was first used by Weston (1963) in chick embryos and subsequently applied by Chibon (1964, 1966) to Amphibians. In 1969, the quail–chick marker system, which provides a stable way to label avian embryonic cells, was devised by one of us (Le Douarin 1969). More recently, transgenic techniques have allowed selective and stable labeling of cells in the mouse embryos (see Jiang et al. 2000).

In this chapter, we first discuss the respective contribution of the ectoderm (via the neural crest) and mesoderm to the craniofacial skeleton. In the latter half of the chapter, we review the present knowledge on the genetic control of the development of this part of the body, which appeared in the transition from protochordates to vertebrates and, together with the brain, was subjected to intensive evolutionary changes within the vertebrate phylum itself.

The Respective Contribution of the NC-derived Mesenchyme and Mesoderm May be Clearly Delineated by Using the Cell-marking Techniques

In a series of studies on *Pleurodeles*, Chibon (1966, 1967) completed the data previously established by the pioneer authors cited earlier by using [3]HTdR-labeled crest cells implanted into a nonlabeled host. He showed that the neurocranium (i.e., the two parachordal cartilages, the basal plate, some *trabeculae cranii* [partly], and the auditory capsules) has a mesodermal origin, except for most of the *trabeculae cranii*, which are derived from the head NC (Chibon 1966). The visceral skeleton, formed of six visceral arches and two basibranchials, appeared to be derived from the mesectoderm.

The avian embryo is particularly suitable for in ovo surgery in the early stages of embryogenesis. This is why our knowledge on cell migrations during the ontogenetic processes has been essentially documented in the chick species, which was commonly used in embryology.

The first experimental study of the role of the NC in head morphogenesis in higher vertebrates is that of Hammond and Yntema (1953, 1964). These authors found that extirpation of the neural fold from the preotic region of 5- to 6-somite chick embryos resulted in the absence of the inter-orbital plate and Meckel's cartilage. Removal of the neural fold from the postotic region in 6- to 12-somite embryos caused total agenesis of the hyoid apparatus. Deletions had to cover a length of about six somites caudal to the otic vesicle for the hyoid skeleton to be totally absent. More thorough studies that entailed the use of cell markers have followed these pioneer investigations.

The early steps of cephalic crest-cell migration were investigated by implanting neural folds labeled with [3]HTdR into unlabelled embryos (Johnston 1966; Noden 1975). Although much more efficient than the previously used techniques, this way of labeling embryonic cells suffered from the dilution from cell divisions.

A stable marker was proposed in 1969 by one of us (Le Douarin 1969, 1973), which consisted of constructing embryonic chimeras between two closely related species, the chick and the quail (*Coturnix coturnix japon-*

ica). The quail marker system was devised from the observation of the structure of the interphase nucleus in embryonic and adult cells of the quail, where heterochromatic DNA is condensed in a large mass, generally located in the center of the nucleus and associated with the nucleolus, making this organelle stainable by the Feulgen reaction. This is not the case in chick cells and in most other animal species (like mouse, man, etc.), where heterochromatic DNA is evenly dispersed in small chromocenters in the nucleoplasm during the interphase. Construction of embryonic chimeras between quail and chick allows the cells of the two species to be recognized during the entire period of embryogenesis and even after birth by applying either the Feulgen staining procedure or by using monoclonal antibodies directed against species-specific antigens. The quail–chick marker system was extensively used to study the migration and fate of the NCC (for reviews, see Le Douarin 1982; Le Douarin and Kalcheim 1999).

In the 1970s, a large series of experiments were carried out in our laboratory to explore the contribution of the NCC to the cephalic region of the embryo. They consisted of exchanging encephalic vesicles between chick and quail embryos at the early stages of their formation and before the NCC had started to leave the neural fold (Le Lièvre 1974, 1976, 1978; Le Lièvre and Le Douarin 1974, 1975). Other authors transplanted more limited fragments of the neural folds, thus excluding the neural tube itself (Johnston et al. 1974; Noden 1978a).

Migration of the NCC starts at the midbrain level as the neural folds join on the dorsal midline. Before spreading, the cephalic crest cells form a clearly distinguishable, tightly packed mass on the dorsal aspect of the neural tube. Their total dispersion at the brain level covers a period of about 9–10 hours (from the 6- to 13-somitic stage [ss]). They first move in a relatively cell-free space beneath the dorsolateral head ectoderm. Later, they meet cells of the paraxial cephalic mesoderm, which migrate from ventral to dorsal on the surface of the developing cephalic vesicles and superficial ectoderm. More posteriorly, the cephalic NCC meet the mesodermal cells of the five first somites of the sclerotome, which is later incorporated into the skull. At these early migration stages, NCC express genes that allow them to be distinguished from the neural epithelium of the encephalic vesicles and from the superficial ectoderm: several transcription factors, such as *Snail-2/Slug* (Nieto et al. 1994), genes of the *Sox* family (*Sox 8, 9,* and *10*) (for review, see Hong and Saint-Jeannet 2005), *FoxD3* (Kos et al. 2001; Sasai et al. 2001), and *Pax3*. During their dispersion, they also acquired a surface marker recognized by the monoclonal antibody (Mab), HNK1 (Tucker et al. 1984). However, none of these markers label all of the NCC. Moreover, their expression is in general transitory.

When midbrain crest cells reach their destination on the ventral side of the head and pharyngeal regions, they are in contact with the pharyngeal endoderm and the superficial ectoderm. Components of the skull are divided into two groups: the neurocranium, surrounding the brain, and the splanchnocranium (or viscerocranium) for face and branchial arches. Moreover, one can also find a subdivision based on the type of calcified tissue: the dermatocranium,, formed by membrane bones, and the chondocranium, in which bones originate from a cartilaginous rudiment (Fig. 1). Thus, bones of both the neurocranium and splanchnocranium develop from cartilaginous rudiments that form the chondrocranium. The dermatocranium is formed by membrane bones that ossify in a noncartilaginous mesenchyme.

The neurocranium forms the base of the skull and the capsules that surround the sense organs: eyes, ears, and olfactory epithelium. The splanchnocranium, also called visceral skeleton, is formed from the branchial arches. Dermatocranial bones enclose neurocranial structures and form the roof of the skull and most of the bones of the splanchnocranium.

In the work originally carried out by using quail–chick transplantations (Le Lièvre 1974, 1976, 1978; Le Lièvre and Le Douarin 1975), large-sized grafts involving entire encephalic vesicles (such as mesencephalon, metencephalon, or myelencephalon) were performed. The goal of such an approach was to determine whether the various skeletal and connective structures of the head and neck were produced by populations of cells of either pure NC or mixed NC, and whether they had a mesodermal origin. *The answer to this question is that the entire facial and hypobranchial skeleton, as well as the cranial vault, were in fact made up by NCC in birds* (Fig. 1).

Although the fate map established about 30 years ago from the work of Johnston et al. (1974), Le Lièvre (1974, 1976, 1978), Le Lièvre and Le Douarin (1975), and Noden (1978a) has been significantly refined since, it is still mostly valid. However, as shown by the work of Couly et al. (1993), the origin of the frontal and parietal bones was found to be entirely of NC origin, a fact that had not been perceived in the previous studies and remains controversial. Using β-galactosidase encoding replication incompetent retrovirus, Evans and Noden (2006) have claimed that most of the frontal and the whole parietal bones of the chick are of mesodermal origin. However, the cellular specificity of this type of labeling remains questionable (for discussion, see Gross and Hanken 2008).

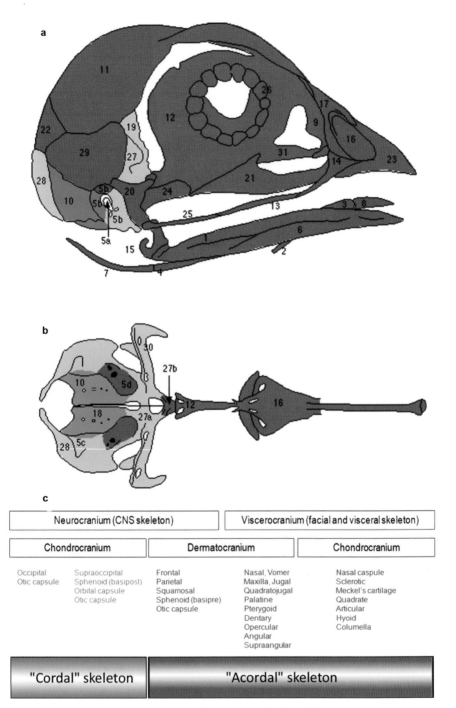

Figure 1. (*See facing page for legend*).

According to this early work, the diencephalic, mesencephalic, and anterior rhombencephalic (also called metencephalic) NCC form the skeleton of the upper and lower jaws, the palate, the tongue (i.e., the entoglossum and basihyal) and, with the posterior rhombencephalic crest, participate in the makeup of the preotic region. The hyoid bone originates from the mesencephalon down to the level of somite 4 (corresponding to the posterior limit of r8). The NC located caudally to the level of somites 4–5 does not yield mesectodermal cells in higher vertebrates (see Le Douarin and Kalcheim 1999).

Relationships Between Brain Development and NCC Migration

As mentioned earlier, the fate of the cephalic NC was established before segmentation of the brain primordium was recognized. When it was demonstrated that, at the critical stage of its specification, a large part of the brain anlage is divided into metameric units characterized by specific sets of gene activities, it appeared essential to establish with precision the contribution of each of these units to the derivatives of the NC. Two methods were used: DiI labeling and the quail–chick chimera system.

The stability of the quail marker in quail–chick chimeras permitted the determination of the precise contribution of each brain segment to the skeleton, the connective components of the head muscles, the peripheral head ganglia and nerves, and finally to the heart and glandular structures associated with the aortic arches and pharynx (carotid body, ultimobranchial bodies, thyroid, parathyroids, and thymus), with a resolution that was not provided by the previously applied experimental designs (see Le Douarin and Kalcheim 1999).

Figure 1. Contribution of the NC (red), the cephalic mesenchyme (blue), and the somitic mesoderm (green) to the vertebrate cranium. (*a*) Lateral view of the cephalic skeleton of a 14-day-old avian embryo. (*b*) Basal view of the chondrocranium of a 10-day-old embryo. (1) Angular, (2) basibranchial, (3) basihyal, (4) ceratobranchial, (5a) columella, (5b) otic capsule, (5c) otic capsule (pars ampullaris), (5d) otic capsule (pars cochlearis), (6) dentary, (7) epibranchial, (8) entoglossum, (9) ethmoid, (10) exoccipital, (11) frontal, (12) inter-orbital septum, (13) jugal, (14) maxilla, (15) Meckel's cartilage, (16) nasal capsule, (17) nasal, (18) basioccipital, (19) postoccipital, (20) quadrate, (21) palatine, (22) parietal, (23) premaxilla, (24) pterygoid, (25) quadratojugal, (26) sclerotic ossicles, (27a) basipostsphenoid, (27b) basipresphenoid, (28) supraoccipital, (29) squamosal, (30) orbital capsule, (31) vomer. (Reprinted, with permission, from Couly et al. 1993 [© Company of Biologists].)

A fate map of the anterior neural plate and neural folds was constructed in the 1980s in our laboratory (Couly and Le Douarin 1985, 1987, 1988) and completed by Cobos et al. in 2001. It appeared that the anteriormost part of the neural fold does not yield NCC, which arise from the mid-diencephalon (the level of the epiphysis anlage) downward. The segmental distribution of crest cells into the branchial arches already observed with larger grafts of neural tube or neural fold and with DiI (Johnston et al. 1974; Le Lièvre and Le Douarin 1975; Noden 1978a,b; Lumdsen et al. 1991) was generally confirmed.

On the basis of experiments carried out at 5–6 ss, the NCC originating from the posterior diencephalic and anterior half of the mesencephalic territory cover the developing telencephalic vesicle as shown in Figure 2a,b. They fill up the fronto-medial nasal bud, hence participate to the upper beak skeleton, the sclera (partly), and the frontal bone.

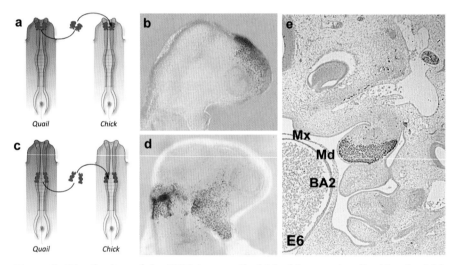

Figure 2. Distribution of the NCC in a quail–chick chimera, visualized by the Mab QCPN (Quail nonChick Peri-Nuclear antigen) (B. Carlson and J. Carlson, University of Michigan). (*a*) Graft of the posterior diencephalic neural fold of a 5-ss quail embryo to a stage-matched chick. (*b*) At E2.5 (25 ss), the NCC exiting from the graft and evidenced with QCPN Mab cover the prosencephalic vesicle. (*c*) Graft similar to *a* but at r1-r2 level. (*d*) NCC invade the maxillary and mandibular buds. Some remain in a dorsal position and will participate in the parietal bone (S. Creuzet, unpubl.). (*e*) Transverse section of an embryo similar to *d* at E6. Note the accumulation of QCPN-positive NCC in the posterior aspect of BA1 (Md). (BA2) 2nd branchial arch, (Md) mandibular bud, (Mx) maxillary bud. (Reprinted, with permission, from Creuzet et al. 2005 [© UBC Press].)

The NCC, exiting from the neural fold, limit the territory of the posterior half of the mesencephalon and migrate to the maxillary-mandibular buds (i.e., the first branchial arch, BA1). A large contribution of cells exiting from r1-r2 invade the posterior part of BA1, while another remains dorsally located and participates in the formation of the parietal bone (Fig. 2c–e).

The cephalic NC largely contributes to the peri-ocular tissues: The sclera is entirely of NC origin. In the anterior segment of the eye, most of the iris is of NC origin, as well as the ciliary process and the corneal endothelium that delineates the anterior optic chamber. In a peri-ocular position, the nictitating membrane and eyelid mesenchyme is derived from the NC (Creuzet et al. 2005). In addition, in birds, NCC give rise to sclerotic ossicles that exert tensile forces upon the intraocular fluid and constrain the proper corneal curvature (Coulombre and Coulombre 1958). Quail NCC emanating from the posterior mesencephalon and r1-r2 region are also found in the squamosal, parietal, pterygoquadrate, quadratojugal, and articular bones, as well as in Meckel's cartilage and membrane bones of the mandible (Couly et al. 1993).

NCC derived from r1 are found throughout the quadrate and pterygoid bones, also called pterygoquadrate, whereas midbrain crest cells occupy only its dorsal margin and articulation. Crest cells from r2 are essentially located in ventral articulations with the lower jaw. The boundaries between these crest-cell populations are not sharp, but r1 and r2-derived cells are always found in the posterior quarter of the jaw in agreement with their localization in the first branchial arch.

The lower jaw skeleton in the embryo is composed of Meckel's cartilage, covered by several membrane bones: the angular, supra-angular opercular, and dentary. The retroarticular process is formed by calcification of connective tissues taking place during the second half of the incubation period. It can thus be identified only in embryos older than E9–10. These tissues originate from cells belonging to the hyoid arch (i.e., BA2, which derives mainly from r4 with also a contribution from r3 and r5; see Köntges and Lumsden 1996).

The hyoid bone was found to have a highly composite origin with the entoglossum, derived from the posterior mesencephalon (Couly et al. 1996), and its other components formed by cells from the successive anteriormost rhombomeres. As a general rule, cells originating from different segmental origins remain coherent and do not mix. This results in sharply defined borders separating the crest-cell populations, originating either from the mesencephalic or the various rhombomeric levels. These borders do not coincide with anatomical entities. The columella (avian homolog of the mammalian stapes) is mostly of NC origin and made up of cells from r4 (i.e., from the hyoid arch, BA2), with a possible contribution from the paraxial mesoderm (Köntges and Lumsden 1996).

Some discrepancies exist between the reports of Couly et al. (1996) and Köntges and Lumsden (1996) as to the exact extent of the contribution of each rhombomere to the hyoid bone, but as a whole their interpretations of the segmental contribution of the hindbrain to the hypobranchial structure are similar. These discrepancies may have arisen from variations in the grafting experiments (for details, see the discussion in Le Douarin and Kalcheim 1999).

Contribution of the Paraxial Mesoderm to the Skull

The paraxial mesoderm contributing to the skull is divided into two domains: cephalic and somitic. Small fragments of quail-cephalic mesoderm were isotopically substituted to their chick counterpart in 3 ss chick embryos. It was found that most of the cephalic mesoderm yields the striated muscles of the head. Their connective components (including muscle tendons) are in contrast to NC origin (Ayer-Le Lièvre and Le Douarin 1982; Noden 1982, 1983, 1986; Couly et al. 1992). The mesoderm also forms the vascular endothelium of the head blood vessels and the meninges of the midbrain and hindbrain, while the prosencephalic meninges are of NC origin (Etchevers et al. 2001).

The bones labeled by quail cells in the experiments involving the cephalic paraxial mesoderm were the corpus sphenoidalis, the orbitosphenoid, and parts of the otic capsule. No cells of mesodermal origin were ever found in the frontal and parietal bones. In contrast, these bones were labeled following the graft of the cephalic NC (Couly et al. 1992, 1993).

The origin of the occipital bone was further documented by grafting the most rostral somites of the quail into chick embryos (and vice versa) and the participation of each of these anterior segments to the skull was analyzed (for details, see Couly et al. 1993).

Origin of the Craniofacial Skeleton in Mammals

In mammals, the migration of the cephalic NCC was studied in mice and rats by using vital dyes, such as DiI. This approach was made possible by the improvement of culture methods, enabling the embryos to survive for a few days in vitro. Apart from the fact that NCC emigration starts from the cephalic neural ridges well before neural tube closure, their migration pattern does not significantly differ from that described in the avian species (Osumi-Yamashita et al. 1996; Le Douarin and Kalcheim 1999). Experiments involving *Cre*-recombinase-mediated transgenesis were successfully carried out to follow the long-term fate of NCC. Two reports have used this method based on expression of the *Wnt1* gene in the dorsal neural ridge of the mouse embryo (Chai et al. 2000; Jiang et al. 2002). In these studies, no β-galactosidase label was found in the parietal and basioccipital bones. It has to be underlined that the *Wnt1* gene is not expressed at the level of r1 (for supplemental information, see Matsuoka et al. 2005), from which NCC forming the parietal bone may arise. More investigations in this point are required to definitively assess the origin of this posterior part of the skull vault. Using *Cre*-recombinase-mediated transgenesis with *Wnt1* and *Sox10* gene promoters, Matsuoka et al. (2005) unraveled NC and mesoderm respective contribution to neck and shoulder in the mouse. The architecture of the neck has been highly modified in birds and specific investigations were needed in mammals and are still needed in various species. In the postotic region, NC and mesoderm do not provide clear-cut boundaries in the intervening neck transition zone since the NCC are not segmentally deployed in this region posterior to r5. Therefore, the respective contribution of the postotic NC (PONC) and mesoderm to bone and muscles, located between the otic capsules and the forelimbs, remained unknown in mammals and was established by this method. The boundaries of embryonic cell populations of mesodermal origins were found to correspond to muscle attachment regions, rather than to ossification modes. Their cellular resolution data led these authors to refute the widely accepted competing "ossification model," which finds dermal (versus endochondral) ossification modes as valid criteria for identifying cellular origins and homologies of neck and shoulder structures. According to these authors, muscle-to-bone connectivity is the critical criteria for determining the cellular origin of the bones because muscles turn out to be connected with skeletal structures that have similar axial origins.

The Notion of Acordal and Cordal Skull: the "New Head" Hypothesis

As seen in Figure 3, the rostralmost limit of the mesodermal contribution to the basis of the skull corresponds to the extreme tip of the notochord and lies in the sella turcica, composed of a basi-presphenoid of NC origin and a basi-postsphenoid, which is mesodermal in nature. This leads to the distinction of a "chordal" skull derived from the NC, the respective bones of which are listed in Figure 1.

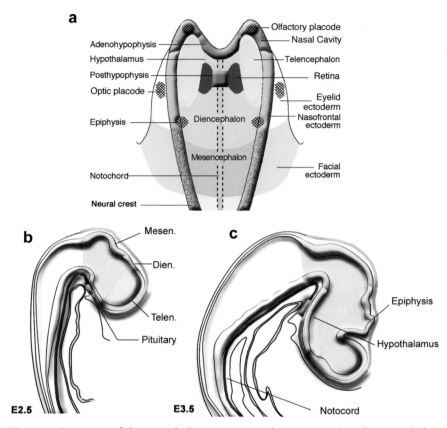

Figure 3. Fate map of the neural plate in avian embryo at 3 ss. (*a*) The neural plate is represented flat with its limiting anterior and lateral neural folds. This map was deduced from substitution experiments of fragments of the neural fold and of the neural plate. (*b,c*) Diagram showing the evolution of the two territories corresponding to the telencephalon join on the dorsal midline and extend more rostrally than the level of the hypothalamus and of the Rathke's pouch (which yields the adenohypophysis). The part of the brain that develops rostrally to the tip of the notochord (No) is later covered by the NC-derived bones. (Di) Diencephalon, (Tel) telencephalon. (Reprinted, with permission, from Le Douarin and Kalcheim 1999 [© Cambridge University Press].)

According to Gans and Northcutt (1983), vertebrates have evolved from the basic body plan of the cordates by addition, rostrally to the notochord, of a "New Head" that developed mostly from the ectoderm. As vertebrates became predators, their life style required the development of sense organs, absent in their hypothetical protocordate-like ancestors (e.g., organisms similar to the extant cephalocordate, the *Amphioxus*).

The presence of sense organs is accompanied by neural structures able to process the sensory information and generate movements. Moreover, the neural folds limiting the cephalic area of the neural plate provide the facial buds with NCC, from which the facial skeleton is entirely derived. The jaw is an important asset of the vertebrates' life style because it provides them with an efficient organ of predation. Thus, considering its mode of formation from the lateral borders of the cephalic neural plate, one can consider that the face is a "product" of the early neural primordium from which it arises.

The results obtained from the quail–chick chimeras have revealed that the entire acordal skull, including the vault formed by the frontal and parietal bones together with the anterior half of the corpus sphenoidalis, is derived from the NC. This can be related to the fact that the differentiation of somitic mesenchyme into cartilage is strictly dependent upon its interaction with the notochord and floor plate (Pourquié et al. 1993). Extirpation of both notochord and floor plate in the chick embryo results in the absence of axial skeleton. The effect of floor plate and notochord on the development of the paraxial mesoderm was shown to be mediated by the product of the gene *Sonic hedgehog* (*Shh*), expressed by both these structures at the critical stages of somitic differentiation (Johnson et al. 1994; Teillet et al. 1998).

It is interesting to discuss these embryological considerations on the origin of the craniofacial skeleton in the frame of fore- and midbrain development. According to the fate map of the anterior neural plate constructed in the chick embryo (Couly and Le Douarin 1985, 1987, 1988), the neural fold corresponding to the diencephalon (down to the level of the epiphysis anlage) does not undergo the epithelio-mesenchymal transition leading to emigration of NCC. In contrast, it yields the adenohypophysis (rostrally), which is in close contact with the notochord tip and the precordal plate. The olfactory placodes and the nasal cavity epithelium arise laterally to the future adenohypophysis with (more caudally) the epithelium of the roof of the mouth and of the upper beak and frontonasal area. At this anterior level, the lateral regions of the neural plate are fated to form the telencephalic vesicles. Figure 3 schematizes the spectacular transformation of a virtually plane structure at the early somitic stages into the encephalic vesicles. It appears that the two anlagen of the cerebral hemispheres join and grow at a considerable rate, which leads them to develop in front of the Rathke's pouch. Protection to this new part of the brain has been ensured by the NCC of the di- and mesencephalic neural folds, which have spread over the forebrain and midbrain to construct the skull vault, the optic and nasal capsules, and the facial skeleton.

Molecular Control of Craniofacial Development

Hox Genes and Craniofacial Skeleton Development

The discovery of the homeotic genes of the Hox family has been critical in the problematics raised by the development of the vertebrate head. The Hox genes, distributed in four linear clusters in the vertebrate genome, obey the same colinearity rule concerning their genomic organization and their expression pattern along the neural axis as seen in Drosophila (Duboule and Dollé 1989; Graham et al. 1989). The genes localized at the 3' end of the DNA molecule are the first to be expressed in the most anterior region of the embryo. The anterior limit of expression of these genes corresponds to the boundary between two rhombomeres. *Hoxa2* is the gene whose expression domain reaches the most rostral level in the body (between r1 and r2). Posteriorly, each rhombomere (or pair of rhombomeres) is characterized by a combination of Hox-gene expression, forming a "Hox code." As a general rule, the NCC exiting from a given level of the neural tube express the same Hox genes as the neural tube epithelium itself. It turns out that the skeletogenic NC of the head is divided into two domains: an anterior one extending from the mid-diencephalon down to r2, in which no Hox genes are expressed, and a posterior domain in which Hox genes of the first four paralog groups are expressed (Fig. 4).

It appears that the anterior, Hox-negative domain is at the origin of the facial skeleton. For this reason, it is designated here as the Facial Skeletogenic Neural Crest (FSNC), whereas the Hox-positive domain yields the hyoid bone (at the exception of the entoglossum and the rostral part of the basihyal) (Fig. 4). Rhombomere 3 is an intermediate zone and provides both BA1 and BA2 with only a small contingent of NCC. Interestingly, when r3 NCC exit the neural fold, they express *Hoxa2*, but lose it as they enter BA1. In contrast, NCC of r3 origin maintain *Hoxa2* expression when they migrate into BA2.

Loss- and Gain-of-function of Hox Genes Strongly Influences Craniofacial Development

The disruption of *Hoxa2* in the mouse results in the homeotic transformation of BA2 into BA1 skeleton (Rijli et al. 1993; Gendron-Maguire et al. 1993). In the chick, the experimental transposition of the NC from the Hox-positive posterior domain to the anterior Hox-negative domain results in the failure of facial structure formation while allowing the differentiation of neural derivatives from the transplanted crest cells (Couly et al. 1998, 2002). Moreover, the complete excision of the FSNC ends up

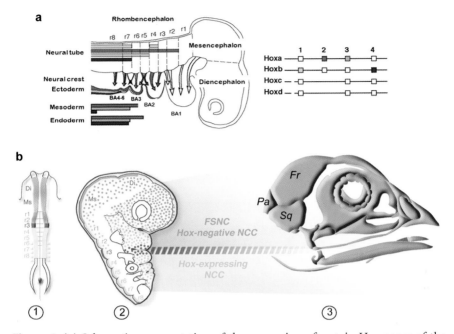

Figure 4. (*a*) Schematic representation of the expression of certain Hox genes of the first paralog groups in chick or quail embryos at E3, when the branchial arches (BA) are being colonized by the NCC originating from the posterior half of the mesencephalon and the rhombomeres (r1-r8). The arrows indicate the anteroposterior origin of the NCC migrating to each BA. Expression of Hox genes is also indicated in the superficial ectoderm. (*b*) Tracing cells from the cephalic neural fold to the craniofacial skeleton. (*1*) The colored areas of the neural fold in a 5-ss embryo correspond to the di- and mesencephalic level (green) of the cephalic vesicles; to the level of r1 and r2 (blue); to r3 (red); and to r4-r8 (yellow). (*2*) Distribution of NCC of the various levels of the cephalic NC in an E2.5 embryo (same color code as in (*1*)). (*3*) The skull vault (Pa, parietal; Fr, frontal; Sq, squamosal) and facial skeleton, both derived from the Hox-negative facial skeletogenic NC (FSNC) (i.e., from the mid-diencephalic level, including r2), are indicated by green and blue spots. In red, retroarticular process and part of the basihyal are partly derived from r3. In yellow, the part of the hyoid cartilage is derived from BA2 (mostly r4 with a modest contribution from r3 and r5). Note that the entoglossum originates from the FSNC, whereas the rest of the hyoid cartilage is derived from the Hox-positive domain of the cephalic NC. (Reprinted, with permission, from Couly et al. 1996 [© Company of Biologists].)

with the total absence of face in the operated embryos associated with anencephaly (Fig. 5a,b).

In strong contrast, any fragment of the Hox-negative NC (whether it belongs to the diencephalon, the mesencephalon, or to r1-r2) is able to regenerate a complete facial skeleton, thus showing that the rostral, Hox-negative, domain of the NC is endowed with large regenerative capacities and behaves as an "equivalence group" since part of it is able to replace the whole territory (Fig. 5c,d) (Couly et al. 2002; Creuzet et al. 2002). This also means that the NCC themselves do not possess the information necessary for patterning the different components of the craniofacial skeleton. Morphogenesis of the latter must therefore be patterned by the environmental cues provided to the NCC in the facial buds and branchial arches.

In the reverse experiment, a fragment of Hox-negative NC transplanted at the r4-r6 level participates in the formation of the hyoid cartilage (Couly et al. 1998). Therefore, it appears that Hox-gene expression has an inhibitory effect on the differentiation of facial cartilages and bones (Couly et al. 1998, 2002), but not of the hyoid bone. This notion was further confirmed by the experiments where gene expression was targeted to limited and well-defined embryonic territories through electroporation of nucleic acids at elected times in development, a major methodological asset of the chick embryo. Forced *Hoxa2* expression in the FSNC prior to its emigration from the neural primordium abolished its capacity to form the facial skeleton and produces a phenotype similar to that of FSNC excision (Creuzet et al. 2002). Experiments similar in principle, using *Hoxa3* and *Hoxb4* as transgenes, prevented partly (but severely) the development of the facial skeleton. Combination of the two constructs (*Hoxa3* + *Hoxb4*) yielded results comparable to those obtained by transfecting *Hoxa2* alone. Different results were obtained by Pasqualetti et al. (2000) in experiments where all the tissues of BA1 were transfected with *Hoxa2* at later developmental stages. In this case, gain-of-function of *Hoxa2* not only by the NCC, but by the ectodermal and mesodermal BA1 components, produced a homeotic transformation of BA1 into BA2 structures. As a whole, these results are in line with the fact that translocation of Hox-positive NCC anteriorly inhibits the development of facial skeletal structures (Couly et al. 1998).

These data emphasize the inability of the cephalic mesoderm to substitute for the absence of NCC in regenerating skeletal elements of any kind, even partly. In NC-deprived embryos, the lack of skeletal regeneration coming from the cephalic mesoderm indicates that molecular skeletogenic signaling at the forepole of the embryos for elaborating a head is

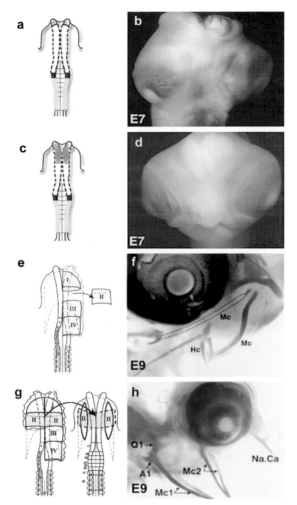

Figure 5. (*a,b*) Effect of excision of the FSNC on face and brain development. (*a*) The neural fold corresponding to the mid-diencephalon down to r2 (included) is removed in a 5-ss chick embryo. (*b*) At E7, the operated embryos had no facial development and exhibited anencephaly. (*c,d*) The phenotype seen in *b* can be rescued by the graft of only a fragment of the FSNC domain, but not of the posterior Hox-positive domain of the cephalic NC (color-coded as in Fig. 4b). (*e,f*) The pharyngeal endoderm plays a critical role in facial skeletogenesis. (*e*) Removal of a stripe of the ventral pharyngeal endoderm, corresponding to the presumptive territory of branchial pouch 1, results in the loss of most of Meckel's cartilage on the operated side. (*f*) Grafting the ventral pharyngeal endoderm in an otherwise intact embryo at the same level induces the development of an extra pair of Meckel's cartilage (Mc2). (Mc1) Endogenous Meckel's cartilage, (NaCa) nasal capsule, (Q1 and A1) quadrate and articular cartilages of the endogenous lower jaw. (Reprinted, with permission, from Couly et al. 2002 [© Company of Biologists].)

strictly NC-specific. This also implies that the somatic and cephalic meso-derm is unable to respond to the molecular determinants specific for NC-derived skeletogenesis. The next question remains to identify the signals that control the morphogenesis of the craniofacial skeleton.

Ectodermal Signaling Is Crucial for Facial Skeleton Development and Patterning

It has long been believed that growth and development of branchial arches and limb buds involve similar signaling pathways. In support of this view is the fact that the superficial ectoderm provides the underly-ing mesenchyme with polarizing and trophic cues. The role of the ecto-derm in the development of the mandibular bud has been documented for some time. For example, in organ cultures of E3 chick mandibular explants, chondrogenesis is not affected by the absence of ectoderm, which, in contrast, is necessary for membrane bone formation (Takahashi et al. 1991). The role of signaling molecules such as *Fibroblast Growth Factor 8* (*Fgf8*), Shh, and the Bone Morphogenetic Proteins were further shown to play a major role in these processes.

A discrete domain of the nasofrontal bud, designated as FEZ (Fronto-nasal Ectodermal Zone), turned out to be critical for the differentiation of NC-derived skeletal elements. This region, where expression of both morphogens *Shh* and *Fgf8* abuts, supports the proliferation of the under-lying mesenchyme (Hu et al. 2003). Moreover, heterotopic FEZ trans-plantation close to its endogenous area results in the duplication of distal upper-beak elements. These data imply that a subset of the NCC popu-lation that can be recruited for skeletogenesis by local ectoderm exhibits a certain amount of plasticity.

The ectoderm of the fronto-nasal process was also shown to play an important role in refining the shape of the beak (Wu et al. 2004). In mammals, *Fgf8* expression by the superficial ectoderm specifies the mesectoderm to form either the mandibular skeleton or the tooth germs, and refines the position of the developing jaw joint (Tucker et al. 1999; Wilson and Tucker 2004).

In turn, NCC, once they have populated the facial processes, pattern the molecular ectodermal landscape. In avian embryos, FSNC ablation is followed by a dramatic decrease of *Fgf8* expression in the anterior neural ridge (ANR) and in the ectoderm of branchial arches. If exogenous recombinant *Fgf8* is provided to the operated embryo after NC excision, a significant rescue of facial structures ensues. The results of these *Fgf8*-

rescue experiments raised the question of the origin of the NCC that regenerate the face. Replacing r3 of the operated chick embryos by their quail counterpart after FSNC removal showed that regeneration of the lower jaw induced by *Fgf8*-soaked beads is due to the strong stimulation of r3-derived NCC. The latter, which massively invade BA1, transiently express *Hoxa2* and become Hox-negative after 24 hours (Creuzet et al. 2004). These experiments demonstrate the strong regeneration capacities of r3-derived NCC and the role of *Fgf8* in regulating NCC proliferation and migration. Hence, a reciprocal relationship links the fate of the NC to the signals produced by the ectoderm.

The influence of the mesectoderm has also been demonstrated by inter-specific combinations of embryonic tissues from quail and duck, which strongly differ in their bill morphology. These experiments revealed that NCC convey the timing for *Shh* and *Pax6* expression to the host ectoderm, and therefore impose a donor- rather than host-type molecular and morphological pattern for the chimeric bill (Schneider and Helms 2003; Tucker and Lumsden 2004).

Therefore, once induced by the endoderm to develop into a particular cartilage, NCC follow a species-specific genetic program involving growth and resulting in the bill shape. These data show that the dermal ossification triggered by ectoderm follows an NC-species-specific timing of growth and differentiation.

Influence of the Endoderm on the Hypobranchial Skeletogenesis

The segmentation of the visceral skeleton in branchial arches depends upon the presence of the endodermal pouches. The segmental pattern of NCC migration into branchial arches is, to a large extent, influenced by the segmental arrangement of the pharyngeal endoderm into pouches as shown in *Danio rerio* by Piotrowski and Nüsslein-Volhard (2000). The zebrafish Van gogh (vgo) mutation, which corresponds to the invalidation of the *Tbx1* gene, results in the absence of endoderm segmentation in the branchial arches posterior to BA2. It secondarily hampers the development and the patterning of the NC-derived skeleton: The hypobranchial skeletal elements are reduced or fused (Piotrowski et al. 2003). The defective crest-derived skeletogenesis of vgo mutants can be significantly restored by implanting wild-type endodermal cells. This indicates that the pharyngeal endoderm is critical for the segregation of migrating NCC streams. In addition to skeletal malformations, other defects generated by *Tbx1* mutations are consistent with the absence of thymus, along with the reduction of aortic arches, and are reminiscent of DiGeorge syndrome.

The initial outpocketing of the pharyngeal endoderm along the anteroposterior axis depends upon dynamic remodeling of endoderm cells (Crump et al. 2004). Actin cables play a critical role in this process and provide a bi-dimensional scaffold network that accompanies the elongation of the endodermal pouches, as shown in chick by Quinlan et al. (2004). Fgfs are essential mediators in these processes.

Signaling pathways regulating the endodermal influence on branchial arch skeletogenesis. In zebrafish, *Fgf8* and *Fgf3* synergize to trigger the lateral progression of the pharyngeal endodermal cells and promote the formation of endodermal pouches (Reifers et al. 1998; Crump et al. 2004). In Fgf8 hypomorphant embryos, altered pouch formation generates rearrangements of branchial cartilages (Abu-Issa et al. 2002). It has been shown in Zebrafish that Fgfs orchestrate foregut segmentation and endodermal pouch formation through a double source of signaling involving the brain and the cephalic mesoderm (Crump et al. 2004). Some studies have suggested that retinoids play a role in patterning the pharyngeal endoderm: Inactivation of the retinoic acid pathway severely affects the proper development of the caudalmost branchial arches (Wendling et al. 2000; Begemann et al. 2001).

Endothelin-1 (Edn1) is a major signaling molecule in the pathway regulating craniofacial development. Embryos, in which the gene encoding the precursor protein of Edn1 has been inactivated, exhibit major craniofacial and cardiac abnormalities. Identical phenotypes are obtained in two other situations: when either the *ECE1* gene (the Edn1 Converting Enzyme 1, which cleaves the active Edn1 peptide from its larger proteic form) or the Edn1 receptor (EDNR-A) is knocked out (Yanagisawa et al. 1998). In chick embryos, the pharmacological inhibition of EDNR-A results in severe hypoplastic mandibular phenotypes (Kempf et al. 1998). Downstream targets of *Edn1*, such as transcription factors *Hand2* and *Bapx1*, are also involved in dorsoventral patterning in the anterior pharyngeal arches (Miller et al. 2003). It turns out that, in zebrafish, *Hand2* plays a role in specifying the ventral pharyngeal cartilages of the mandible, while *Bapx1* is critical for specifying the jaw joint (Wilson and Tucker 2004).

Classical organotypic culture experiments have indicated that intimate contact between pharyngeal endoderm and NCC is required for cartilage differentiation (Epperlein and Lehman 1975). Questioning this mechanism, our experiments carried out in the avian embryo have established the capacity of defined regions of the pharyngeal endoderm to convey

instructive patterning cues to the NCC during facial skeleton morphogenesis. Extirpation of endodermal stripes at neurula stage results in the
disruption of pieces of the facial skeleton, which vary according to the
level of the ablation (see Fig. 5e,f) (Couly et al. 2002). Reciprocally, grafting similar endodermal stripes into otherwise intact host embryos within
the migration pathway of cephalic NCC induces predicted duplications of
facial skeletal components (see Fig. 5g,h). Further experiments showed
that, in addition to being essential for shaping cartilage rudiments, signals
from the ventral foregut endoderm also dictate the position adopted by
the mandibular skeleton with respect to the body axes (Couly et al. 2002).
Hox-expressing NCC are similarly responsive to endodermal cues arising
from the more caudal part of the foregut endoderm (Ruhin et al. 2003).
Taken together, these data demonstrated that patterning and orientation
of the pharyngeal arch skeleton is dependent upon the endoderm.

Among the critical cues provided by the pharyngeal endoderm, Shh
plays a key role in the survival of the NCC and the development of BA1
derivatives. In $Shh^{-/-}$ mutant embryos, the expression of several genes
fails to be activated in the developing BA1. Such is the case for $Fgf8$,
$Bmp4$, $Barx1$, Gsc, $Sox9$, and $Twist$ (Moore-Scott and Manley 2005;
Washington Smoak et al. 2005; Yamagishi et al. 2006). It was shown that
Shh produced by the precordal plate (Pcp) and the anterior foregut endoderm is critical for lower jaw development (Wada et al. 2005; Brito et al.
2006). In chick embryos, the expression of Shh in the foregut endoderm
starts in the early neurula in the rostralmost part of the endoderm, which
is in continuity with the Pcp. Excision of the forehead (including the
complete area of the anterior foregut endoderm producing Shh) prevents
the expression of Shh in the first branchial pouch endoderm. However,
the NCC, destined to form the mandible, migrate and normally populate
BA1, but they fail to survive, and the molecular patterning of the oral
ectoderm does not occur. This results in the absence of lower jaw (Brito
et al. 2006). In decapitated embryos, supplementation with an exogenous
source of Shh restores the expression of Shh in the ventral pharyngeal
endoderm of the branchial pouches and the survival of NCC, which col-
onize BA1 and thus rescue the lower jaw development.

NC and Brain Development

Excision of the FSNC in the chick embryo at 5–6 ss resulted in multiple
consequences. In addition to the absence of facial skeleton formation, it
ends up with the unpredicted impairment of brain development, consisting in an extended anencephaly with the complete absence of the cranial

vault. A similar phenotype is obtained when Hox genes of the first paralog group are experimentally expressed in the preotic NC. In humans,
neural tube defects (exencephaly and anencephaly) are congenital anomalies that encompass failure of neural tube closure to total or partial
absence of cranial vault and cerebral hemispheres. They represent the second most common type of birth congenital abnormalities after those that
affect the heart. In NC-deprived anencephalic embryos, the neural tube
defect coincides with the loss of gene activities normally present in the
prosencephalon and mesencephalon. Accordingly, long-term structural
deficiencies in the development of the telencephalic vesicles ensue together
with agenesis of the thalamic and pretectal nuclei (Creuzet et al. 2006).

Depriving the developing head of FSNC also results in severe perturbations in the signaling centers, considered as "secondary brain organizers": NC ablation ends up with the loss of *Fgf8* expression in the ANR
(corresponding to the prosencephalic organizer) (Crossley and Martin
1995; Houart et al. 1998), and of *Wnt* (McMahon and Bradley 1990)
expression along the dorsal midline. In addition, *Shh* expression in the
prosencephalic basal plate is expanded at the expense of the diencephalic
alar plate.

If supplemented with an exogenous source of *Fgf8* placed on the
ANR, brain morphogenesis is rescued. In this context, exogenous *Fgf8*
stimulates the progression of NCC (from r3, at the edge of the excised
territory) to the forehead. Our data show that migrating NCC are necessary for the maintenance of neural tube closure in the mid- and forebrain. The deployment of NCC restores the formation of the roof plate,
concurs to the patterning of the prosencephalic alar plate, and cooperates with *Fgf8* for repressing the ventralizing signals from the basal plate
(Creuzet et al. 2006).

All together, these data support the notion that the cephalic NC,
while forming the facial and cranial vault skeleton, also controls the
development of the preotic brain. We show that NCC are at the origin
of signals crucial for (1) growth and patterning of the lateral regions of
the anterior neural plate, thus ensuring the closure of the neural tube,
(2) for dorsalizing the preotic neuroepithelium, and (3) for controling
telencephalic development through the regulation of *Fgf8* expression in
the ANR (S. Creuzet, in prep.).

CONCLUDING REMARKS

Most of the data reviewed in this chapter have been obtained from experiments carried out in the avian embryo. Their extrapolation to other ver-

tebrate forms must be taken with caution. More research is needed in other groups of vertebrates. However, general features arise from the results presently available:

1. By using the stable cell marking technique provided by the quail–chick marker system, the origin of the complete facial and hypobranchial skeleton could be demonstrated, and this notion has been confirmed in the other experimental models used so far. The origin of the entire cranial vault from the NC, although the subject of some controversy, is according to our results of NC origin.

2. The facial and cranial vault skeleton arise from the part of the neural primordium in which the homeotic genes of the Hox-family are not expressed. Moreover, their experimental expression in this domain prevents skeletal differentiation from this part of the NC, here designated as the FSNC.

3. The NCC of this anterior region (FSNC) exerts a crucial influence on the development of the preotic brain via signaling pathways that are presently under scrutiny (Creuzet et al. 2006; S. Creuzet, in prep.)

This paramount influence of the NC on head development supports Gans and Northcutt's (1983) contention that this structure has been an important asset in the evolution of vertebrate, which, as far as brain is concerned, has reached its most spectacular complexity in Humans.

ACKNOWLEDGMENTS

The authors thank Christine Martin for preparation of the manuscript. Work in the authors' laboratory is supported by the Centre National de la Recherche Scientifique, the Fondation Bettencourt-Schueller (N.L.D.), and the Association pour la Recherche sur le Cancer (grant n°3929, S.C.).

REFERENCES

Abu-Issa, R., Smyth, G., Smoak, I., Yamamura, K., and Meyers, E.N. 2002. Fgf8 is required for pharyngeal arc hand cardiovascular development in the mouse. *Development* **129**: 4613–4625.

Ayer-le Lièvre, C.S. and Le Douarin, N.M. 1982. The early development of cranial sensory ganglia and the potentialities of their component cells studied in quail-chick chimeras. *Dev. Biol.* **94**: 291–310.

Begemann, G., Schilling, T.F., Rauch, G.J., Geisler, R., and Ingham, P.W. 2001. The zebrafish neckless mutation reveals a requirement for raldh2 in mesodermal signals that pattern the hindbrain. *Development* **128**: 3081–3094.

Brito, J.M., Teillet, M.A., and Le Douarin, N.M. 2006. An early role for sonic hedgehog from foregut endoderm in jaw development: Ensuring neural crest cell survival. *Proc. Natl. Acad. Sci.* **103:** 11607–11612.

Chai, Y., Jiang, X., Ito, Y., Bringas Jr., P., Han, J., Rowitch, D.H., Soriano, P., McMahon, A.P., and Sucov, H.M. 2000. Fate of the mammalian cranial neural crest during tooth and mandibular morphogenesis. *Development* **127:** 1671–1679.

Chibon, P. 1964. Analyse par la méthode de marquage nucléaire à la thymidine tritiée des dérivés de la crête neurale céphalique chez l'Urodèle *Pleurodèles waltii*. *C.R. Acad. Sci.* **259:** 3624–3627.

Chibon, P. 1966. Analyse expérimentale de la régionalisation et des capacités morphogénétiques de la crête neurale chez l'Amphibien Urodèle *Pleurodèles waltii* Michah. *Mem. Soc. Zool. Fr.* **36:** 1–107

Chibon, P. 1967. Nuclear labelling by tritiated thymidine of neural crest derivatives in the amphibian Urodele *Pleurodeles waltlii* Michah. *J. Embryol. Exp. Morphol.* **18:** 343–358.

Cobos, I., Shimamura, K., Rubenstein, J.L., Martínez, S., and Puelles, L. 2001. Fate map of the avian anterior forebrain at the four-somite stage, based on the analysis of quail-chick chimeras. *Dev Biol.* **239:** 46–67.

Coulombre, A.J. and Coulombre, J.L. 1958. The role of the intraocular pressure in the development of the chick eye. IV. Corneal curvature. *A.M.A. Arch. Ophthalmol.* **59:** 502–506.

Couly, G.F. and Le Douarin, N.M. 1985. Mapping of the early neural primordium in quail-chick chimeras. I. Developmental relationships between placodes, facial ectoderm, and prosencephalon. *Dev. Biol.* **110:** 422–439

Couly, G.F. and Le Douarin, N.M. 1987. Mapping of the early neural primordium in quail-chick chimeras. II. The prosencephalic neural plate and neural folds: Implications for the genesis of cephalic human congenital abnormalities. *Dev. Biol.* **120:** 198–214.

Couly, G.F. and Le Douarin, N.M. 1988. The fate map of the cephalic neural primordium at the presomitic to the 3-somite stage in the avian embryo. *Development* (suppl.) **103:** 101–103.

Couly, G.F., Coltey, P. M., and Le Douarin, N.M. 1992. The developmental fate of the cephalic mesoderm in quail-chick chimeras. *Development* **114:** 1–15.

Couly, G.F., Coltey, P.M., and Le Douarin, N.M. 1993. The triple origin of skull in higher vertebrates: A study in quail-chick chimeras. *Development* **117:** 409–429.

Couly, G., Grapin-Botton, A., Coltey, P., and Le Douarin, N.M. 1996. The regeneration of the cephalic neural crest, a problem revisited: The regenerating cells originate from the contralateral or from the anterior and posterior neural fold. *Development* **122:** 3393–3407.

Couly, G., Grapin-Botton, A., Coltey, P., Ruhin, B., and Le Douarin, N.M. 1998. Determination of the identity of the derivatives of the cephalic neural crest: Incompatibility between Hox gene expression and lower jaw development. *Development* **125:** 3445–3459.

Couly, G., Creuzet, S., Bennaceur, S., Vincent, C., and Le Douarin, N.M. 2002. Interactions between Hox-negative cephalic neural crest cells and the foregut endoderm in patterning the facial skeleton in the vertebrate head. *Development* **129:** 1061–1073.

Creuzet, S., Couly, G., Vincent, C., and Le Douarin, N.M. 2002. Negative effect of *Hox* gene expression on the development of neural crest derived facial skeleton. *Development* **129:** 4301–4313.

Creuzet, S., Schuler, B., Couly, G., and Le Douarin, N.M. 2004. Reciprocal relationships between *Fgf8* and neural crest cells in facial and forebrain development. *Proc. Natl. Acad. Sci.* **101:** 4843–4847.

Creuzet, S., Vincent, C., and Couly, G. 2005. Neural crest derivatives in ocular and peri-ocular structures. *Int. J. Dev. Biol.* **49:** 161–171.

Creuzet, S., Martinez, S., and Le Douarin, N.M. 2006. The cephalic neural crest exerts a critical effect on forebrain and midbrain development. *Proc. Natl. Acad. Sci.* **103:** 14033–14038.

Crossley, P.H. and Martin, G.R. 1995. The mouse *Fgf8* gene encodes a family of polypeptides and is expressed in regions that direct outgrowth and patterning in the developing embryo. *Development* **121:** 439–451.

Crump, J.G., Maves, L., Lawson, N.D., Weinstein, B.M., and Kimmel, C.B. 2004. An essential role for Fgfs in endodermal pouch formation influences later craniofacial skeletal patterning. *Development* **131:** 5703–5716.

Duboule, D. and Dollé, P. 1989. The structural and functional organization of the murine HOX gene family resembles that of *Drosophila* homeotic genes. *EMBO J.* **8:** 1497–1505.

Epperlein, H.H. and Lehmann, R. 1975. The ectomesenchymal-endodermal interaction-system (EEIS) of *Triturus alpestris* in tissue culture. I. Observations of visceral cartilage. *Differentiation* **4:** 159–174.

Etchevers, H.C., Vincent, C., Le Douarin, N.M., and Couly, G.F. 2001. The cephalic neural crest provides pericytes and smooth muscle cells to all blood vessels of the face and forebrain. *Development* **128:** 1059–1068.

Evans, D.J. and Noden, D.M. 2006. Spatial relations between avian craniofacial neural crest and paraxial mesoderm cells. *Dev. Dyn.* **235:** 1310–1325.

Gans, C. and Northcutt, R. 1983. Neural crest and the origin of vertebrates: A new head. *Science* **220:** 268–274.

Gendron-Maguire, M., Mallo, M., Zhang, M., and Gridley, T. 1993. Hoxa-2 mutant mice exhibit homeotic transformation of skeletal elements derived from cranial neural crest. *Cell* **75:** 1317–1331.

Goronowitsch, N. 1892. Die axiale and die laterale Kopfmentamerie des Vögelembryonen. Die Rolle der sog. "Ganglienleisten" im aufbaue der Nervenstämme. *Anat. Anz.* **7:** 454–464.

Goronowitsch, N. 1893. Weiters über die ektodermale Entstehung von Skeletanlagen im Kopfe der Wirbeltiere. *Morphol. Jahrb.* **20:** 425–428.

Graham, A., Papalopulu, N., and Krumlauf, R. 1989. The murine and *Drosophila* homeobox gene complexes have common features of organization and expression. *Cell* **57:** 367–378.

Gross, J.B. and Hanken, J. 2008. Review of fate-mapping studies of osteogenic cranial neural crest in vertebrates. *Dev Biol.* **317:** 389–400.

Haeckel, E. 1874. Die Gastrea-Theorie, die Phylogenetische Klasification des Thierreiches und die Homologie der Keimblätter. *Jena Z. Naturwiss.* **8:** 1–55.

Hammond, W.S. and Yntema, C.L. 1953. Deficiencies in visceral skeleton of the chick after removal of cranial neural crest. *Anat. Rec.* **115:** 393–394.

Hammond, W.S. and Yntema, C.L. 1964. Depletions of pharyngeal arch cartilages following extirpation of cranial neural crest in chick embryos. *Acta Anat.* **56:** 21–34

His, W. 1868. *Untersuchungen über die erste Anlage des Wirbeltierleibes. Die erste Entwicklung des Hühnchens im Ei.* F.C.W. Vogel, Leipzig, Germany.

Hong, C.-H. and Saint-Jeannet, J.-P., 2005. Sox proteins and neural crest development. *Semin. Cell Dev. Biol.* **16:** 694–703.

Houart, C., Westerfield, M., and Wilson, S.W. 1998. A small population of anterior cells patterns the forebrain during zebrafish gastrulation. *Nature* **391:** 788–792.

Hörstadius S. 1950. *The neural crest: Its properties and derivatives in the light of experimental research.* Oxford University Press, London.

Hörstadius, S. and Sellman, S. 1941. Experimental studies in the determination of the chondrocranium in *Amblystoma mexicanum. Ark. Zool. Stockholm* **33A:** 1–8.

Hörstadius, S. and Sellman, S. 1946. Experimentelle Untersuchungen über die Determination des Knorpeligen Kopfskelettes bei Urodelen. *Nova Acta Reg. Soc. Scient. Ups.* **13:** 1–170.

Hu, D., Marcucio, R.S., and Helms, J.A. 2003. A zone of frontonasal ectoderm regulates patterning and growth in the face. *Development* **130:** 1749–1758.

Huxley, T.H. 1849. On the anatomy and the affinities of the family of the Medusæ. *Philos. Trans. R. Soc.* **139:** 413–434.

Jiang, X., Iseki, S., Maxson, R.E., Sucov, H.M., and Morriss-Kay, G.M. 2002. Tissue origins and interactions in the mammalian skull vault. *Dev. Biol.* **241:** 106–116.

Johnson, R.L., Laufer, E., Riddle, R.D., and Tabin, C. 1994. Ectopic expression of *Sonic hedgehog* alters dorsal-ventral patterning of somites. *Cell* **79:** 1165–1173.

Johnston, M.C. 1966. An autoradiographic study of the migration and fate of cranial neural crest cells in the chick embryo. *Anat. Rec.* **156:** 143–156.

Johnston, M.C., Bhakdinaronk, A., and Reid, Y.C. 1974. An expanded role of the neural crest in oral and pharyngeal development. In *Oral sensation and perception development in the fetus and infant* (ed. J.F. Bosma), pp. 37–52. U.S. Government Printing Service, Washington, D.C.

Katschenko, N. 1888. Zur Entwicklungsgeschichte des selachierembryos. *Anat. Anz.* **3:** 445–467.

Kempf, H., Linares, C., Corvol P., and Gasc, J.M.. 1998. Pharmacological inactivation of the endothelin type A receptor in the early chick embryo: A model of mispatterning of the branchial arch derivatives. *Development* **125:** 4931–4941.

Köntges, G. and Lumsden, A. 1996. Rhombencephalic neural crest segmentation is preserved throughout craniofacial ontogeny. *Development* **122:** 3229–3242.

Kos, R., Reedy, M., Johnson, R.L., and Erickson, C. 2001. The winged-helix transcription factor Foxd3 is important for establishing the neural crest lineage and repressing melanogenesis in avian embryos. *Development* **128:** 1467–1479.

Le Douarin, N.M. 1969. Particularités du noyau interphasique chez la Caille japonaise (*Coturnix coturnix japonica*). Utilisation de ces particularités comme "marquage biologique" dans les recherches sur les interactions tissulaires et les migrations cellulaires au cours de l'ontogenèse. *Bull. Biol. Fr. Belg.* **103:** 435–452.

Le Douarin, N.M. 1973. A biological cell labelling technique and its use in experimental embryology. *Dev. Biol.* **30:** 217–222.

Le Douarin, N. 1982. *The neural crest.* Cambridge University Press, Cambridge.

Le Douarin, N.M. and Kalcheim, C. 1999. *The neural crest,* 2nd edition. Cambridge University Press, New York.

Le Lièvre, C. 1974. Rôle des cellules mésectodermiques issues des crêtes neurales céphaliques dans la formation des arcs branchiaux et du squelette viscéral. *J. Embryol. Exp. Morphol.* **31:** 453–477.

Le Lièvre, C. 1976. "Contribution des crêtes neurales à la genèse des structures céphaliques et cervicales chez les oiseaux." Ph.D. thesis, University of Nantes, France.

Le Lièvre, C.S. 1978. Participation of the neural crest-derived cells in the genensis of the skull in birds. *J. Embryol. Exp. Morphol.* **47**: 17–37.

Le Lièvre, C.S. and Le Douarin, N.M. 1974. Ectodermic origin of the derma of the face and neck, demonstrated by interspecific combinations in the bird embryo. *C.R. Acad. Sci. Ser. III Life Sci.* **278**: 517–520.

Le Lièvre, C.S. and Le Douarin, N.M. 1975. Mesenchymal derivatives of the neural crest: Analysis of chimaeric quail and chick embryos. *J. Embryol. Exp. Morphol.* **34**: 125–154.

Lumsden, A., Sprawson, N., and Graham, A. 1991. Segmental origin and migration of neural crest cells in the hindbrain region of the chick embryo. *Development* **113**: 1281–1291.

Matsuoka, T., Ahlberg, P., Kessaris, N., and Koentges, G. 2005. Neural crest origins of the neck and shoulder. *Nature* **436**: 347–355.

McMahon, A.P. and Bradley, A. 1990. *Wnt*-1 (*int*-1) proto-oncogene is required for development of a large region of the mouse brain. *Cell* **62**: 1073–1085.

Miller, C.T., Yelon, D., Stainier, D.Y., and Kimmel, C.B. 2003. Two endothelin 1 effectors, hand2 and bapx1, pattern ventral pharyngeal cartilage and the jaw joint. *Development* **130**: 1353–1365.

Moore-Scott, B.A. and Manley, N.R. 2005. Differential expression of Sonic hedgehog along the anterior-posterior axis regulates patterning of pharyngeal pouch endoderm and pharyngeal endoderm-derived organs. *Dev. Biol.* **278**: 323–335.

Nieto, M.A., Sargent, M.G., Wilkinson, D.G., and Cooke, J. 1994. Control of cell behavior during vertebrate development by Slug, a zinc finger gene. *Science* **264**: 835–839.

Noden, D.M. 1975. An analysis of migratory behavior of avian cephalic neural crest cells. *Dev. Biol.* **42**: 106–130.

Noden, D.M. 1978a. The control of avian cephalic neural crest cytodifferentiation. I. Skeletal and connective tissues. *Dev. Biol.* **67**: 296–312.

Noden, D.M. 1978b. The control of avian cephalic neural crest cells cytodifferentiation. II. Neural tissues. *Dev. Biol.* **67**: 313–329.

Noden, D.M. 1982. Patterns and organization of craniofacial skeletogenic and myogenic mesenchyme: A perspective. *Prog. Clin. Biol. Res.* **101**: 167–203.

Noden, D.M. 1983. The role of the neural crest in patterning of avian cranial skeletal, connective, and muscle tissues. *Dev. Biol.* **96**: 144–165.

Noden, D.M. 1986. Origins and patterning of craniofacial mesenchymal tissues. *J. Craniofac. Genet. Dev. Biol.* (Suppl.) **2**: 15–31.

Osumi-Yamashita, N., Ninomiya, Y., Doi, H., and Eto, K. 1996. Rhombomere formation and hind-brain crest cell migration from prorhombomeric origins in mouse embryos. *Dev. Growth Differ.* **38**: 107–118.

Pander, C.H. 1817. *Beiträge zur Entwicklungsgeschichte. des Hühnchens im Ei.* Stahel, Würzburg, Germany.

Pasqualetti, M., Ori, M., Nardi, I., and Rijli, F.M. 2000. Ectopic Hoxa2 induction after neural crest migration results in homeosis of jaw elements in *Xenopus. Development* **127**: 5367–5378.

Piotrowski, T., and Nüsslein-Volhard, C. 2000. The endoderm plays an important role in patterning the segmented pharyngeal region in zebrafish (*Danio rerio*). *Dev. Biol.* **225**: 339–356.

Piotrowski, T., Ahn, D.G., Schilling, T.F., Nair, S., Ruvinsky, I., Geisler, R., Rauch, G.J., Haffter, P., Zon, L.I., Zhou, Y., et al. 2003. The zebrafish *van gogh* mutation disrupts *tbx1*, which is involved in the DiGeorge deletion syndrome in humans. *Development* **130**: 5043–5052.

Platt, J. 1893. Ectodermic origin of the cartilage of the head. *Anat. Anz.* **8:** 506–509.

Pourquié, O., Coltey, M., Teillet, M.A., Ordahl, C. and Le Douarin, N.M. 1993. Control of dorsoventral patterning of somitic derivatives by notochord and floor plate. *Proc. Natl Acad. Sci.* **90:** 5242–5246.

Quinlan, R., Martin, P., and Graham, A. 2004. The role of actin cables in directing the morphogenesis of the pharyngeal pouches. *Development* **131:** 593–599.

Reifers, F., Bohil, H., Walsh, E.C., Crossley, P.H., Stainer, D.Y., and Brand, M. 1998. *Fgf8* is mutated in zebrafish *acerebellar* (*ace*) mutants and is required for maintenance of midbrain-hindbrain boundary development and somitogenesis. *Development* **125:** 2381–2395.

Rijli, F.M., Mark, M., Lakkaraju, S., Dierich, A., Dollé, P., and Chambon, P. 1993. A homeotic transformation is generated in the rostral branchial region of the head by disruption of *Hoxa-2*, which acts as a selector gene. *Cell* **75:** 1333–1349.

Ruhin, B., Creuzet, S., Vincent, C., Le Douarin, N.M., and Couly, G. 2003. Patterning of the hyoid cartilage depends upon signals arising from the ventral foregut endoderm. *Dev. Dyn.* **228:** 239–246.

Sasai, N., Mizuseki, K., and Sasai, Y. 2001. Requirement of FoxD3-class signaling for neural crest determination in *Xenopus*. *Development* **128:** 2525–2536.

Schneider, R.A. and Helms, J.A. 2003. The cellular and molecular origins of beak morphology. *Science* **299:** 565–568.

Takahashi Y., Bontoux, M., and Le Douarin, N.M. 1991. Epithelio-mesenchymal interactions are critical for Quox 7 expression and membrane bone differentiation in the neural crest derived mandibular mesenchyme. *EMBO J.* **10:** 2387–2393.

Teillet M., Watanabe Y., Jeffs P., Duprez D., Lapointe F., and Le Douarin N.M. 1998. Sonic hedgehog is required for survival of both myogenic and chondrogenic somitic lineages. *Development* **125:** 2019–2030.

Tucker, A.S. and Lumsden, A. 2004. Neural crest cells provide species-specific patterning information in the developing branchial skeleton. *Evol. Dev.* **6:** 32–40.

Tucker, G.C., Aoyama, H., Lipinski, M., Tursz, T., and Thiery, J.P. 1984. Identical reactivity of monoclonal antibodies HNK-1 and NC-1: Conservation in vertebrates on cells derived from the neural primordium and some leukocytes. *Cell Differ.* **14:** 223–230.

Tucker, A.S., Yamada, G., Grigoriou, M., Pachnis, V., and Sharpe, P.T. 1999. Fgf-8 determines rostral-caudal polarity in the first branchial arch. *Development* **126:** 51–61.

von Baer, K.E. 1828. *Ueber Entwicklungsgeschichte der Tiere.* Bornträger, Köningsberg, Germany.

Wada, N., Javidan, Y., Nelson, S., Carney, T.J., Kelsh, R.N., and Schilling, T.F. 2005. Hedgehog signaling is required for cranial neural crest morphogenesis and chondrogenesis at the midline in the zebrafish skull. *Development* **132:** 3977–3988.

Washington Smoak, I., Byrd, N.A., Abu-Issa, R., Goddeeris, M.M., Anderson, R., Morris, J., Yamamura, K., Klingensmith, J., and Meyers, E.N. 2005. *Sonic hedgehog* is required for cardiac outflow tract and neural crest cell development. *Dev. Biol.* **283:** 357–372.

Wilson, J. and Tucker, A.S. 2004. Fgf and Bmp signals repress the expression of *Bapx1* in the mandibular mesenchyme and control the position of the developing jaw joint. *Dev. Biol.* **266:** 138–150.

Wendling, O., Dennefeld, C., Chambon, P., and Mark, M. 2000. Retinoid signaling is essential for patterning the endoderm of the third and fourth pharyngeal arches. *Development* **127:** 1553–1562.

Weston, J. 1963. A radiographic analysis of the migration and localization of trunk neural crest cells in the chick. *Dev. Biol.* **6:** 279–310.

Wilson, J. and Tucker, A.S.2004. Fgf and Bmp signals repress the expression of Baxpx1 in the mandibular mesenchyme and control the position of the developing jaw joint. *Dev. Biol.* **266:** 138–150.

Wu, P., Jiang, T.X., Suksaweang, S., and Chuong, C.-M. 2004. Molecular shaping of the beak. *Science* **305:** 1465–1466.

Yamagishi, C., Yamagishi, H., Maeda, J., Tsuchihashi, T., Ivey, K., Hu, T., and Srivastava, D. 2006. Sonic hedgehog is essential for first pharyngeal arch development. *Pediatr. Res.* **59:** 349–335.

Yanagisawa, H., Yanagisawa, M., Kapur, R.P., Richardson, J.A., Williams, S.C., Clouthier, D.E., de Wit, D., Emoto, N., and Hammer, R.E. 1998. Dual genetic pathways of endothelin-mediated intercellular signaling revealed by targeted disruption of endothelin converting enzyme-1 gene. *Development* **125:** 825–836.

5

Transcriptional Control of Chondrocyte Differentiation

Benoit de Crombrugghe
Department of Molecular Genetics
The University of Texas
M.D. Anderson Cancer Center
Houston, Texas 77030

Haruhiko Akiyama
Department of Orthopaedics
Kyoto University
Sakyo, Kyoto 606-8507, Japan

Bone formation occurs through two distinct processes. Most skeletal elements form by endochondral ossification, which involves a cartilage intermediate. The other skeletal elements, which mainly include craniofacial bones, are formed by a process of intramembranous ossification, whereby bones form directly from mesenchymal condensations without involvement of a cartilage intermediate. In addition to forming the templates for the development of endochondral bones, cartilage is also present as a permanent connective tissue at the ends of bones (articular cartilages) and in ear, nose, and throat tissues.

Chondrogenesis is a multistep process that begins with the commitment of mesenchymal cells to a chondrogenic cell lineage (Fig. 1). These cells then aggregate into condensations that prefigure the future shape of endochondral bones. Cells in these mesenchymal condensations overtly differentiate into chondrocytes and produce a characteristic cartilage extracellular matrix (ECM). These cells then undergo several more changes. The first is a unidirectional proliferation that results in parallel columns of dividing cells that fuel the longitudinal growth of bones. In contrast to the overtly differentiated chondrocytes, which are round cells,

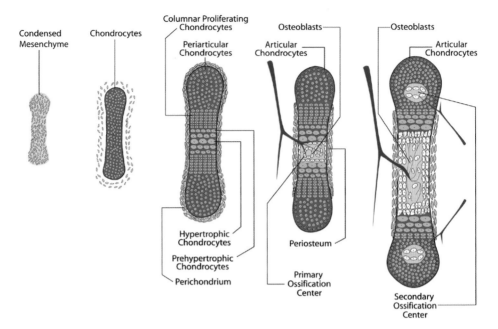

Figure 1. Schematic representation of the various steps involved in endochondral bone formation. (Reprinted from Murakami et al. 2003 [© Oxford University Press].)

the proliferating chondrocytes in these parallel columns have a flat morphology. These cells then exit the cell cycle, gradually change their genetic program, and become prehypertrophic and then hypertrophic chondrocytes. The most mature hypertrophic chondrocytes, which acquire the ability to mineralize their ECM, later die by apoptosis.

In endochondral skeletal elements, first a thin layer of mesenchymal cells on the periphery of the condensations forms the perichondrium, which subsequently develops into the periosteum. Cells in the periosteum invade the zone of hypertrophic chondrocytes, initially localized in the center of the endochondral skeletal elements, together with blood vessels and cells of hematopoietic origin that develop into multinucleated osteoclasts. The mesenchymal cells differentiate into osteoblasts, which deposit a bone-specific ECM while the cartilage template is simultaneously degraded. The pathways of chondrocyte and osteoblast differentiation are thus interconnected during endochondral bone formation and must be coordinated. In membranous skeletal elements, cells in mesenchymal condensations directly differentiate into osteoblasts.

Chondrogenic mesenchymal cells and chondrocytes derive from three different embryonic tissues—the cranial neural crest for craniofacial cartilages, the paraxial mesoderm for the axial skeleton, and the lateral mesoderm for limb skeletal elements—but the morphological and molecular characteristics of these cells are essentially identical regardless of tissue of origin.

Key transcription factors control chondrocyte differentiation, and these—in particular, three members of the SOX family of transcription factors—are an important focus of this chapter. SOX transcription factors were originally identified because of their homologies with SRY, the male sex-determination transcription factor. They contain a high mobility group (HMG)-box domain that binds DNA in the minor groove and also bends DNA. This domain also includes sequences needed for nuclear import and export. Twenty SRY-related HMG-box (*SOX*) genes have been identified in the human and mouse genomes, and these are divided into eight subgroups (Schepers et al. 2002; Lefebvre et al. 2007). Along with *SOX8* and *SOX10, SOX9* belongs to group E, whose members have a well-conserved HMG-box domain, a *trans*-activation domain, and a dimerization domain. In chondrocytes, Sox9 binds as a dimer to pairs of recognition sites in target genes. Dimer binding is a requirement for activating expression of these genes in these cells (Bernard et al. 2003; Sock et al. 2003). Another group of SOX proteins in group D includes SOX5, SOX6, and SOX13. Members of this group have an HMG-box domain but no clear *trans*-activation domain, suggesting that these proteins act as architectural organizers. As discussed later in this chapter, SOX5 and SOX6 also have an essential role in chondrogenesis.

In addition, critical signaling molecules have an essential role in coordinating the successive steps of endochondral bone formation. These signaling molecules include bone morphogenetic proteins (BMPs), Indian hedgehog (Ihh), Wnt polypeptides, insulin-like growth factors (IGFs), parathyroid hormone-related peptide (PTHrP), fibroblast growth factors (Fgfs), and others.

The ECM of cartilage contains a unique combination of secreted proteins. A major component of this ECM is type II collagen, which, together with collagen types IX and XI, forms fibrils. The cartilage ECM also contains different types of proteoglycans (including aggrecan and link protein, as well as several members of the family of small leucine-rich proteoglycans) and other glycoproteins such as cartilage oligomeric proteins and matrilins. Many skeletal birth defects are caused by mutations that disrupt the complex process of chondrogenesis and endochondral bone formation. Not surprisingly, a number of these mutations map to the genes for ECM components of cartilage. In this chapter, we discuss the role of transcription factors in the sequential steps of chondrogenesis during endochondral bone formation.

FORMATION OF CHONDROGENIC MESENCHYMAL CONDENSATIONS

Several lines of evidence have led to the conclusion that commitment of cells to chondrogenic mesenchymal condensations and to chondrocytic differentiation requires the transcription factor SOX9. First, a rare human skeletal dysplasia, called campomelic dysplasia, was shown in 1994 to be due to mutations in the gene for SOX9 (Foster et al. 1994; Wagner et al. 1994). Heterozygous mutations in and around *SOX9* produce distinct clinical features, including a disproportionately short stature, bowing of limbs, low ears, a depressed nasal bridge, talipes equinovarus, a long philtrum, and micrognathia. Radiological findings show bowing of the long bones, hypoplasia of the scapula, narrow iliac wings, and a small thorax with hypoplastic ribs. In most cases, the disease leads to perinatal lethality due to respiratory complications. These clinical features suggest that campomelic dysplasia is a generalized disease of the endochondral bones and, because the disease is caused by haploinsufficiency of *SOX9*, imply a key role for SOX9 in cartilage development. In addition to the major skeletal symptoms, the disease is often associated with sex reversal in XY individuals and occasionally with milder defects in the heart and kidney.

SOX9 mutations identified in campomelic dysplasia include missense or deletion mutations in the HMG-box DNA-binding domain, in the dimerization domain, or in HMG-box sequences involved in nuclear translocation, as well as nonsense mutations that lead to truncation of the SOX9 polypeptide and chromosomal translocations that involve regulatory DNA segments upstream of the *SOX9* gene (Gimovsky et al. 2008).

A second line of evidence that suggests that SOX9 has a crucial role in cartilage development is the pattern of expression of *Sox9* during chondrogenesis. During mouse embryogenesis, *Sox9* is expressed in all chondroprogenitors and chondrocytes, but this expression is abolished in hypertrophic chondrocytes (Fig. 2) (Wright et al. 1995; Ng et al. 1997; Zhao et al. 1997). *Sox9* is also expressed in the male gonad, otic vesicle, heart, kidney, pancreas, intestine, and neural crest.

The expression of *Sox9* in endochondral skeletal elements overlaps largely with that of *Col2a1*, the gene for type II collagen, a major cartilage matrix protein, suggesting that SOX9 controls *Col2a1* expression.

Figure 2. Expression of *Sox9* in a mouse embryo at embryonic day 12.5. A *lacZ* gene was inserted in the *Sox9* gene by homologous recombination in mouse embryonic stem cells. X-gal staining shows *lacZ* expression reproducing expression of *Sox9* in all cartilages and cartilage precursors.

This possibility is supported by the results of experiments involving cell transfections and transgenic mice, which identified chondrocyte-specific regulatory sequences in the first intron of the *Col2a1* gene. These sequences include four imperfect single Sox-binding sites (Lefebvre et al. 1997; Ng et al. 1997; Zhou et al. 1998). SOX9 activates this enhancer, and mutations in several of the SOX9 DNA-binding sites, which abrogate the binding of SOX9 to these sites, abolish the activity of this chondrocyte-specific enhancer both in DNA transfection experiments in vitro and in transgenic mice (Lefebvre et al. 1997; Zhou et al. 1998). Similar SOX9-activated chondrocyte enhancers are present in several other genes, including *Col11a2, Col9a2, Col27a1, Cdrap, matrilin1,* and *aggrecan* (Bridgewater et al. 1998, 2003; Xie et al. 1999; Liu et al. 2000; Sekiya et al. 2000; Jenkins et al. 2005; Rentsendorj et al. 2005).

Definite evidence of the essential role of SOX9 in chondrogenesis came from mouse genetic studies that resulted in deletion of the *Sox9* gene. Heterozygous *Sox9* mouse mutants phenocopy the skeletal manifestations of campomelic dysplasia (Bi et al. 2001; Kist et al. 2002). Hypoplasia of endochondral bones in these mutant mice is due to smaller chondrogenic mesenchymal condensations, suggesting that fewer *Sox9*-expressing cells are recruited into these condensations. The mutants also show an enlargement of the zone of hypertrophic chondrocytes and premature mineralization of endochondral bones. These results provide evidence that an adequate level of SOX9 is required both for the proper formation of mesenchymal condensations and for the proper control of hypertrophic chondrocyte differentiation. However, the perinatal lethality of *Sox9* heterozygous mouse mutants precluded the use of conventional crosses of heterozygous animals to generate *Sox9* homozygous mutants in order to elucidate the role of SOX9. This obstacle was overcome first by the generation of mouse embryo chimeras using *Sox9* homozygous-mutant embryonic stem cells in which the *LacZ* gene was knocked in at both *Sox9* loci. These chimeras were generated by inject-ing *Sox9*-null embryonic stem cells into wild-type blastocysts so that the cellular composition of the resulting embryos consisted of both wild types and *Sox9*-null cells. Remarkably, in these chimeric embryos, the *Sox9*-null cells are excluded from the wild-type chondrogenic mesenchy-mal condensations and are found outside these condensations, even though before these condensations form, the mutant and wild-type cells are intermingled (Fig. 3) (Bi et al. 1999). The *Sox9*-mutant cells in these chimeras are unable to express typical chondrogenic markers such as *Col2a1, Col9a2, Col11a2,* and *aggrecan*. The mutant *Sox9*-null cells are, however, specified and migrate to their proper locations in the embryo. Thus, the analysis of these chimeras strongly suggests that SOX9 is needed for commitment to the chondrogenic lineage and for formation of chondrogenic mesenchymal condensations. The exclusion of *Sox9*-null cells from these condensations raises the possibility that SOX9 may be

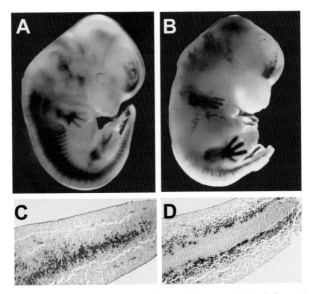

Figure 3. Embryonic day 12.5 mouse embryo chimeras derived from $Sox9^{+/-}$ (A, C) and from $Sox9^{-/-}$ (B,D) embryonic stem ES cells injected in wild-type blastocysts. Limb bud sections are shown (B, D). In $Sox9^{+/-}$ and $Sox9^{-/-}$ ES cells, a *lacZ* gene was inserted to replace the deleted portion of the *Sox9* gene. (Modified from Bi et al. 1999 [© Nature Publishing Group].)

required for the expression or the activity of cell-surface proteins that may be needed for mesenchymal condensations. At later stages in these chimeras, the mutant *Sox9*-null cells are also found outside the periosteum, implying that these cells are also unable to participate in the formation of osteogenic tissues (Bi et al. 1999).

In separate genetic experiments, the requirement of SOX9 for the commitment of mesenchymal cells to a chondrogenic cell fate was examined by using the Cre recombinase–*LoxP* system. When *Sox9* has been inactivated using the *Prx1-Cre* transgene in limb mesenchyme prior to formation of mesenchymal condensations, the condensations are totally absent, and chondrogenic marker genes are not expressed (Akiyama et al. 2002). In addition, in these cells, the expression of *Runx2*, an osteogenic marker gene required for osteoblast differentiation (Ducy et al. 1997; Komori et al. 1997; Otto et al. 1997), is missing, suggesting that SOX9 is needed at an early step for the commitment of osteochondroprogenitors. In the limbs of these embryos, patterning markers for the three different axes of limb development display a normal pattern of expression, indicating that the expression of these markers is independent of SOX9 and strongly suggesting that, in the skeletal system, SOX9 controls cell-fate determination but not patterning processes (Akiyama et al. 2002). In a related experiment, *Sox9*-null embryos were generated by mating *Sox9* chimeras with mice with oocytes in which *Sox9* had been inactivated. In these embryos, which die around day 12 of embryonic development due to cardiac defects, the skeletal progenitor cells are correctly specified but are arrested in their differentiation at a stage preceding the formation of chondrogenic mesenchymal condensations (Fig. 4). This finding again indicates that lack of SOX9 does not affect the specification and patterning of presumptive *Sox9*-expressing cells. It only affects their differentiation.

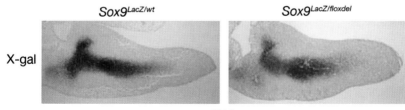

Figure 4. Embryonic day 12 heterozygous and homozygous *Sox9* mutant limb buds. The pattern of expression of *lacZ* (inserted in the mutant *Sox9* allele) reproduces the expression of *Sox9*, and is very similar in heterozygous and homozygous *Sox9* mutant limb buds.

Although SOX9 is the only transcription factor as yet identified that is required for formation of all chondrogenic mesenchymal condensations, other transcription factors, including members of the HOX, FORKHEAD, PAX, and HOMEODOMAIN families, influence the migration, survival, and shape of subsets of chondrogenic condensations (Karsenty 1999; Olsen et al. 2000). Among these are the paired box-containing transcription factors PAX1 and PAX9, which have overlapping patterns of expression in sclerotomal mesenchymal cells and are expressed in these cells up to the stage of mesenchymal condensation. Whereas *PAX9*-null mouse mutants show no detectable change in phenotype, *PAX1*-null mutants display markedly deformed vertebral bodies, and the double mutants almost completely lack vertebrae, indicating that PAX1 and PAX9 have partially overlapping roles. In these double-mutant embryos, *SOX9*-expressing sclerotomal mesenchymal cells migrate to the notochord but do not form proper condensations, and undergo apoptosis (Wallin et al. 1994; Peters et al. 1998). Thus, the role of PAX1 and PAX9 together consists of allowing *SOX9*-expressing sclerotomal mesenchymal cells to form and maintain appropriate chondrogenic condensations in the axial skeleton. Two other homeodomain transcription factors, NKX3.1 and NKX3.2, are also expressed in sclerotomal mesenchymal cells and have a role that is analogous to that of PAX1 and PAX9 in the maintenance of the SOX9-dependent chondrogenic cell fate of these cells (Lettice et al. 1999; Tribioli and Lufkin 1999; Herbrand et al. 2002). Thus, these four transcription factors are needed in the axial skeleton to organize *Sox9*-expressing mesenchymal cells to form correctly shaped and functioning condensations, which prefigure the future skeletal elements. These examples illustrate how a cell-fate and differentiation factor, SOX9, cooperates with so-called patterning factors—PAX1, PAX9, NKX3.1, and NKX3.2—to generate specific functioning cellular entities.

OVERT CHONDROCYTE DIFFERENTIATION

L-SOX5, a larger isoform of SOX5 that was initially identified in the testis, and SOX6 form homo- and heterodimers because of a highly conserved coiled-coil domain (Lefebvre et al. 1998). During chondrogenesis, *Sox5* and *Sox6* are coexpressed with *Sox9*, although somewhat later than *Sox9* in mesenchymal condensations. As with *Sox9*, their expression is completely shut off in hypertrophic chondrocytes. Moreover, the overall high degree of sequence identity of L-SOX5 and SOX6 suggests that they may have redundant functions in tissues in which they are coexpressed.

In developing mouse cartilage, this view is supported by the relatively mild skeletal phenotypes of both *Sox5*-null and *Sox6*-null mutants. In contrast, *Sox5*; *Sox6* double mutants die in utero with very severe defects in cartilage formation (Smits et al. 2001). In these mutants, expression of chondrocyte markers, including *Col2a1*, *Col11a2*, *Col9a1*, *aggrecan*, *comp*, and *matrilin*, occurs at very low levels. Moreover, in the absence of L-SOX5 and SOX6, chondrocyte proliferation is severely inhibited. However, in these *Sox5*; *Sox6* double mutants, *Sox9* expression is normal, and mesenchymal condensations form normally. These observations indicate that L-SOX5 and SOX6 are required at the step that follows the formation of mesenchymal condensations during chondrogenesis: the step of overt chondrocyte differentiation. L-SOX5 and SOX6 bind to the same pairs of HMG-like sites to which SOX9 binds in the *Col2a1* and *Col11a2* genes (Bridgewater et al. 1998; Lefebvre et al. 1998; Smits et al. 2001), and in vitro transfection experiments indicate that SOX9 cooperates with L-SOX5 and SOX6 to activate the *Col2a1* and *aggrecan* genes in mesenchymal cells (Lefebvre et al. 1998). It is therefore likely that these two types of SOX proteins act together to activate expression of cartilage ECM genes and the chondrocyte genetic program. However, the precise mechanism by which SOX9 and L-SOX5 and SOX6 cooperate in the activation of downstream genes in chondrocytes is not yet understood.

When the Cre-*LoxP* system is used to inactivate *Sox9* in mouse embryos at the stage of mesenchymal condensations using a *Col2a1-Cre* transgene, these cells are arrested in their differentiation as condensed mesenchymal cells and do not undergo overt differentiation into chondrocytes (Akiyama et al. 2002). The phenotype of these embryos shows important similarities with that of *Sox5*; *Sox6* double mutants. Moreover, in these embryos, proliferation and maturation of differentiated chondrocytes into hypertrophic chondrocytes are also severely defective. In these embryos and in embryos in which *Sox9* is inactivated in limb mesenchyme prior to mesenchymal condensation formation, expression of *Sox5* and *Sox6* is abolished, indicating that SOX9 is needed for the activation of *Sox5* and *Sox6* in chondrogenic cells.

In summary, SOX9 is required for the commitment of osteochondroprogenitors, for the formation of chondrogenic mesenchymal condensations, and, largely because SOX9 is required for the expression of *Sox5* and *Sox6*, also for the overt differentiation of chondrocytes. L-SOX5 and SOX6 are not required for formation of mesenchymal condensations but are needed, presumably together with SOX9, for the overt differentiation of chondrocytes and the formation of cartilage (Fig. 5). It should be obvious that the control of chondrogenesis by SOX9 and L-SOX5 and SOX6 is just one layer of control. Other layers are provided by signaling molecules, which

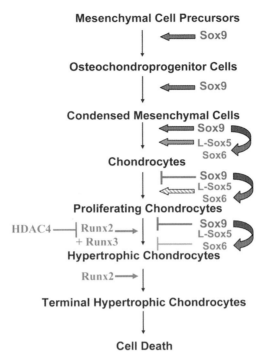

Figure 5. Role of SOX9, SOX5 and SOX6, and RUNX2 and RUNX3 in the different steps of chondrocyte differentiation.

have been shown to control and coordinate various steps of chondrogenesis, and also by chromatin proteins that regulate the activity of SOX9, L-SOX5, and SOX6 on the target genes of these transcription factors.

Indeed, several transcription factors that control the activity of SOX9 have been identified. The coactivator protein CBP/p300 interacts with SOX9 and regulates its activity, presumably through its ability to acetylate histones (Tsuda et al. 2003). During chondrogenesis, SMAD3, a signaling molecule in the transforming growth factor-β (TGF-β) family, stimulates SOX9-dependent transcriptional activation by modulating the interactions between SOX9 and CBF/p300 (Furumatsu et al. 2005). SOX9 also interacts with a component of the Mediator complex, the thyroid hormone receptor-associated protein TRAP230 (Zhou et al. 2002). The multicomponent Mediator complex connects DNA-bound transcription factors with the general transcriptional machinery. Another factor that enhances the activity of SOX9 during chondrogenesis is the coactivator peroxisome proliferation-activated receptor-gamma (PGC1-γ) (Kawakami et al. 2005). Recently, the histone acetyl transferase TIP60 was also shown to interact with SOX9 and to increase its activity. TIP60, together with SOX9 and L-SOX5, is present in the chromatin of the chondrocyte-specific enhancer in intron 1 of the *Col2a1* gene (Hattori et al. 2008). One can anticipate that still other transcription factors that interact with and regulate the activity of SOX9, L-SOX5, and SOX6 will be identified. Given that SOX9 also has an essential role in determination of the fate of other cell types, some of these factors should provide added specificity to the control by SOX9 of the chondrocyte genetic program.

Control of Chondrocyte Proliferation

Several signaling molecules and the pathways that they activate control the proliferation of chondrocytes. These include IHH, FGFs, PTHrP, WNT/β-catenin, and BMPs (Kronenberg 2003; Pogue and Lyons 2006). L-SOX5 and SOX6 also have an essential function in the establishment of the parallel columns of proliferating chondrocytes. Indeed, in the mouse model, in *Sox5; Sox6* double mutants, in which proliferation of chondrogenic mesenchymal cells is severely inhibited, these columnar proliferating chondrocytes are completely missing (Smits et al. 2001). Similarly, in embryos in which *Sox9* is inactivated at the stage of mesenchymal condensation formation, the typical columnar proliferating chondrocytes are also missing, most likely because *Sox5* and *Sox6* are not expressed (Akiyama et al. 2002). However, the mechanism by which L-SOX5 and SOX6 control chondrocyte proliferation is not understood. In mouse embryos, it is possible that the almost complete absence of an organized cartilage ECM in *Sox5; Sox6* double mutants prevents the relevant signaling molecules from exerting their control.

Both in vitro and in vivo experiments indicate that SOX9 is a powerful inhibitor of cell proliferation. Indeed, mouse embryos in which the expression of *Sox9* is increased in chondrogenic cells by knock-in of a *Sox9* cDNA in the *Col2a1* gene are dwarfed and show striking inhibition of chondrocyte proliferation (Akiyama et al. 2004). In addition, in cultured cells, FGFs increase the expression of *Sox9* via the mitogen-activated protein (MAP) kinase pathway, as well as the activity of SOX9 through the P38 pathway (Murakami et al. 2000; Zhang et al. 2006). Since the most common forms of human dwarfism are due to activating mutations in FGF receptor 3 (Horton et al. 2007), it is likely that SOX9 has a role in these diseases. One mechanism by which SOX9 could inhibit cell proliferation is by inhibiting the transcriptional activity of β-catenin. Indeed, by interacting with β-catenin, SOX9 targets β-catenin for proteosome degradation and markedly decreases its transcriptional activity.

REGULATION OF PREHYPERTROPHIC AND HYPERTROPHIC CHONDROCYTE DIFFERENTIATION

When the chondrocytes at the bottom of the characteristic columnar structures of proliferating chondrocytes exit the cell cycle, they gradually change both their morphology and their genetic program. The cells increase in size, and this enlargement of hypertrophic chondrocytes has an important role in the longitudinal growth of bones. At an early prehypertrophic stage, expression of *Sox9,* as well as expression of the typical chondrocyte markers such as *Col2a1, aggrecan,* and others, increases compared with their level of expression in proliferating chondrocytes. Prehypertrophic chondrocytes also activate genes that are not expressed in earlier chondrocytes, including those for the signaling molecules IHH and parathyroid hormone receptor, whereas the gene for the ECM protein type X collagen (*Col10a1*) is expressed when the cells become more hypertrophic.

The premature mineralization of endochondral skeletal elements in mouse *Sox9* heterozygous mutants and the delay in chondrocytic maturation in embryos that overexpress *Sox9* in chondrocytes suggest that SOX9 plays an important role in delaying hypertrophic chondrocyte differentiation (Bi et al. 2001; Akiyama et al. 2004). In this role, SOX9 is also a mediator of the effect of PTHrP, whose major function in chondrogenesis is to delay hypertrophic chondrocyte maturation. Indeed, by stimulating phosphorylation of two serine residues in SOX9, PTHrP signaling increases the DNA binding and the activity of SOX9 (Huang et al. 2001). This increased SOX9 activity is mediated by the high level of expression of the PTHrP receptor in prehypertrophic chondrocytes.

In contrast, in hypertrophic chondrocytes in mice, expression of *Sox9* and of *Sox5* and *Sox6* is shut off completely, leading to a parallel decrease in expression of the characteristic chondrocyte markers, including *Col2a1, Col11a2,* and *Col9a2* (Zhao et al. 1997).

It is possible that the abrupt loss of *Sox9* expression is necessary for differentiation into hypertrophic chondrocytes. Two other transcription factors, SNAIL and SLUG, both members of the SNAIL family of zinc finger proteins, could also play a role in the inhibition of the program of overt chondrocyte differentiation in hypertrophic chondrocytes (Seki et al. 2003). Their genes are expressed in hypertrophic chondrocytes, and both SNAIL and SLUG were shown to repress expression of *Col2a1* and *aggrecan* in transfection experiments.

In contrast to these negative regulators, two other transcription factors, RUNX2 and RUNX3, are needed for prehypertrophic and hypertrophic chondrocyte differentiation. Expression of RUNX2, which is present at low levels in chondrogenic mesenchymal condensations, increases in prehypertrophic chondrocytes and hypertrophic chondrocytes. RUNX2 and RUNX3 are members of the small family of Runt-domain transcription factors. RUNX2 is present in all preosteoblasts and osteoblasts, and is required for osteoblast differentiation and bone formation in both membranous and endochondral skeletal elements (Ducy et al. 1997; Komori et al. 1997; Otto et al. 1997). In *Runx2*-null mouse embryos, which show a complete absence of endochondral and membranous bones, the maturation of chondrocytes into prehypertrophic and hypertrophic chondrocytes is strongly inhibited, although not in all endochondral skeletal elements (Inada et al. 1999; Kim et al. 1999). In transfection experiments, RUNX2 binds to several sites in the *Col10a1* promoter and activates this promoter (Zheng et al. 2003). Since RUNX2 also interacts with SOX9 and inhibits the activity of SOX9, it may promote hypertrophic differentiation both by directly stimulating the program of hypertrophic chondrocytes and by inhibiting SOX9 (Zhou et al. 2006). *Runx3* is also expressed in prehypertrophic chondrocytes. Whereas in *Runx3*-null mouse embryos, there is a delay in endochondral bone formation, in *Runx2*-null; *Runx3*-null double mutants, there is a complete absence of prehypertrophic and hypertrophic chondrocytes and no expression of *Ihh* and *Col10a1* (Yoshida et al. 2004). Thus, RUNX2 has a dual role—it is needed for osteoblast differentiation and, together with RUNX3, is also needed for prehypertrophic and hypertrophic chondrocyte differentiation. Another protein, CBFB, constitutes a common subunit that interacts with all three Runt family members and is needed for their activity. Hence, *CbfB* mouse mutants show defective hypertrophic chondrocyte differentiation (Kundu et al. 2002; Yoshida et al. 2002).

Finally, the chromatin protein histone deactylase 4 (HDAC4), which is expressed in prehypertrophic chondrocytes, negatively regulates chondrocyte hypertrophy and endochondral bone formation by interacting with RUNX2 and inhibiting the activity of RUNX2 in prehypertrophic chondrocytes. Indeed, *Hdac4*-null mice show premature ossification due to early onset of chondrocyte hypertrophy, whereas overexpression of HDAC4 in chondrocytes in vivo inhibits chondrocyte hypertrophic differentiation (Vega et al. 2004).

Hypertrophic chondrocytes undergo a terminal step in differentiation when they no longer express *Col10a1*, acquire the ability to mineralize their ECM, and activate the expression of several other genes, including those for metalloproteinase 13, osteopontin, and alkaline phosphatase (Adams and Shapiro 2002). The expression of these genes and the ability to mineralize are properties that these cells share with osteoblasts. It is likely that the presence of RUNX2 in these terminally differentiated hypertrophic chondrocytes, together with other still unknown factors, have an important role in establishing this particular phenotype.

The growth-plate cartilage is an avascular tissue that has been shown to be hypoxic during development. Adaptation of cells to hypoxia involves the transcription factor hypoxia inducible factor (HIF)-1α and the Von Hippel–Lindau (VHL) tumor suppressor protein VHL, a ubiquitin ligase that promotes the degradation of HIF-1α. This adaptation also leads to the stimulation of vascular endothelial growth factor (VEGF) expression (Schipani 2005). Mouse genetic experiments have shown that HIF-1α and VHL have a critical role in the survival of hypoxic chondrocytes and protect the cells in the growth plate from undergoing cell death (Schipani et al. 2001). Thus, a broadly available system has a physiological role in adapting chondrocytes to their naturally occurring hypoxic state.

TRANSCRIPTION FACTORS IN ARTICULAR CHONDROCYTES

Chondrocytes are present in several layers of cells in articular cartilage at the ends of bones, providing joint surfaces. These cells express high levels of *aggrecan* but lower levels of *Col2a1*. They also express *Col11a2* and some other markers of growth-plate chondrocytes. In addition, the most superficial layer of articular chondrocytes and the synovial cells express high levels of lubricin (Rhee et al. 2005). Articular chondrocytes are metabolically hypoactive and only rarely undergo cell divisions and differentiation into hypertrophic chondrocytes. However, cells below the so-called tidemark that are located in the immediate vicinity of the bones become hypertrophic and express *Col10a1* (Archer et al. 2003). Articular chondrocytes express *Sox9, Sox5,* and *Sox6* (Davies et al. 2002; Lefebvre and Smits 2005), and mouse embryos that lack SOX9 in limb buds or are null for *Sox5* and *Sox6* show defective joint development (Smits et al. 2001; Akiyama et al. 2002). However, the embryonic lethality of these embryos precludes using these mutants to study the role of these transcription factors in the maintenance of articular cartilage after it fully forms postnatally. A variant form of the chicken ERG transcription factor containing an internal deletion is preferentially expressed in articular chondrocytes. This ERG variant, based on its ability to inhibit hypertrophic chondrocyte differentiation when expressed in a viral vector in chick embryo limb buds, has been postulated to prevent hypertrophic differentiation of articular chondrocytes (Iwamoto et al. 2000). However, no precise mechanism has been suggested for this function.

Several experiments using articular chondrocytes in culture support the view that SOX9 stimulates the chondrocytic phenotype both in normal chondrocytes and in chondrocytes of patients with osteoarthritis, suggesting that SOX9 has an important role in these cells, and raising the possibility that SOX9 could be a target for drugs in osteoarthritis (Li et al. 2004; Tew et al. 2005, 2007; Cucchiarini et al. 2007).

ROLE OF TRANSCRIPTION FACTORS IN THE SEGREGATION OF THE CHONDROCYTE AND OSTEOBLAST LINEAGES FROM A COMMON OSTEOCHONDROPROGENITOR

Cell-fate mapping has provided strong evidence that all osteoblasts in both endochondral and membranous bones derive from *Sox9*-expressing pluripotent mesenchymal cells (Akiyama et al. 2005). In addition, there is evidence, as discussed earlier, that SOX9 is required for the commitment of osteochondroprogenitors in limb buds (Akiyama et al. 2002). Because SOX9 is not expressed in osteoblasts, it is likely that a mechanism is in place that silences its expression and the activity of SOX9 in these cells. When *Sox9* is inactivated in *Wnt1*-expressing delaminating cranial neural crest cells, the *Sox9*-null cells migrate to their correct locations in different craniofacial presumptive cartilages, but instead of expressing chondrocyte-specific genes, these cells express the osteoblast markers *Osterix*, a transcription factor required for osteoblast differentiation (Nakashima et al. 2002), and *Col1a1* (Mori-Akiyama et al. 2003). This finding suggests that at the time of *SOX9* inactivation, the delaminating neural crest cells are bipotential and that normally the presence of SOX9 prevents the activation of an osteoblast program in these cells.

Conversely, in both endochondral and membranous skeletal elements, inactivation of the β-*catenin* gene (which is required for osteoblast differentiation at all stages of bone development) in osteoblast precursor cells causes expression of chondrocyte markers such as *Sox9* and *Col2a1*, instead of osteoblast markers (Day et al. 2005; Hill et al. 2005; Hu et al. 2005; Rodda and McMahon 2006). Similarly, in null mutants for *Osterix*, expression of chondrocyte markers is also activated in osteoblast precursor cells in both types of skeletal elements (Nakashima et al. 2002). Thus, in the absence of either β-CATENIN or OSTERIX, the SOX9-dependent chondrocyte program fails to be silenced in osteoblast precursors. Both factors are needed to fully commit osteoblast precursors to the osteoblast phenotype and to prevent expression of SOX9 and the chondrocyte program. Thus, transcription factors that are required for the respective commitment and differentiation of either the chondrogenic or the osteogenic lineage have a role in silencing the genetic program of the reciprocal lineage. It is likely that this role is critical, especially at the time when the two lineages segregate from a common progenitor.

The essential role of SOX9 to commit cells to a chondrogenic cell fate at the time a choice is made between the chondrocyte and osteoblast lineage is not unique. Indeed, SOX9 is required for the differentiation of glial cells when they and neuronal cells segregate from a common proneural progenitor. Inactivation of *Sox9* in proneural cells of the spinal cord in mice prevents differentiation of the glial cells into oligodendrocytes and astrocytes, and at least temporarily, increases the number of motoneuronal cells (Stolt et al. 2003). Also, in primitive gonads, SOX9 has an essential role in mediating a switch from the ovarian to the testicular pathway and in the differentiation of Sertoli cells. Inactivation of *Sox9* in XY gonads prevents sex cord development and activation of male-specific markers, and instead causes the expression of female-specific markers (Chaboissier et al. 2004). Furthermore, in the outer epithelial cell layer of the hair follicle in mice, inactivation of *Sox9* causes the ectopic appearance of epidermal characteristics in this cell layer and also prevents the formation of the stem-cell niche in the bulge of the hair follicle (Vidal et al. 2005). Still another example of *Sox9*'s role in cell-fate determination is that inactivation of *Sox9* in epithelial cells of the intestine in mice prevents the differentiation of Paneth cells, located in the bottom of the intestinal crypt (Mori-Akiyama et al. 2007). Paneth cells are one of four epithelial cell types of the intestine. It is obvious that the genetic programs controlled by SOX9 in chondrocytes, glial cells, Sertoli cells of male gonads, cells of the outer sheath of the hair follicle, and Paneth cells of the intestine are very different. In addition, in each of these cell-fate "decisions"—between chondrocytes and osteoblasts, glial cells and neuronal cells in the spinal cord, cells of the male and female gonads, etc.—SOX9 specifies one cell type and at the same time essentially silences the program of the other cell type. One can speculate that additional factors are needed to enhance the specificity of SOX9 in each of the different cell types whose differentiation is dependent on SOX9.

RELATIONSHIP BETWEEN SOX9 AND BMP SIGNALING DURING CHONDROGENESIS

The fact that BMPs were initially identified by their ability to induce ectopic cartilage and bone formation suggests that they act at an early stage of chondrogenesis (Wozney et al. 1988). This view is supported by experiments that have shown that cartilage formation is blocked when viral vectors that express noggin, a specific BMP antagonist, are added to chick limb-bud mesenchyme (Capdevila and Johnson 1998; Pizette and Niswander 2000). Furthermore, in vitro, BMPs induce chondrogenic differentiation in pluripotent mesenchymal cells (Pogue and Lyons 2006). Moreover, in vivo, implantation of BMP2 beads near mesenchymal condensation in chick embryo limb buds causes up-regulation of *Sox9* expression (Chimal-Monroy et al. 2003). However, genetic analysis in mouse embryos on the effects of BMP signaling in initiating chondrogenesis is rendered difficult by the essential role of BMP signaling for early mouse embryonic development, by the multiplicity of BMP ligands, and by the existence of three BMP receptors. Mouse embryos that lack BMP receptor 1A (*Bmpr1a*) die early during embryonic development with gastrulation defects, indicating that the lack of this receptor cannot be compensated by other BMP receptors in the early embryo (Mishina et al. 1995). In contrast, mice defective in BMP receptor 1B (*Bmpr1b*) have a fairly mild skeletal phenotype that is mainly restricted to phalanges. In one recent experiment (Yoon et al. 2005), embryos were generated in which *Bmpr1b* was ablated in the germ line and in which *Bmpr1a* was conditionally inactivated in chondrogenic mesenchymal condensations using the *Col2a1-Cre* transgene. These mutants showed a loss of chondrogenic mesenchymal condensations in limb buds and a complete absence of *Sox9* expression. Given that *Sox9* is expressed in skeletal progenitors prior to formation of mesenchymal condensations, these results indicate that BMP signaling is needed for the maintenance of *Sox9* expression and the maintenance of mesenchymal condensations, but the results do not demonstrate that BMP is required for the initiation of *Sox9* expression in skeletal progenitor cells. SOX9 and BMP signaling might thus act together in the formation of chondrogenic mesenchymal condensations (Yoon et al. 2005). A somewhat similar conclusion was reached regarding the relationship between BMP signaling and SOX9 in chondrogenesis by experiments using chick embryo somite explants (Zeng et al. 2002). These experiments suggest a regulatory cascade whereby signaling by the sonic hedgehog protein first induces expression of the transcriptional repressor *Nkx3.2*, which then derepresses *Sox9* expression in somitic tissues, maybe by inhibiting expression of a negative regulator of *Sox9*. Only in the presence of BMP signals does the mutual induction by SOX9 and NKX3.2 of each other's expression result in robust chondrogenesis. BMP is needed for this autoregulatory loop to be maintained and hence for activation of the chondrocyte differentiation program by SOX9 to proceed (Zeng et al. 2002).

CONCLUSIONS AND PERSPECTIVE

This chapter has summarized the evidence that has elucidated the central roles of SOX9 and of L-SOX5 and SOX6 as master transcription factors of chondrogenesis. We have also outlined the critical function of RUNX2 and RUNX3 in hypertrophic chondrocyte differentiation.

A number of crucial questions regarding the functions of SOX9 and of L-SOX5 and SOX6 remain. It will be important to identify the factors that add to the specificity of SOX9 in the activation of chondrogenesis given that SOX9 has similar essential roles in cell-fate determination in several very different cell types. It is also possible that SOX9 itself may be endowed with particular properties that allow it to initiate specific genetic programs of differentiation in the precursors of these different cell types. It will also be important to identify the signaling molecules that activate *Sox9* in multipotent mesenchymal cells and also more clearly identify those that stimulate the expression of *Sox9* at the time of chondrogenic mesenchymal condensation. To better comprehend the mechanisms whereby SOX9, L-SOX5, and SOX6 control chondrocyte differentiation, it will also be critical to examine the extent of the genetic program of these SOX proteins in the genome of chondrocytes and to determine whether SOX9 and L-SOX5 and SOX6 have mainly common direct target genes or sometimes have separate target genes in these cells. If they share common target genes, whether these transcription factors bind to adjacent sites in the chromatin of those genes will also be important to examine. Also, given that SOX9, L-SOX5, and SOX6 are expressed in articular cartilage, it will be important to determine, preferably by genetic experiments in vivo, whether their role is essential in the formation and especially in the maintenance of this cartilage postnatally.

ACKNOWLEDGMENTS

The work from the authors' laboratory described in this chapter was supported by grant numbers 2P01 AR42919 and 5R01 AR053568 from the National Institutes of Health. We are grateful to Janie Finch for editorial assistance.

REFERENCES

Adams, C.S. and Shapiro, I.M. 2002. The fate of the terminally differentiated chondrocyte: Evidence for microenvironmental regulation of chondrocyte apoptosis. *Crit. Rev. Oral Biol. Med.* **13:** 465–473.

Akiyama, H., Chaboissier, M.C., Martin, J.F., Schedl, A., and de Crombrugghe, B. 2002. The transcription factor Sox9 has essential roles in successive steps of the chondrocyte differentiation pathway and is required for expression of Sox5 and Sox6. *Genes Dev.* **16:** 2813–2828.

Akiyama, H., Lyons, J.P., Mori-Akiyama, Y., Yang, X., Zhang, R., Zhang, Z., Deng, J.M., Taketo, M.M., Nakamura, T., Behringer, R.R., et al. 2004. Interactions between Sox9 and β-catenin control chondrocyte differentiation. *Genes Dev.* **18:** 1072–1087.

Akiyama, H., Kim, J.E., Nakashima, K., Balmes, G., Iwai, N., Deng, J.M., Zhang, Z., Martin, J.F., Behringer, R.R., Nakamura, T., et al. 2005. Osteo-chondroprogenitor cells are derived from Sox9 expressing precursors. *Proc. Natl. Acad. Sci.* **102:** 14665–14670.

Archer, C.W., Dowthwaite, G.P., and Francis-West, P. 2003. Development of synovial joints. *Birth Defects Res. C Embryo Today* **69:** 144–155.

Bernard, P., Tang, P., Liu, S., Dewing, P., Harley, V.R., and Vilain, E. 2003. Dimerization of SOX9 is required for chondrogenesis, but not for sex determination. *Hum. Mol. Genet.* **12:** 1755–1765.

Bi, W., Deng, J.M., Zhang, Z., Behringer, R.R., and de Crombrugghe, B. 1999. Sox9 is required for cartilage formation. *Nat. Genet.* **22:** 85–89.

Bi, W., Huang, W., Whitworth, D.J., Deng, J.M., Zhang, Z., Behringer, R.R., and de Crombrugghe, B. 2001. Haploinsufficiency of Sox9 results in defective cartilage primordia and premature skeletal mineralization. *Proc. Natl. Acad. Sci.* **98:** 6698–6703.

Bridgewater, L.C., Lefebvre, V., and de Crombrugghe, B. 1998. Chondrocyte-specific enhancer elements in the *Col11a2* gene resemble the *Col2a1* tissue-specific enhancer. *J. Biol. Chem.* **273:** 14998–15006.

Bridgewater, L.C., Walker, M.D., Miller, G.C., Ellison, T.A., Holsinger, L.D., Potter, J.L., Jackson, T.L., Chen, R.K., Winkel, V.L., Zhang, Z., et al. 2003. Adjacent DNA sequences modulate Sox9 transcriptional activation at paired Sox sites in three chondrocyte-specific enhancer elements. *Nucleic Acids Res.* **31:** 1541–1553.

Capdevila, J. and Johnson, R.L. 1998. Endogenous and ectopic expression of *noggin* suggests a conserved mechanism for regulation of BMP function during limb and somite patterning. *Dev. Biol.* **197:** 205–217.

Chaboissier, M.C., Kobayashi, A., Vidal, V.I., Lutzkendorf, S., van de Kant, H.J., Wegner, M., de Rooij, D.G., Behringer, R.R., and Schedl, A. 2004. Functional analysis of Sox8 and Sox9 during sex determination in the mouse. *Development* **131:** 1891–1901.

Chimal-Monroy, J., Rodriguez-Leon, J., Montero, J.A., Ganan, Y., Macias, D., Merino, R., and Hurle, J.M. 2003. Analysis of the molecular cascade responsible for mesodermal limb chondrogenesis: *sox* genes and BMP signaling. *Dev. Biol.* **257:** 292–301.

Cucchiarini, M., Thurn, T., Weimer, A., Kohn, D., Terwilliger, E.F., and Madry, H. 2007. Restoration of the extracellular matrix in human osteoarthritic articular cartilage by overexpression of the transcription factor SOX9. *Arthritis Rheum.* **56:** 158–167.

Davies, S.R., Sakano, S., Zhu, Y., and Sandell, L.J. 2002. Distribution of the transcription factors Sox9, AP-2, and δEF1 in adult murine articular and meniscal cartilage and growth plate. *J. Histochem. Cytochem.* **50:** 1059–1065.

Day, T.F., Guo, X., Garrett-Beal, L., and Yang, Y. 2005. Wnt/β-catenin signaling in mesenchymal progenitors controls osteoblast and chondrocyte differentiation during vertebrate skeletogenesis. *Dev. Cell* **8:** 739–750.

Ducy, P., Zhang, R., Geoffroy, V., Ridall, A.L., and Karsenty, G. 1997. Osf2/Cbfa1: A transcriptional activator of osteoblast differentiation. *Cell* **89:** 747–754.

Foster, J.W., Dominguez-Steglich, M.A., Guioli, S., Kowk, G., Weller, P.A., Stevanovic, M.,

Weissenbach, J., Mansour, S., Young, I.D., Goodfellow, P.N., et al. 1994. Campomelic dysplasia and autosomal sex reversal caused by mutations in an SRY-related gene. *Nature* **372**: 525–530.

Furumatsu, T., Tsuda, M., Yoshida, K., Taniguchi, N., Ito, T., Hashimoto, M., and Asahara, H. 2005. Sox9 and p300 cooperatively regulate chromatin-mediated transcription. *J. Biol. Chem.* **280**: 35203–35208.

Gimovsky, M., Rosa, E., Tolbert, T., Guzman, G., Nazir, M., and Koscica, K. 2008. Campomelic dysplasia: Case report and review. *J. Perinatol.* **28**: 71–73.

Hattori, T., Coustry, F., Stephens, S., Eberspaecher, H., Takigawa, M., Yasuda, H., and de Crombrugghe, B. 2008. Transcriptional regulation of chondrogenesis by coactivator Tip60 via chromatin association with Sox9 and Sox5. *Nucleic Acids Res.* **36**: 3011–3024.

Herbrand, H., Pabst, O., Hill, R., and Arnold, H.H. 2002. Transcription factors Nkx3.1 and Nkx3.2 (Bapx1) play an overlapping role in sclerotomal development of the mouse. *Mech. Dev.* **117**: 217–224.

Hill, T.P., Spater, D., Taketo, M.M., Birchmeier, W., and Hartmann, C. 2005. Canonical Wnt/β-catenin signaling prevents osteoblasts from differentiating into chondrocytes. *Dev. Cell* **8**: 727–738.

Horton, W.A., Hall, J.G., and Hecht, J.T. 2007. Achondroplasia. *Lancet* **370**: 162–172.

Hu, H., Hilton, M.J., Tu, X., Yu, K., Ornitz, D.M., and Long, F. 2005. Sequential roles of Hedgehog and Wnt signaling in osteoblast development. *Development* **132**: 49–60.

Huang, W., Chung, U.I., Kronenberg, H.M., and de Crombrugghe, B. 2001. The chondrogenic transcription factor Sox9 is a target of signaling by the parathyroid hormone-related peptide in the growth plate of endochondral bones. *Proc. Natl. Acad. Sci.* **98**: 160–165.

Inada, M., Yasui, T., Nomura, S., Miyake, S., Deguchi, K., Himeno, M., Sato, M., Yamagiwa, H., Kimura, T., Yasui, N., et al. 1999. Maturational disturbance of chondrocytes in *Cbfa1*-deficient mice. *Dev. Dyn* **214**: 279–290.

Iwamoto, M., Higuchi, Y., Koyama, E., Enomoto-Iwamoto, M., Kurisu, K., Yeh, H., Abrams, W.R., Rosenbloom, J., and Pacifici, M. 2000. Transcription factor ERG variants and functional diversification of chondrocytes during limb long bone development. *J. Cell Biol.* **150**: 27–40.

Jenkins, E., Moss, J.B., Pace, J.M., and Bridgewater, L.C. 2005. The new collagen gene COL27A1 contains SOX9-responsive enhancer elements. *Matrix Biol.* **24**: 177–184.

Karsenty, G. 1999. The genetic transformation of bone biology. *Genes Dev.* **13**: 3037–3051.

Kawakami, Y., Tsuda, M., Takahashi, S., Taniguchi, N., Esteban, C.R., Zemmyo, M., Furumatsu, T., Lotz, M., Belmonte, J.C., and Asahara, H. 2005. Transcriptional coactivator PGC-1α regulates chondrogenesis via association with Sox9. *Proc. Natl. Acad. Sci.* **102**: 2414–2419.

Kim, I.S., Otto, F., Zabel, B., and Mundlos, S. 1999. Regulation of chondrocyte differentiation by *Cbfa1*. *Mech. Dev.* **80**: 159–170.

Kist, R., Schrewe, H., Balling, R., and Scherer, G. 2002. Conditional inactivation of Sox9: A mouse model for campomelic dysplasia. *Genesis* **32**: 121–123.

Komori, T., Yagi, H., Nomura, S., Yamaguchi, A., Sasaki, K., Deguchi, K., Shimizu, Y., Bronson, R.T., Gao, Y.H., Inada, M., et al. 1997. Targeted disruption of Cbfa1 results in a complete lack of bone formation owing to maturational arrest of osteoblasts. *Cell* **89**: 755–764.

Kronenberg, H.M. 2003. Developmental regulation of the growth plate. *Nature* **423**: 332–336.

Kundu, M., Javed, A., Jeon, J.P., Horner, A., Shum, L., Eckhaus, M., Muenke, M., Lian, J.B., Yang, Y., Nuckolls, G.H., et al. 2002. Cbfβ interacts with Runx2 and has a critical role in bone development. *Nat. Genet.* **32:** 639–644.

Lefebvre, V. and Smits, P. 2005. Transcriptional control of chondrocyte fate and differentiation. *Birth Defects Res. C Embryo Today* **75:** 200–212.

Lefebvre, V., Huang, W., Harley, V.R., Goodfellow, P.N., and de Crombrugghe, B. 1997. SOX9 is a potent activator of the chondrocyte-specific enhancer of the pro α1(II) collagen gene. *Mol. Cell. Biol.* **17:** 2336–2346.

Lefebvre, V., Li, P., and de Crombrugghe, B. 1998. A new long form of Sox5 (L-Sox5), Sox6 and Sox9 are coexpressed in chondrogenesis and cooperatively activate the type II collagen gene. *EMBO J.* **17:** 5718–5733.

Lefebvre, V., Dumitriu, B., Penzo-Mendez, A., Han, Y., and Pallavi, B. 2007. Control of cell fate and differentiation by Sry-related high-mobility-group box (Sox) transcription factors. *Int. J. Biochem. Cell Biol.* **39:** 2195–2214.

Lettice, L.A., Purdie, L.A., Carlson, G.J., Kilanowski, F., Dorin, J., and Hill, R.E. 1999. The mouse bagpipe gene controls development of axial skeleton, skull, and spleen. *Proc. Natl. Acad. Sci.* **96:** 9695–9700.

Li, Y., Tew, S.R., Russell, A.M., Gonzalez, K.R., Hardingham, T.E., and Hawkins, R.E. 2004. Transduction of passaged human articular chondrocytes with adenoviral, retroviral, and lentiviral vectors and the effects of enhanced expression of SOX9. *Tissue Eng.* **10:** 575–584.

Liu, Y., Li, H., Tanaka, K., Tsumaki, N., and Yamada, Y. 2000. Identification of an enhancer sequence within the first intron required for cartilage-specific transcription of the α2(XI) collagen gene. *J. Biol. Chem.* **275:** 12712–12718.

Mishina, Y., Suzuki, A., Ueno, N., and Behringer, R.R. 1995. *Bmpr* encodes a type I bone morphogenetic protein receptor that is essential for gastrulation during mouse embryogenesis. *Genes Dev.* **9:** 3027–3037.

Mori-Akiyama, Y., Akiyama, H., Rowitch, D.H., and de Crombrugghe, B. 2003. Sox9 is required for determination of the chondrogenic cell lineage in the cranial neural crest. *Proc. Natl. Acad. Sci.* **100:** 9360–9365.

Mori-Akiyama, Y., van den Born, M., van Es, J.H., Hamilton, S.R., Adams, H.P., Zhang, J., Clevers, H., and de Crombrugghe, B. 2007. SOX9 is required for the differentiation of paneth cells in the intestinal epithelium. *Gastroenterology* **133:** 539–546.

Murakami, S., Kan, M., McKeehan, W.L., and de Crombrugghe, B. 2000. Up-regulation of the chondrogenic *Sox9* gene by fibroblast growth factors is mediated by the mitogen-activated protein kinase pathway. *Proc. Natl. Acad. Sci.* **97:** 1113–1118.

Murakami, S., Akiyama, H., and de Crombrugghe, B. 2003. Development of bone and cartilage. In *Inborn errors of development: The molecular basis of clinical disorders of morphogenesis. II. Patterns of development,* 2nd ed. (Oxford Monographs on Medical Genetics) (ed. C.J. Epstein et al.), vol. 49, pp. 162–181. Oxford University Press, New York.

Nakashima, K., Zhou, X., Kunkel, G., Zhang, Z., Deng, J.M., Behringer, R.R., and de Crombrugghe, B. 2002. The novel zinc finger-containing transcription factor Osterix is required for osteoblast differentiation and bone formation. *Cell* **108:** 17–29.

Ng, L.J., Wheatley, S., Muscat, G.E., Conway-Campbell, J., Bowles, J., Wright, E., Bell, D.M., Tam, P.P., Cheah, K.S., and Koopman, P. 1997. SOX9 binds DNA, activates transcription, and coexpresses with type II collagen during chondrogenesis in the mouse. *Dev. Biol.* **183:** 108–121.

Olsen, B.R., Reginato, A.M., and Wang, W. 2000. Bone development. *Annu. Rev. Cell Dev. Biol.* **16**: 191–220.

Otto, F., Thornell, A.P., Crompton, T., Denzel, A., Gilmour, K.C., Rosewell, I.R., Stamp, G.W., Beddington, R.S., Mundlos, S., Olsen, B.R., et al. 1997. *Cbfa1*, a candidate gene for cleidocranial dysplasia syndrome, is essential for osteoblast differentiation and bone development. *Cell* **89**: 765–771.

Peters, H., Neubuser, A., Kratochwil, K., and Balling, R. 1998. Pax9-deficient mice lack pharyngeal pouch derivatives and teeth and exhibit craniofacial and limb abnormalities. *Genes Dev.* **12**: 2735–2747.

Pizette, S. and Niswander, L. 2000. BMPs are required at two steps of limb chondrogenesis: Formation of prechondrogenic condensations and their differentiation into chondrocytes. *Dev. Biol.* **219**: 237–249.

Pogue, R. and Lyons, K. 2006. BMP signaling in the cartilage growth plate. *Curr. Top. Dev. Biol.* **76**: 1–48.

Rentsendorj, O., Nagy, A., Sinko, I., Daraba, A., Barta, E., and Kiss, I. 2005. Highly conserved proximal promoter element harbouring paired Sox9-binding sites contributes to the tissue- and developmental stage-specific activity of the matrilin-1 gene. *Biochem. J.* **389**: 705–716.

Rhee, D.K., Marcelino, J., Baker, M., Gong, Y., Smits, P., Lefebvre, V., Jay, G.D., Stewart, M., Wang, H., Warman, M.L., et al. 2005. The secreted glycoprotein lubricin protects cartilage surfaces and inhibits synovial cell overgrowth. *J. Clin. Invest.* **115**: 622–631.

Rodda, S.J. and McMahon, A.P. 2006. Distinct roles for Hedgehog and canonical Wnt signaling in specification, differentiation and maintenance of osteoblast progenitors. *Development* **133**: 3231–3244.

Schepers, G.E., Teasdale, R.D., and Koopman, P. 2002. Twenty pairs of sox: Extent, homology, and nomenclature of the mouse and human sox transcription factor gene families. *Dev. Cell* **3**: 167–170.

Schipani, E. 2005. Hypoxia and HIF-1α in chondrogenesis. *Semin. Cell Dev. Biol.* **16**: 539–546.

Schipani, E., Ryan, H.E., Didrickson, S., Kobayashi, T., Knight, M., and Johnson, R.S. 2001. Hypoxia in cartilage: HIF-1α is essential for chondrocyte growth arrest and survival. *Genes Dev.* **15**: 2865–2876.

Seki, K., Fujimori, T., Savagner, P., Hata, A., Aikawa, T., Ogata, N., Nabeshima, Y., and Kaechoong, L. 2003. Mouse *snail* family transcription repressors regulate chondrocyte, extracellular matrix, type II collagen, and aggrecan. *J. Biol. Chem.* **278**: 41862–41870.

Sekiya, I., Tsuji, K., Koopman, P., Watanabe, H., Yamada, Y., Shinomiya, K., Nifuji, A., and Noda, M. 2000. SOX9 enhances aggrecan gene promoter/enhancer activity and is upregulated by retinoic acid in a cartilage-derived cell line, TC6. *J. Biol. Chem.* **275**: 10738–10744.

Smits, P., Li, P., Mandel, J., Zhang, Z., Deng, J.M., Behringer, R.R., de Crombrugghe, B., and Lefebvre, V. 2001. The transcription factors L-Sox5 and Sox6 are essential for cartilage formation. *Dev. Cell* **1**: 277–290.

Sock, E., Pagon, R.A., Keymolen, K., Lissens, W., Wegner, M., and Scherer, G. 2003. Loss of DNA-dependent dimerization of the transcription factor SOX9 as a cause for campomelic dysplasia. *Hum. Mol. Genet.* **12**: 1439–1447.

Stolt, C.C., Lommes, P., Sock, E., Chaboissier, M.C., Schedl, A., and Wegner, M. 2003. The Sox9 transcription factor determines glial fate choice in the developing spinal cord. *Genes Dev.* **17**: 1677–1689.

Tew, S.R., Li, Y., Pothacharoen, P., Tweats, L.M., Hawkins, R.E., and Hardingham, T.E. 2005. Retroviral transduction with SOX9 enhances re-expression of the chondrocyte phenotype in passaged osteoarthritic human articular chondrocytes. *Osteoarthritis Cartilage* **13**: 80–89.

Tew, S.R., Clegg, P.D., Brew, C.J., Redmond, C.M., and Hardingham, T.E. 2007. SOX9 transduction of a human chondrocytic cell line identifies novel genes regulated in primary human chondrocytes and in osteoarthritis. *Arthritis Res. Ther.* **9**: R107.

Tribioli, C. and Lufkin, T. 1999. The murine *Bapx1* homeobox gene plays a critical role in embryonic development of the axial skeleton and spleen. *Development* **126**: 5699–5711.

Tsuda, M., Takahashi, S., Takahashi, Y., and Asahara, H. 2003. Transcriptional co-activators CREB-binding protein and p300 regulate chondrocyte-specific gene expression via association with Sox9. *J. Biol. Chem.* **278**: 27224–27229.

Vega, R.B., Matsuda, K., Oh, J., Barbosa, A.C., Yang, X., Meadows, E., McAnally, J., Pomajzl, C., Shelton, J.M., Richardson, J.A., et al. 2004. Histone deacetylase 4 controls chondrocyte hypertrophy during skeletogenesis. *Cell* **119**: 555–566.

Vidal, V.P., Chaboissier, M.C., Lutzkendorf, S., Cotsarelis, G., Mill, P., Hui, C.C., Ortonne, N., Ortonne, J.P., and Schedl, A. 2005. Sox9 is essential for outer root sheath differentiation and the formation of the hair stem cell compartment. *Curr. Biol.* **15**: 1340–1351.

Wagner, T., Wirth, J., Meyer, J., Zabel, B., Held, M., Zimmer, J., Pasantes, J., Bricarelli, F.D., Keutel, J., Hustert, E., et al. 1994. Autosomal sex reversal and campomelic dysplasia are caused by mutations in and around the SRY-related gene SOX9. *Cell* **79**: 1111–1120.

Wallin, J., Wilting, J., Koseki, H., Fritsch, R., Christ, B., and Balling, R. 1994. The role of Pax-1 in axial skeleton development. *Development* **120**: 1109–1121.

Wozney, J.M., Rosen, V., Celeste, A.J., Mitsock, L.M., Whitters, M.J., Kriz, R.W., Hewick, R.M., and Wang, E.A. 1988. Novel regulators of bone formation: Molecular clones and activities. *Science* **242**: 1528–1534.

Wright, E., Hargrave, M.R., Christiansen, J., Cooper, L., Kun, J., Evans, T., Gangadharan, U., Greenfield, A., and Koopman, P. 1995. The Sry-related gene Sox9 is expressed during chondrogenesis in mouse embryos. *Nat. Genet.* **9**: 15–20.

Xie, W.F., Zhang, X., Sakano, S., Lefebvre, V., and Sandell, L.J. 1999. Trans-activation of the mouse cartilage-derived retinoic acid-sensitive protein gene by Sox9. *J. Bone Miner. Res.* **14**: 757–763.

Yoon, B.S., Ovchinnikov, D.A., Yoshii, I., Mishina, Y., Behringer, R.R., and Lyons, K.M. 2005. *Bmpr1a* and *Bmpr1b* have overlapping functions and are essential for chondrogenesis in vivo. *Proc. Natl. Acad. Sci.* **102**: 5062–5067.

Yoshida, C.A., Furuichi, T., Fujita, T., Fukuyama, R., Kanatani, N., Kobayashi, S., Satake, M., Takada, K., and Komori, T. 2002. Core-binding factor β interacts with Runx2 and is required for skeletal development. *Nat. Genet.* **32**: 633–638.

Yoshida, C.A., Yamamoto, H., Fujita, T., Furuichi, T., Ito, K., Inoue, K., Yamana, K., Zanma, A., Takada, K., Ito, Y., et al. 2004. Runx2 and Runx3 are essential for chondrocyte maturation, and Runx2 regulates limb growth through induction of *Indian hedgehog*. *Genes Dev.* **18**: 952–963.

Zeng, L., Kempf, H., Murtaugh, L.C., Sato, M.E., and Lassar, A.B. 2002. Shh establishes an Nkx3.2/Sox9 autoregulatory loop that is maintained by BMP signals to induce somitic chondrogenesis. *Genes Dev.* **16**: 1990–2005.

Zhang, R., Murakami, S., Coustry, F., Wang, Y., and de Crombrugghe, B. 2006. Constitutive activation of MKK6 in chondrocytes of transgenic mice inhibits proliferation and delays endochondral bone formation. *Proc. Natl. Acad. Sci.* **103:** 365–370.

Zhao, Q., Eberspaecher, H., Lefebvre, V., and De Crombrugghe, B. 1997. Parallel expression of Sox9 and Col2a1 in cells undergoing chondrogenesis. *Dev. Dyn.* **209:** 377–386.

Zheng, Q., Zhou, G., Morello, R., Chen, Y., Garcia-Rojas, X., and Lee, B. 2003. Type X collagen gene regulation by Runx2 contributes directly to its hypertrophic chondrocyte-specific expression in vivo. *J. Cell Biol.* **162:** 833–842.

Zhou, G., Lefebvre, V., Zhang, Z., Eberspaecher, H., and de Crombrugghe, B. 1998. Three high mobility group-like sequences within a 48-base pair enhancer of the *Col2a1* gene are required for cartilage-specific expression in vivo. *J. Biol. Chem.* **273:** 14989–14997.

Zhou, R., Bonneaud, N., Yuan, C.X., de Santa Barbara, P., Boizet, B., Schomber, T., Scherer, G., Roeder, R.G., Poulat, F., and Berta, P. 2002. SOX9 interacts with a component of the human thyroid hormone receptor-associated protein complex. *Nucleic Acids Res.* **30:** 3245–3252.

Zhou, G., Zheng, Q., Engin, F., Munivez, E., Chen, Y., Sebald, E., Krakow, D., and Lee, B. 2006. Dominance of SOX9 function over RUNX2 during skeletogenesis. *Proc. Natl. Acad. Sci.* **103:** 19004–19009.

6

Growth Factors and Chondrogenesis

Henry M. Kronenberg
Endocrine Unit, Massachusetts General Hospital and
Harvard Medical School
Boston, Massachusets 02114

Andrew P. McMahon
Department of Molecular and Cellular Biology and
Harvard Stem Cell Institute, Harvard University
Cambridge, Massachusetts 02138

Clifford J. Tabin
Department of Genetics
Harvard Medical School
Boston, Massachusetts 02115

CHONDROGENESIS, WHETHER DURING THE FORMATION of cartilage models during endochondral bone formation, longitudinal and apositional growth, maintenance of the articular surfaces, or during repair and healing, is a carefully orchestrated multistep process. As such, regulation of this process requires the interplay of a large number of factors, including inductive cues from surrounding tissues, intercellular signals emanating from within the cartilage itself, and intrinsic factors within the chondrocytes. Indeed, intrinsic and extrinsic regulation are intimately related to one another. In response to specific sets of growth factors, cells at various stages of chondrogenic differentiation activate expression of unique sets of transcription factors committing them to, and defining, particular cell states. In turn, a consequence of the expression of these transcription factors is the regulated production of stage-specific secreted proteins that feedback on other chondrogenic cells. Of equal importance, the transcriptional state of cells in the chondrogenic pathway determines their ability to respond to specific factors and the nature of their response. These intrinsic and extrinsic factors are thus components of a complex integrated network, a fact that must be borne in mind while considering any one aspect of chondrogenic regulation. Nonetheless, given the incomplete nature of our current understanding of chondrogenesis, and the complexity of the problem, it is perhaps easiest to organize a discussion of chondrogenesis by considering inductive factors and intrinsic regulation individually.

This review focuses on the secreted proteins and signal transduction systems that regulate various aspects of chondrogenesis, while the transcription factors upstream and downstream of these activities are discussed in a separate chapter. Chondrogenesis itself is a multistep process and we will endeavor to draw together what is known about intercellular signaling at each step. As described in more detail elsewhere in this volume, endochondral skeletal development, typified by the process of bone formation in the limb, begins with the condensation of mesenchymal cells at the sites of the future skeletal elements (see Fig. 1A,B). In the limbs, these condensations initiate proximally and then extend distally as a single, continuous branching structure. The condensed mesenchymal cells then differentiate into round chondrocytes, secreting the fibrillar type II collagen and aggrecan (Fig. 1C). These cells then proliferate, except at the locations of the future joints, where the cells are diverted to a distinct "interzone" pathway. As the cartilage elements enlarge, mesenchymal cells directly abutting the cartilage flatten and differentiate into a distinct tissue, the perichondrium. The perichondrium will subsequently become an important source of signals guiding the future growth and the differentiation of the cartilage, as well as a source of cells for appositional growth. Next, the cells in the center of the growing cartilage elements drop out of the cell cycle (Fig. 1D). Through a series of distinct steps, these cells undergo dramatic hypertrophy, and alter the cellular matrix they produce, activating production of the sheet-forming type X collagen and cause mineralization of the surrounding matrix. In addition, the hypertrophic cells produce signals that initiate osteogenesis in the adjacent perichondrium (here after called the periosteum) (Fig. 1E). Finally, the hypertrophic cells undergo apoptosis, leaving the mineralized matrix as a scaffold for future osteogenesis (Fig. 1F) (considered separately in this volume).

Even as this process of terminal differentiation is continuing in the center of the nascent skeletal elements, chondrocytes proximal and distal to this zone continue to proliferate (Fig. 1G). Those closest to the hypertrophic zone flatten while maintaining a high proliferation rate. As these rapidly growing cells divide, daughter cells are displaced directly above

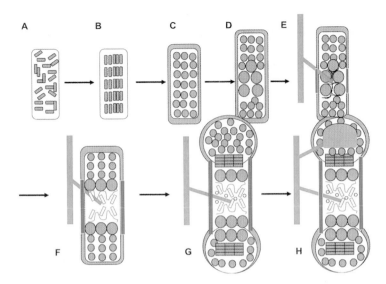

Figure 1. Endochondral bone formation in long bones of the limb. (*A, B*) Mesenchymal cells condense. (*C*) Condensed cells become chondrocytes, surrounded by perichondrium. (*D*) After proliferation, chondrocytes in the center of bone anlage stop proliferating and hypertrophy. (*E*) Hypertrophic chondrocytes mineralize surrounding matrix, induce adjacent bone collar, and attract blood vessels. (*F*) Hypertrophic chondrocytes die, leaving behind a matrix upon which osteoblasts secrete a bone matrix. (*G*) Further chondrocyte proliferation is associated with formation of long columns; hematopoiesis begins in marrow. (*H*) Secondary ossification center forms in epiphyseal region, establishing true growth-plate cartilage between the primary and secondary ossification centers.

and below one another, forming ordered columns. The chondrocytes situated further towards the ends of the cartilage elements remain round and proliferate more slowly, forming the resting zone, a reserve of chondroprogenitors that continuously generate additional flattened columns of growing and then differentiating chondrocytes. Thus, the columns of proliferating chondrocytes continue to expand the skeletal elements. Hence, the growth of new cartilage balanced by a wave of hypertrophic differentiation spreading from the center towards the ends keeps the length of the region of growth cartilage relatively constant. As morphogenesis of the skeletal elements progresses, there is a second site of hypertrophic differentiation at the very ends of the bones (Fig. 1H). These domains of hypertrophy give way to osteogenesis as in the medial portion of the skeletal elements, forming the secondary ossification centers. The resting, flattened proliferating and hypertrophic cartilage caught between the primary and secondary ossification centers constitute the growth plate. Continued steady-state signaling maintains the growth and differentiation of the growth plates, thereby lengthening the skeletal elements, until closure of the epiphysis (or growth plate fusion) at puberty. Thus, the signaling systems involved in establishing and regulating chondrogenesis in the context of the embryo are of additional relevance in the continued growth of the postnatal skeleton. Even the earliest steps of mesenchymal condensation and differentiation are of postnatal relevance since they are recapitulated during bone healing. We therefore consider what is known about signaling during each of these steps. We discuss some of the best-understood signaling systems separately, but should emphasize that these systems communicate with each other to coordinate orderly growth.

BONE MORPHOGENETIC PROTEINS

Bone morphogenetic proteins (BMPs) were discovered as components of bone matrix capable of inducing endochondral bone formation after subcutaneous injection (Urist 1965). This assay was subsequently used to clone the first identified BMP molecules (Wozney et al. 1988). More than 20 family members have been identified by structural homology, alternatively called BMPs or growth and differentiation factors (GDFs) by their discoverers. Subsequent genetic studies have shown that BMPs are required for chondrogenesis and also have effects on chondrocyte proliferation and differentiation.

BMPs are members of the larger TGF-β family of secreted proteins. They are cysteine-rich dimers, generated by cleavage of larger preproBMP precursors. They bind to and activate type I and type II receptors that are membrane-bound serine-threonine protein kinases. The three type I receptors are all expressed on chondrocytes and include ALK2/ActRI, ALK3/BMPRIA, and/or ALK6/BMPRIB. The type II BMP receptors include BMP RII, Act RII, and Act IIb. When BMPs bind to type I receptors, they recruit type II receptors that then phosphorylate the type I receptor (for review, see Rosen 2006). The type I receptor, in turn, phosphorylates Smad 1, 5, and/or 8. These proteins move with Smad 4 to the nucleus where they act as transcription factors (for review, see Massagué et al. 2005). BMP receptor activation also activates MAP kinases, though the possible roles of activation of this pathway by BMPs in chondrocytes is unclear (Pogue and Lyons 2006). BMP signaling is constrained by the action of a number of extracellular inhibitors, including noggin, chordin, follistatin, and gremlin, that bind BMPs.

BMP signaling is important during the formation of mesenchymal condensations and subsequent chondrogenesis. A dominant-negative BMPR1 blocked the formation of cartilage condensations in infected chick limbs (Zou et al. 1997). With similar effect, overexpression of noggin in early chick limbs also blocked the formation of mesenchymal condensations, and interfered with subsequent chondrogenesis (Capdevila and Johnson 1998; Pizette and Niswander 2000). Formation of chondrocytes in mice was blocked by ablation of both BMPRI and BMPRII, using a conditional strategy that used a collagen II-driven Cre to remove BMPRI from condensations of BMPRII knockout mice (Yoon et al. 2005). In these mice, SOX9 expression in the condensations, needed for chondrocyte formation, did not occur. While BMP signaling is thus necessary for forming skeletal elements, it remains unclear why the elements form in their proper locations. Neither the expression patterns of the various BMP ligands, nor those of their known antagonists, give a clue to the specific locations of the condensations. It may be that the integration of these various activities provides such a cue, but it is also possible that BMP signaling is permissive but not instructive in this regard.

After chondrocyte form, BMPs regulate chondrocyte proliferation and differentiation. Noggin-knockout mice have enlarged regions of growth cartilage (Brunet et al. 1998). This enlargement may well involve recruitment of mesenchymal cells by unopposed BMP signaling, but probably also reflects the actions of noggin to decrease chondrocyte proliferation, as shown when noggin is added to murine fetal limb explants (Minina et al. 2001). The effects of BMPs on chondrocyte differentiation have been variable, probably reflecting multiple actions of BMPs, depending on the state of differentiation of the target chondrocytes and the effects of BMPs on adjacent signaling cells such as perichondrial cells (for review, see Kobayashi et al. 2005a; Pogue and Lyons 2006). BMP signaling also interacts with other signaling systems to assure coordination of the multiple signaling systems, determining chondrocyte behavior in development. For example, BMP signaling increases expression of Indian hedgehog (Minina et al. 2001).

BMP signaling also has a major influence on the formation of the joints. Most of the joints between the endochondral bones originate by segmentation of an initially continuous precartilaginous condensation. Stripes of high cell density perpendicular to the long axis of the skeletal elements form at the location of the future joints (Haines 1947). These stripes, called interzones, subdivide into three layers, with apoptosis leading to decreased cell density in the middle layer and ultimately cavitation, while the outer layers develop into the articular cartilage of the mature joint (Mitrovic 1978). The signals and/or cellular mechanisms that trigger the formation of joints at specific locations are still unclear. As discussed below, one of the first genes activated in the prospective joint is *Wnt9a* (*Wnt14*). It is capable of inducing expression of other joint markers when activated ectopically in developing cartilage (Hartmann and Tabin 2001), acting through the β-catenin pathway (Guo et al. 2004). Among the genes induced downstream of *Wnt9a* are a number of markers of the *BMP* family, including *GDF5*, *GDF6*, *GDF7*, *BMP2*, and *BMP4* (Storm and Kingsley 1996; Wolfman et al. 1997; Francis-West et al. 1999; Settle et al. 2003). The roles of BMP signaling are quite complex during the time when joints are specified and initially formed. This has particularly been well studied in the distal limb. Limb-specific conditional mutations of *BMP2* and *BMP4*, or *BMP2* and *BMP7*, do not result in alteration of digit pattern (Bandyopadhyay et al. 2006). This may reflect redundancy between family members; alternatively, BMP signaling may be essential for growth of distal phalanges, but not be involved in patterning them. A localized distal region of the growing digital ray expresses phosphorylated Smads indicative of active BMP signaling, which can be shown to be necessary for continued distal skeletal condensation (Suzuki et al. 2008). One candidate for the ligand responsible for this activity, in addition to BMP2, -4, and -7, is GDF5. GDF5 functions in several different phases. In the first phase, it is expressed more broadly than in just the joint interzone, promotes prechondrogenic condensation, and stimulates proliferation of the chondrocytes adjacent to the forming joint interzone (Francis-West et al. 1999). Because of these activities, ectopic application of GDF results in an increase in cartilage growth, to an extent inhibiting synovial joint formation, with resulting localized fusion of adjacent skeletal elements (Storm and Kingsley 1996; Francis-West et al. 1999). Similarly, mutations in the gene encoding the BMP/GDF antagonist Noggin also result in an inhibition of joint formation (Brunet et al. 1998). Nonetheless, the GDF family members play critical roles in potentiating joint formation. GDF5 is expressed in the location of the future joints 1–1.5 days before they become morphologically visible in mouse embryos (Storm and Kingsley 1996). Mice with mutations in GDF5 display completely normal cartilage condensations, indistinguishable from wild-type littermates up until the time segmentation normally occurs. However, specific joints fail to form, resulting in fusion of proximal and medial phalanges, wrist and ankle bones, and certain other joints (Storm and Kingsley 1996). Mutations in GDF6 similarly block formation of joints at specific sites (Settle et al. 2003). Moreover, the GDF subfamily can induce the differentiation of tissues associated with the joints. For example, GDF7 expression is found at the distal tips of skeletal elements where tendons will attach, and ectopic GDF5, GDF6, and particularly GDF7 applied in conjunction with BMP2 result in de novo formation of tendons and joints as well as cartilage and bone (Wolfman et al. 1997).

Consistent with the role of GDFs in promoting proliferation of articular chondrocytes, this signaling system continues to be required for maintenance of joints after they have formed. When BMPR1A, a receptor for BMP and GDF signaling, is conditionally removed using a Cre-recombinase allele driven by *GDF5* regulatory elements, most joints nonetheless form normally. However, articular cartilage subsequently wears off and is not replaced, in a process resembling osteoarthritis (Rountree et al. 2004).

TGF-βS

TGF-βs are closely related to the BMPs in structure and signaling mechanisms. Their patterns of expression and in vitro actions suggest that they probably have roles in regulating chondrocyte proliferation and differentiation. However, perhaps because of overlapping functions of the multiple TGF-β ligands and receptors, and the consequent challenges in designing informative genetically altered mice, in addition to early fetal lethality of knockout of type I and type II TGF-β receptors, few studies in vivo have clarified the roles of TGF-β signaling in the growth plate (for review, see Pogue and Lyons 2006).

Three TGF-β isoforms exist in mammals. TGF-β1, -2, and -3 are all found in the perichondrium of developing bones and in hypertrophic chondrocytes (Minina et al. 2005). The type II TGF-β receptor (TGF-βRII) and the type I TGF-β receptor (ALK5/Tβ RI), as well as Smad2 and Smad3, the intracellular transcriptional mediators of TGF-β action, are expressed in all growth plate chondrocytes and perichondrium. Thus, the machinery for TGF-β action is widely expressed in growth cartilage and the surrounding perichondrium.

In vitro, TGF-β directs mesenchymal cells to become chondrocytes (Pittenger et al. 1999). When exposed to proliferating chondrocytes, TGF-β delays hypertrophic differentiation (Serra et al. 1999; Ferguson et al. 2000). Knockout of *TGF-β2* leads to widespread skeletal patterning defects (Sanford et al. 1997), while knockout of *TGF-β1* or *-3* leads to no skeletal defects (Shull et al. 1992; Proetzel et al. 1995). Despite the in vitro overexpression studies suggesting an important role for TGF-β signaling in the early differentiation of mesenchymal cells into chondrocytes, knockout of the *TGF-βRII* early in limb development, using the *Prx1*-Cre and a floxed *TGF-βRII* gene in vivo, had no apparent effect in the appearance of early chondrocytes of the limb (Seo and Serra 2007). Perhaps the activation of Smad2 and/or Smad3 by activins in this setting could explain the apparently normal formation of chondrocytes in these limbs. Alternatively, despite the apparent success in eliminating *TGF-βRII* gene expression in the limbs of these mice, low levels of expression may have obscured the potential early role of TGF-β in condensations. However, subsequent limb development was not normal in the *Prx1*-Cre; floxed *TGF-βRII* mice, demonstrating roles for TGF-β signaling after chondrocyte formation. Chondrocyte proliferation was decreased and the relative proportion of hypertrophic chondrocytes was increased, consistent with in vitro studies, suggesting a role for TGF-β in slowing hypertrophic differentiation. Maintenance of the interzone and subsequent joint formation were also defective in these mice.

In contrast to the abnormal limbs in the *Prx1*-Cre; floxed *TGF-βRII* mice, the limbs of *collagen II*-Cre; floxed *TGF-βRII* mice are normal (Baffi et al. 2004). Seo and Serra (2007) suggest that this contrast may reflect the importance of perichondrial TGF-β signaling in regulation of chondrocyte proliferation and differentiation, since the *collagen II*-Cre is probably less active in the perichondrium than is the *Prx1*-Cre. Such a role for TGF-β signaling in the perichondrium had been suggested by earlier studies of bone explants (Alvarez et al. 2001).

The mechanisms whereby activation of the TGF-βRII receptor leads to chondrocyte proliferation and slowed differentiation are not known. Nevertheless, the clarification that TFG-β signaling is needed for normal bone development should lead to further instructive studies.

PARATHYROID HORMONE-RELATED PROTEIN

Parathyroid hormone-related protein (PTHrP) is named because of its resemblance in sequence to parathyroid hormone (PTH), the major regulator of blood calcium in birds and mammals. PTHrP was discovered because cancers frequently overexpress PTHrP and, when they do, the PTH-like actions of PTHrP result in elevation of calcium in the blood (Stewart 2005). The normal functions of PTHrP, however, overlap very little with those of PTH. PTHrP is normally present only in very low amounts in the circulation. Instead, it is produced locally in a broad variety of tissues and acts on receptors on nearby cells (Philbrick et al. 1996). Mammalian PTHrP is 141 residues long and only the initial 36 residues resemble those of PTH. Nevertheless, all the actions of PTHrP during development that have been characterized so far appear to reflect activation of a G protein-coupled receptor that mediates actions of both PTH and PTHrP. Not surprisingly, this PTH/PTHrP receptor (PPR) is expressed both in PTH and PTHrP target cells, usually close to cells expressing PTHrP.

As bones develop, PTHrP expression is first seen at the ends of the early cartilage molds of long bones, both in perichondrial cells and in some chondrocytes (Lee et al. 1995). This expression is absolutely dependent on the expression of Indian hedgehog (see below) (St-Jacques et al. 1999). Further, if Indian hedgehog is expressed in higher levels than normal or in cells closer to PTHrP-producing cells than normal, PTHrP expression increases, though only the cells at the ends of the bones remain competent to express PTHrP, even with dramatic increase in Indian hedgehog expression (Vortkamp et al. 1996; Kobayashi et al. 2002). Indian hedgehog probably acts directly on PTHrP-producing cells, because removal of Smoothened from these cells stops PTHrP expression (Hilton et al. 2007). Suppression of Gli3 repressor activity by Indian hedgehog action is part of the pathway used to generate PTHrP in response to Indian hedgehog, at least in perichondrial cells (Hilton et al. 2005; Koziel et al. 2005). In these cells, no PTHrP is produced in the *Indian hedgehog*$^{-/-}$ mouse. PTHrP expression is restored, however, in *Indian hedgehog*$^{-/-}$; *Gli3*$^{-/-}$ fetuses.

The PPR is expressed at low levels on flat proliferating chondrocytes and at high levels just as chondrocytes stop proliferating and become prehypertrophic cells (Lee et al. 1996). In the absence of either PTHrP or the PPR, the columns of flat proliferating chondrocytes are much shorter than normal or virtually undetectable (in the universal or chondrocyte-specific *PPR*$^{-/-}$ mice) (Karaplis et al. 1994; Lanske et al. 1996; Kobayashi et al. 2002). Conversely, when PTHrP is overexpressed in chondrocytes through a transgene driven by the collagen II promoter, differentiation of chondrocytes into hypertrophic chondrocytes is suppressed until after birth (when the promoter probably decreases its level of expression) (Weir et al. 1996). Thus, PTHrP acts to keep chondrocytes proliferating and delay their further differentiation. The delayed differentiation includes a delay in the expression of Indian hedgehog. Thus, PTHrP and Indian hedgehog together regulate a negative feedback loop: Indian hedgehog stimulates the expression of PTHrP and also stimulates the proliferation of chondrocytes and the conversion of round-proliferating to flat-proliferating chondrocytes (see p. 184). PTHrP then delays the termination of chondrocyte proliferation and expression of Indian hedgehog. Because PTHrP is expressed only at the top of the growth plate and Indian hedgehog only by prehypertrophic chondrocytes, the actions of PTHrP and Indian hedgehog lead to a separation of the cells synthesizing each ligand, thus limiting the actions of each signaling system (Fig. 2). The steady state distance between

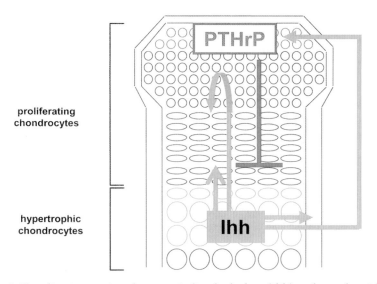

proliferating
chondrocytes

hypertrophic
chondrocytes

Figure 2. Signaling interactions between Indian hedgehog (Ihh) and parathyroid hormone-related protein (PTHrP). PTHrP is synthesized by chondrocytes and perichondrial cells at the ends of bones, and acts to slow the differentiation of prehypertrophic and hypertrophic chondrocytes, thus delaying the synthesis of Ihh. Ihh, synthesized by early hypertrophic chondrocytes, stimulates the production of PTHrP, increases the proliferation of chondrocytes, directs perichondrial cells to become bone collar osteoblasts, and accelerates the conversion of round chondrocytes to flat-proliferating chondrocytes.

the chondrocytes synthesizing these two ligands is thus determined by the program that determines synthesis of each ligand.

The mechanisms by which activation of the PPR accomplishes these actions are incompletely understood. The PPR activates multiple heterotrimeric G proteins. Chondrocyte-specific knockout of $Gs\alpha$ leads to a phenotype that resembles that of knockout of the PPR (Sakamoto et al. 2005). In contrast, the growth plates of mice with PPRs mutated so that they cannot activate $G_{q/11}$ (but can activate Gs normally) resemble those with overexpression of PTHrP (Guo et al. 2002). This result suggests that activation of $G_{q/11}$ by the PPR in the growth plate acts to oppose the effects of activation of $Gs\alpha$ and that the latter effects predominate after activation of the PPR.

PTHrP suppresses expression of the cdk inhibitor p57 in fetal chondrocytes (MacLean et al. 2004). Some (but not all) growth plates of the $PTHrP^{-/-}$; $p57^{-/-}$ mouse are restored close to normal with respect to morphology and patterns of gene expression. Consequently, the suppression of p57 expression normally found as chondrocytes stop proliferating explains part of the actions of PTHrP. Transcription factors mediating the actions of PTHrP are incompletely understood. SOX9 actions serve to slow the differentiation of proliferating chondrocytes to hypertrophic chondrocytes. PTHrP signaling leads to phosphorylation of SOX9 by protein kinase A (Huang et al. 2001). This activating phosphorylation thus probably contributes to the actions of PTHrP in the growth plate. PTHrP also lowers the levels of Runx2 mRNA (Hilton et al. 2007). Runx2 is required for chondrocyte hypertrophy, so the suppression of Runx2 expression plausibly contributes to the actions of PTHrP. As predicted by this hypothesis, $PTHrP^{-/-}$; $Runx2^{+/-}$ mice have partial correction of the short proliferative columns of the $PTHrP^{-/-}$ mouse. However, in bones from transgenic mice engineered to produce Runx2 in response to a *collagen II* promoter, PTHrP still slows hypertrophy. As expected, PTHrP does not decrease Runx 2 protein in these mice. Consequently, suppression of Runx2 expression only partly explains the action of PTHrP on the growth plate. Nkx3.2 is a hox-family transcription factor expressed in proliferating chondrocytes. When overexpressed in the chicken growth plate, expression of Runx2 and hypertrophy are suppressed (Provot et al. 2006). In that system, PTHrP overexpression increases the expression of Nkx3.2. Thus, Nkx3.2 may mediate actions of PTHrP as well. In the murine system, knockout of Nkx3.2 (called Bapx in mammals) does not have a growth plate phenotype (perhaps because of actions of other family members); consequently, the physiologic role of *Nkx3.2* in mammals remains an hypothesis requiring further testing. Thus, multiple potential mediators of PTHrP action on the growth plate, involving regulators of the cell cycle and differentiation, have been identified, but a more complete understanding of essential pathways downstream of PTHrP will require further analysis.

In postnatal mice, secondary centers of ossification form and PTHrP expression then occurs in reserve chondrocytes at the top of the growth plate. $PTHrP^{-/-}$ mice engineered to live postnatally through expression of a constitutively active PPR in chondrocytes fuse their growth plates 3 weeks after birth (Schipani et al. 1997). Since the collagen II-driven transgene probably decreases its expression postnatally, this growth plate fusion may reflect the absence of PTHrP signaling. Preliminary experiments in which the PPR is specifically deleted in postnatal growth plates confirm this observation (T. Hirai and H. Kronenberg, unpubl.). Thus, signaling by the PPR is required for normal growth plate function postnatally. In addition to the actions to slow the differentiation of chondrocytes, in this setting, the actions of PTHrP are required to preserve the growth plate as an entity. In humans, heterozygous inactivating mutations in Gsα lead to short stature associated with early fusion of growth plates. This observation suggests that PTHrP signaling through Gsα may be required for normal growth plate function postnatally in humans as well.

INDIAN HEDGEHOG

The mammalian Hedgehog family comprises three members, two of which, Indian hedgehog (Ihh) and Sonic hedgehog (Shh), regulate skeletal development (for review, see McMahon et al. 2003). However, Shh acts upstream of the active process of skeletogenesis in patterning of neural crest, somitic, and lateral plate mesenchymal derivatives. Consequently, this review focuses on Ihh action.

In Hedgehog signaling, ligands bind to Ptch1, a multipass membrane protein and primary receptor for all Hedgehog ligands. As a result, Ptch1 inhibition of a second membrane protein, Smoothened, is lost. Derepression of Smoothened activates an intracellular pathway targeting Gli transcriptional effectors. Production of Gli repressor (chiefly Gli3) is attenuated, derepressing a subset of targets. In addition, Gli activator forms (chiefly Gli2) directly activate other targets (for review, see Varjosalo and Taipale 2008).

Ihh and the Endochondral Skeleton

The role of Ihh has been most extensively analyzed in the context of long bone development, in which Hedgehog signaling coordinates growth and differentiation within the long bone. Ihh is first expressed in, and its expression restricted to, differentiating, postmitotic prehypertrophic chondrocytes (Bitgood and McMahon 1995; Vortkamp et al. 1996). The Ihh mouse mutant provides a good starting point for understanding Ihh function (St-Jacques et al. 1999). Skeletal elements form but exhibit a severe growth deficiency that is evident as early as E13.5. Moreover, the undifferentiated chondrocytes fail to undergo normal stacking in the shaft of the growing long bone. Most likely, as a result of the growth deficiency, chondrocyte maturation is delayed. When hypertrophic chondrocytes do appear, these are not separated from the presumptive articular, epiphyseal surfaces by a large population of undifferentiated chondrocytes as in wild-type embryos. Rather, hypertrophic chondrocytes extend to the ends of these bones. Finally, no osteoblast development is observed, and consequently no bone forms. Genetic removal of Ihh activity in the postnatal animal (Maeda et al. 2007) and inhibition of Ihh action by pharmacological pathway inhibitors (Kimura et al. 2008) indicates that Ihh most likely acts in a similar way throughout the extensive period of bone development in mammals.

Collectively, studies of *Ihh* mutants, mutants in other critical Hedgehog pathway components, bone explants, and cultured cells indicate that Ihh plays four critical roles in development of long bones and likely more generally in the endochondral skeleton, proliferation of chondrocyte progenitors, organization of chondrocyte columns, inhibition of chondrocyte differentiation, and osteoblast induction (see Fig. 2).

Ihh and Chondrocyte Proliferation

Ihh is a key mitogen for chondrocytes, most likely acting directly on proliferating chondrocytes to regulate their proliferative program chiefly through the inhibition of Gli3 repressor action (Long et al. 2001; Hilton et al. 2005; Koziel et al. 2005). Hedgehog ligands are lipid-modified; cholesterol is covalently attached at the carboxyl terminus of the bioactive amino-terminal-signaling peptide in an autocatalytic processing step (Porter et al. 1996). Reducing cholesterol levels can decrease skeletal growth rate, but it is not clear whether altered cholesterol levels directly influence Ihh production (Wu and De Luca 2004). Ihh's proliferative activity is modulated by a surface-bound proteoglycan, Syndecan-3 (Shimo et al. 2004). As proteoglycans regulate cellular response and trafficking of Hedgehog ligands, this has led to the suggestion that Syndecan-3 may serve as receptor for Ihh in driving chondrocyte expansion in the Ihh target field (Shimo et al. 2004).

In its role as a major organ-specific mitogen, Ihh action mirrors that of Shh in driving the expansion of granule cell precursors in the cerebellum (Wechsler-Reya and Scott 1999), and more distantly, cell-cycle control regulated by the Hedgehog pathway in growth of imaginal disc structures in *Drosophila melanogaster* (Baker 2007). Transcriptional control of G1 cyclin activity, in particular, expression of cyclin D1, provides a common, conserved mechanistic link amongst these examples, but the demonstration that loss of all G1 cyclin activity (Kozar and Sicinski 2005) is less severe than the loss of Ihh function (St-Jacques et al. 1999) indicates that other mechanisms must underpin Ihh's proliferative activity. Growth of the entire embryo is regulated by insulin-like growth factor (IGF) signaling. Reduction of IGF1/2 levels leads to mice that are dramatically reduced in size. Surprisingly, Ihh and IGF act independently, as compound mutants in both pathways exhibit a more marked reduction in long-bone growth than either single mutant (Long et al. 2006).

Chondrocyte proliferation is also linked to mechanical sensing, and some evidence suggests Ihh production may be an output of mechano-sensing in the condylar cartilage (Tang et al. 2004). Finally, although all endochondral bones utilize Ihh signaling as a primary driver of chondrocyte proliferation, the final sizes of these bones are dramatically different within an organism, and a homologous bone very different in size across species. Thus, production and interpretation of an Ihh-mediated proliferative program are likely targets of element-specific growth programs. *Shox2* encodes a transcriptional regulator restricted to proximal limb long bones (stylopod). Interestingly, *Shox2* mutants exhibit a marked reduction in stylopod development that correlates with reduced Ihh levels, providing a potential molecular insight into this process (Yu et al. 2007).

Ihh and Growth Plate Organization

In the growth plate of long bones, proliferating chondrocytes organize into longitudinal columns prior to entering a postmitotic, hypertrophic pathway. This organization is thought to underlie longitudinal versus radial growth of the skeletal element. Ihh accelerates the conversion of round to flat columnar chondrocytes. This organization also appears to be governed directly through Ihh modulation of Gli3 activity (Hilton et al. 2005; Kobayashi et al. 2005b; Koziel et al. 2005). Further, columnar development is disrupted in cilial-deficient *Kif3a* mutants, where intraflagellar transport is disrupted (Koyama et al. 2007; Song et al. 2007). Recent work demonstrates a functional connection between the primary cilium and Hedgehog signaling at multiple levels in the pathway (Varjosalo and Taipale 2008).

Ihh and Cartilage Differentiation

Proliferating columnar chondrocytes transition to postmitotic prehypertrophic chondrocytes. Subsequently, these undergo hypertrophic development and finally death in conjunction with the bone-forming process. As discussed earlier (see "PTHrP"), Ihh controls the flow of cells into the hypertrophic program by regulating expression of PTHrP, most likely directly through the modulation of Gli3 repressor levels (Hilton et al. 2005; Kobayashi et al. 2005b; Koziel et al. 2005). PTHrP blocks chondrocyte differentiation. Thus, the positive regulation of PTHrP by Ihh signaling is predicted to lower Ihh levels as Ihh is itself a product of prehypertrophic chondrocytes. As a consequence, PTHrP production drops, new prehypertrophic cells form, and Ihh levels rise. A prediction of this model is that reducing the levels of either PTHrP or Ihh is expected to reduce the size of the growth plate, as the system accommodates to the levels of a new regulatory input, establishing a new balance point between the opposing actions of these signals in stimulating proliferation (Ihh) and differentiation (PTHrP). This prediction is borne out by analysis of mutants in Disp1. Disp1 participates in cellular export of cholesterol-modified Hedgehog ligands, and Disp1^{C829F}- point mutants (Caspary et al. 2002) show reduced Ihh within the target field (Tsiairis and McMahon 2008). As a result, the active proliferative zone is reduced in size, accommodating to reduced Ihh levels, and long bone growth is diminished (Tsiairis and McMahon 2008). The distinct roles that Ihh plays in regulating chondrocyte proliferation and differentiation raises the question of whether blocking differentiation and cell-cycle exit of chondrocytes can functionally restore skeletal growth in Ihh mutants by potentially enabling a larger pool of chondrocytes to respond to other proliferative signals (e.g., IGF signals). However, this does not appear to be the case. When differentiation is blocked by expression of constitutive PPR receptor in an *Ihh* mutant background, chondrocyte differentiation is blocked, but skeletal growth is not improved (Karp et al. 2000).

Ihh and the Osteoblast

Osteoblasts first appear in the perichondrial region of long bones adjacent to Ihh-producing prehypertrophic chondrocytes. Several lines of evidence indicate that Ihh initiates the osteoblast program in this region. First, these cells are actively responding to Ihh (St-Jacques et al. 1999). Second, no osteoblasts form in an Ihh mutant (St-Jacques et al. 1999). Third, if Smoothened activity is removed from the perichondrial region so that cells there cannot respond to Ihh, osteoblasts fail to form (Long et al. 2004). Interestingly, mutant cells form ectopic chondrocytes (Long et al. 2004). Fourth, when PPR is removed in chimeric mice, ectopic osteoblast formation is observed adjacent to the ectopic source of Ihh signal in an Ihh-dependent process (Chung et al. 2001). Fifth, Ihh triggers osteoblast differentiation in a number of cell lines that retain osteoblast potential (Enomoto-Iwamoto et al. 2000; Chung et al. 2001; Jemtland et al. 2003; van der Horst et al. 2003).

Analysis of *Ihh/Gli3* compound mutants indicates that loss of Gli3 repressor, while sufficient to rescue Ihh's other regulatory actions, is not sufficient to fully restore normal osteoblast formation, implicating distinct Gli-activator functions in the bone-lineage program (Hilton et al. 2005; Koziel et al. 2005). This program is initiated on induction of the transcriptional regulator Runx2, and Runx2 mutants lack most osteoblasts (Komori et al. 1997; Lee et al. 1997; Otto et al. 1997). No perichondrial activation of Runx2 is observed in Ihh mutants, suggesting that Ihh signaling is required at the earliest stage of osteoblast development (St-Jacques et al. 1999; Long et al. 2004). In contrast, removal of Ihh responsiveness after Runx2-dependent activation of Osx1 in the osteoblast lineage has no pronounced effect on subsequent development of osteoblasts (Rodda and McMahon 2006). Thus, Ihh's action is limited to induction of osteoblasts, and Runx2 is a potential target of Ihh action. Consistent with this view, Gli2 activator is associated with Runx2 activation in vitro (Shimoyama et al. 2007). Runx factors are themselves regulators of Ihh synthesis in chondrocytes (Yoshida et al. 2004). Analysis of Ihh function in the context of other osteogenic factors suggests a model in which Ihh enables an osteogenic program by modulating the response of a bipotential progenitor to BMP signaling (Long et al. 2004). In the absence of Ihh, this progenitor adopts a chondrocyte fate (Long et al. 2004).

Whereas evidence for Ihh action in specifying osteoblasts in the bone collar is compelling, a role in osteogenesis within the primary spongiosa is not clear-cut. A layer of HH-responding cells lies at the interface of the hypertrophic chondrocytes and primary spongiosa, which may potentially represent a continued target of Ihh-mediated osteogenesis (St-Jacques et al. 1999). However, the identity and function of these cells has not been established. Formation of the primary spongiosa is closely associated with vascularization of an acellular region, formed on apoptosis of hypertrophic chondrocytes, leading to the suggestion that osteoprogenitors or osteoblasts move from the periosteal region with the invading vasculature. Vascular ingrowth is defective in Ihh mutants—why is unclear—but Hedgehog action has been reported to promote vascularization elsewhere in the mouse embryo (Vokes et al. 2004; Astorga and Carlsson 2007). Consequently, defective vascular development may underpin the failure in osteogenesis (Colnot et al. 2005).

Cranial Bone Development

Much of the head skeleton is generated by a process of membranous bone formation in the absence of a cartilage template. In Ihh mutants, cranial osteoblasts form, but the ossification fronts are retarded (St-Jacques et al. 1999). Thus, either another Hedgehog member can functionally substitute for Ihh or there is no absolute requirement for a Hedgehog signal in these neural crest-derived osteoblasts. Initial reports indicated Shh was expressed in conjunction with cranial osteogenesis (Kim et al. 1998), but this may reflect a cross-hybridization of probes. Smo mutant cells form osteoblasts in the mandible of $WT;smo^{-/-}$ chimeric mice, indicating that an osteoblast can form in the absence of a Hedgehog response (Long et al. 2004). More recently, an extensive characterization of the intramembranous bone forming process suggests that definitive osteoblasts emerge from CLO cells, cells with chondrocyte (collagen II), and osteoblast (osteopontin) features (Abzhanov et al. 2007). In this, Ihh and PTHrP are suggested to act in parallel to block CLO differentiation, promoting expansion of mitotically active progenitors and consequently enabling normal bone growth.

WNTS

As with Hedgehog ligands, Wnt signals play early roles in the growth, specification, movement, and organization of preskeletal cell types. We restrict this review to the actions of Wnt signaling at a later stage after skeletal progenitors are formed. Only a brief overview of Wnt signaling is provided here. The reader is directed to recent reviews for a thorough discussion of the pathway (Gordon and Nusse 2006; Huang and He 2008).

There are 19 members of the mammalian family of lipid-modified Wnt glycoproteins. These are thought to interact with at least three families of signal-transducing receptors, the Frizzled, Ror, and Ryk receptors, as well as Lrp5/6 coreceptors. Wnt signaling activates several intracellular signaling pathways. The two best characterized are the canonical Wnt pathway and planar polarity pathway. The former results in stabilization of a complex of β-catenin with members of the Tcf/Lef family of DNA-binding proteins and transcriptional activation of targets through binding to *cis*-regulatory regions. The latter principally alters cell behaviors, such as movement and polarity, and may not have a direct transcriptional component. In addition, several other pathways have been invoked that may mediate Wnt action in distinct cellular contexts, including developing bones (see later discussion).

Canonical Wnt Action in Osteoblast Development

Studies in the mouse indicate that canonical Wnt signaling regulates the osteogenic program (Day et al. 2005; Hill et al. 2005; Hu et al. 2005; Rodda and McMahon 2006). These analyses have used a common genetic approach, removal of β-catenin activity, but with different regulatory systems that vary the cell type in which canonical Wnt signaling is lost. The model emerging from this work is one of an extended period of Wnt signaling following initial specification of osteoblast progenitors under Ihh regulation in the endochondral skeleton. In addition, β-catenin is also essential for membranous bone formation.

Removal of β-catenin activity in preosteoblast cells demonstrates that β-catenin is not essential for Ihh-mediated specification of a Runx2[+] osteoblast progenitor. In contrast, mutant cells fail to progress to Osx1[+] osteoblast progenitors (Hill et al. 2005; Hu et al. 2005). Though these experiments removed β-catenin throughout much of the developing long bone, the absence of an apparent phenotype in chondrocytes argues for a specific role in osteoblast cells. In a study that focused more narrowly on Osx1[+] osteoblast progenitors, β-catenin was shown to be essential for maintenance of expression of the bone determinants, Runx2 and Osx1, in the osteoblast lineage; consequently, mature Osteocalcin[+] osteoblasts were not formed (Rodda and McMahon 2006). Interestingly, failure of osteogenesis was accompanied by ectopic chondrogenesis. In contrast to β-catenin removal, activation of a stabilized form of β-catenin in Osx1[+] cells dramatically enhanced the bone forming process, resulting in premature synthesis of a bone matrix and expansion of osteoblasts (Rodda and McMahon 2006). However, these bone-synthesizing cells did not progress to become mature Osteocalcin[+] cells. Together, these data demonstrate that β-catenin promotes an osteoblast program at the expense of an alternative chondrogenic program, most likely by positively regulating Runx2 and Osx1, and potentially other osteogenic factors. At the end of the osteoblast pathway, β-catenin activity may be attenuated to establish a mature osteoblast. Indeed, Lef/b-catenin action is reported to prevent Runx2 activation of the osteocalcin promoter (Kahler and Westendorf 2003) and terminal differentiation of osteoblasts (Kahler et al. 2006; Rodda and McMahon 2006). The recent finding that Osx1 transcriptionally activates the promoter of the secreted Wnt antagonist Dkk1 suggests a possible mechanism whereby Wnt signaling may be down-regulated (Zhang et al. 2008). The identity of Wnt ligand(s) regulating the osteoblast program remains unclear. Wnt7b is activated in early osteoblast progenitors, but bone forms following Wnt7b removal from progenitor cells (Rodda and McMahon 2006). Further, Wnt7b appears to act through a noncanonical G-protein coupled PKC delta pathway (Tu et al. 2007).

Canonical Wnt Signaling and Bone Mass

The study of mouse and human mutants in the Wnt coreceptor Lrp5 indicates that Wnt signaling functions in the adult to control bone mass. Mutants in the osteoblast-expressed Lrp5 antagonists, Dkk1 and Sost, enhance bone mass (Ellies et al. 2006; Morvan et al. 2006; Semenov and He 2006). Point mutants that prevent antagonist binding to Lrp5 also increase bone mass, while mutants that remove Lrp5 activity decrease bone mass, resulting in osteoporosis (Gong et al. 2001; Kato et al. 2002; Little et al. 2002). Further, Lrp5 is reported to be a target of both thyroid-stimulating hormone receptor (TSHR) (Abe et al. 2007), and the vitamin D receptor (Fretz et al. 2007), two regulatory pathways that control bone mass. Analysis of anabolic (PTH-triggered) and mechanosensory (load-generated) stimulation of bone mass has implicated Lrp5 function in the latter only (Sawakami et al. 2006). The osteocyte, the osteoblast derivative that senses mechanical stress, has been shown to undergo a rapid induction of a canonical Wnt reporter when load is applied to the ulna. Further, the osteocyte-like cell line, MLO-Y4, activates a canonical Wnt pathway in response to mechanical loading (Bonewald and Johnson 2008).

How osteocyte load sensing through a Wnt pathway results in altered bone mass is unclear. In one model, this stimulates osteoblast induction, proliferation, or bone matrix synthesis, or blocks osteoblast or osteocyte apoptosis. Alternatively, or in addition, Wnt signaling in the osteoblast may regulate the interplay between osteoblasts and bone matrix-digesting osteoclasts. Evidence in support of the latter model comes from removal or stabilization of β-catenin at late stages of osteoblast development, resulting in osteopenia or increased bone mass, respectively, through the regulation of osteoprotegerin (Glass et al. 2005). Osteoprotegerin, a RANK ligand decoy receptor produced by osteoblasts in response to canonical Wnt action, inhibits RANK ligand-dependent osteoclast production and hence bone resorption. Clearly, modulating the canonical pathway provides a clinical opportunity for treatment of bone disease and age-related loss of bone mass.

Wnts in Joint Formation

Individual skeletal elements are segmented from a continuous condensation by the formation of synovial joints. Wnt4, Wnt9a (formally Wnt14), and Wnt16 are expressed in condensates prior to cartilage segmentation (Hartmann and Tabin 2001; Guo et al. 2004). Functional analysis demonstrates that at least one of these can initiate a joint-forming program at ectopic sites (Hartmann and Tabin 2001). Joints are induced by stabilized β-catenin (Guo et al. 2004), though β-catenin is apparently not essential for initiating the process (Später et al. 2006a). Further analysis of Wnt4/Wnt9a mutants, however, indicates a semiredundant role for these factors not in induction but in maintenance of joint interzone cells

(Später et al. 2006b). The molecular pathway through which these signals act, and the interface between Wnt and other joint-associated pathways, remain to be determined.

FIBROBLAST GROWTH FACTORS

The fibroblast growth factor (FGF) family comprises 23 ligands that signal through a surface receptor complex comprising heparin/heparin sulfate and a receptor tyrosine kinase derived from one of four *FGF receptor* (*FGFR*) genes (for review, see Ornitz 2005; Su et al. 2008). Ligand binding to the extracellular domain of the FGFR results in intracellular tyrosine kinase activation. The common result is the triggering of a MAP kinase-based signaling cascade that results in the regulation of gene activity. Unlike some pathways in which all signaling intersects with a small family of transcriptional regulators (e.g., Glis in Hedgehog signaling), FGF signaling likely uses several distinct regulators depending on the cell context.

The first functional evidence for FGF regulation of bone development came from the striking finding that distinct mutations in FGFR3 lead to two types of human dwarfism, a severe achondroplasia (Rousseau et al. 1994; Shiang et al. 1994) and less severe hypochondroplasia (Bellus et al. 1995). Since this time, a large number of FGFR mutants have been described, encompassing FGFR1, -2, and -3, that lead to a spectrum of skeletal pathologies (Ornitz 2005). Cell culture and genetic studies, predominantly in the mouse, have complemented the analysis of human skeletal disorders in this pathway.

FGF Signaling in Chondrocytes

FGFR2 expression is one of the earliest identifiers of the developing skeletal condensate, suggesting that FGF signaling may play an early role in initiating cartilage development (Peters et al. 1992). Consistent with this view, FGF signaling up-regulates expression of the chondrocyte determinant, Sox9 (Murakami et al. 2000).

FGFR3 negatively regulates chondrocyte growth. Null mutants have enlarged skeletons (Deng et al. 1996), while activating mutants are dwarfed (Chen et al. 1999; Iwata et al. 2001). One likely regulator of FGFR3 action is perichondrial-derived FGF18, as FGF18 mutants also exhibit enhanced skeletal growth (Ellsworth et al. 2002; Liu et al. 2002; Ohbayashi et al. 2002). FGFR3 appears to regulate skeletal growth by controlling both chondrocyte proliferation and chondrocyte differentiation, utilizing MAP kinase and STAT-based regulation. In the normal skeleton, ectopic activation of MAP kinase signaling generates a dwarf phenotype, in which differentiation rather than cell proliferation is modified (Murakami et al. 2004). Expression of this transgene rescues dwarfism in FGFR3 mutants (Murakami et al. 2004). While an FGF18–FGFR3 pathway regulates chondrocyte maturation throughout the endochondral skeleton, an FGF9–FGFR1 pathway appears to act more specifically in proximal limb structures (Hung et al. 2007). Nuclear STAT accumulation is observed in patients and mice with FGFR3-mutant related dwarfism (Su et al. 1997; Iwata et al. 2001) and removal of STAT1 action can suppress dwarfism in an FGFR gain-of-function model (Sahni et al. 2001). STAT1 action is linked to direct regulation of cell-cycle inhibitors, and FGF fails to induce cell-cycle arrest of STAT1 mutant chondrocytes (Sahni et al. 1999), suggesting that STAT1 mediates cell-cycle control downstream of FGFR3. In addition to STAT-based growth regulatory mechanisms, FGFR3 activity modulates expression of Ihh and BMPs, indicating regulatory cross talk amongst growth-controlling pathways, though the regulatory details have not been determined (Naski et al. 1998).

FGF Signaling in Osteoblasts and Vasculogenesis

Reduced FGFR2 activity has been shown to decrease osteoblast proliferation and function in the developing skeleton (Eswarakumar et al. 2002; Yu et al. 2003), while mice deficient in FGF2 exhibit reduced bone mass in the adult (Montero et al. 2000). Thus, FGF signaling may act in both developmental and physiological control of the bone program. FGF18 mutant mice have decreased endochondral and membranous bone formation (Liu et al. 2002). FGF18 is itself a product of osteoblasts where it is a likely target of STAT1-mediated inhibition of osteoblast proliferation (Xiao et al. 2004). Thus, FGF18 may have an auto-stimulatory activity in osteoblasts, acting through FGFR2 in this cellular context. As discussed earlier, there is close correlation between vascularization and osteoblast development. In both FGF9 and FGF18 mutants, delayed osteoblast development correlates with reduced VEGF levels and delayed vascularization (Hung et al. 2007; Liu et al. 2007).

VASCULAR ENDOTHELIAL GROWTH FACTOR IN BONE DEVELOPMENT

Though the initial condensations of bone and initial generation of chondrocytes in these condensations occur in an avascular setting, blood vessels surround the growing condensations (from E13.5 in the mouse tibia), and vascularization is an early event in the formation of the primary spongiosa. Not surprisingly, Vascular Endothelial Growth Factor (VEGF)-A, a protein secreted during vascularization of most tissues, has multiple roles in early bone development. VEGF-A is produced in three isoforms, VEGF 120, VEGF 164, and VEGF 188, generated by alternative spicing of mRNA. VEFG 120 is the only isoform that does not bind heparan sulfate. All isoforms bind the tyrosine kinase receptors, VEGFR1 (Flt 1) and VEGFR2 (Flk 1), but only VEGF 164 binds the coreceptors, neuropilin 1 and neuropilin 2. The *VEGF* gene is expressed in perichondrial cells and hypertrophic chondrocytes and, somewhat later in development, in epiphyseal chondrocytes (for review, see Zelzer and Olsen 2005).

The roles of VEGF in early bone development have been shown through either chondrocyte-specific knockout of the gene using the Cre-lox approach, or through universal expression of one or another of the splicing isoforms of VEGF. When only VEGF 120 is produced, the appearance of perichondrial blood vessels is delayed, suggesting that one or both of the heparan sulfate binding isoforms of VEGF is needed for timely attraction of these blood vessels (Maes et al. 2002; Zelzer et al. 2002). Shortly after the synthesis of VEGF by hypertrophic chondrocytes, the vascular invasion of the primary ossification center occurs. This invasion is curtailed in mice with chondrocyte-specific knockout of VEGF, with concomitant expansion of the hypertrophic chondrocyte layer. A similar phenotype is seen in mice with only expression VEGF 120 (Maes et al. 2002; Zelzer et al. 2002, 2004). An independent action of VEGF is revealed by the phenotype of mice expressing only VEGF 188 (Maes et al. 2004). In these mice, epiphyseal chondrocytes exhibit increased hypoxia, with deficits in the vascularity surrounding the cartilage. Chondrocytes in the center of the epiphyseal region die. This phenotype is also seen in mice with chondrocyte-specific knockout of VEGF (Zelzer et al. 2004). A similar but somewhat more severe phenotype is exhibited by mice that are missing the transcription factor HIF1α in chondrocytes (Schipani et al. 2001). Studies in these mice show that VEGF expression in the epiphyseal region, but not in the hypertrophic chondrocytes, is regulated by HIF1α. In the hypertrophic region, Runx2 is a major regulator of VEGF expression (Amarilio et al. 2007). Thus, VEGF expression in chondrocytes is crucial for vascularization of bone and associated chondrocyte differentiation and survival.

CONCLUSIONS

In this chapter, we attempted to summarize the current understanding of the regulation of chondrogenesis by inductive signaling at various stages of differentiation. We have seen that a large number of classical signaling systems converge to regulate this process, including Hedgehog, canonical and noncanonical Wnt, FGF, BMP, TGF-β, PTHrP, and VEGF signaling. Understanding the exact roles each of these play, the steps of chondrogenesis at which they act, and the way they are integrated with one another is critical for understanding the developmental process of chondrogenesis and also for the prospect of regenerative medicine. Already, tissue engineering has provided a framework for incorporating chondrogenic progenitor cells and scaffolding materials with embedded signaling molecules, such as BMPs, to facilitate proper growth and differentiation. As more is learned about the signaling pathways involved, there will be an opportunity for increased precision in directive tissue engineering. Yet, in spite of the very rapid progress in this area over the last decade that has resulted in the picture summarized here, a large number of critical questions remain open. We list here a few of the most obvious and important steps for which the regulatory signals still remain elusive:

1. While we know a number of factors, most notably the BMPs that are necessary for initiating chondrogenesis and sufficient to trigger the process when provided exogenously, we still do not know the specific signals and/or signal antagonists responsible for the location, size, and shape of the initial chondrogenic mesenchymal condensations, or how such signals relate to upstream global patterning information.

2. As the condensations elongate from proximal to distal within the limb by recruitment of additional mesenchyme at the distal ends, we do not know the signals that trigger branching, producing (for example) two bones in the zeugopod (radius and ulna, or tibia and fibula) as opposed to one in the stylopod (humerus or femur).

3. We also do not know the signaling systems responsible for the precise placement of the joints. While a number of factors (most notably Wnt9a and GDF5) have been implicated as playing critical roles in the process of forming a joint interzone, the trigger for forming the joints with the right spacing remains elusive.

4. The signals regulating the later stages of joint formation are also obscure. Of particular importance are the still unknown signals directing the formation of articular cartilage.

5. Of equal mystery to the inductive processes forming joints at the ends of the chondrogenic condensation is the induction process in the middle that triggers the initial hypertrophic differentiation of the cartilage.

6. While a great deal has been learned about the control of proliferation and differentiation of the chondrocytes once the process has been initiated, there are many cellular details for which the signaling remains obscure. For example, what orients the columns of proliferating chondrocytes, and what signals direct them to flatten out and form linear columns?

7. Once the growth plates are mature, why do some produce more longitudinal growth than others? For example, the phalanx and femur start out as very similarly sized condensations. What signals are differentially modulated, and by what means, to yield the differences in their growth rates?

8. In the context of the fairly well-understood influence of estrogens at puberty in humans, what are the local interactions that lead to growth-plate fusion and the termination of longitudinal growth?

While a great deal remains to be discovered in understanding the growth factors orchestrating chondrogenesis, there is also a reason for optimism that many of these questions will be answered in the near future. There are now very powerful tools for profiling genes expressed at particular times and locations, such as DNA microarrays and high-throughput sequencing, suggesting that candidate signaling molecules will soon be identified for the remaining elusive activities. Equally important, transgenic technologies have become exceedingly sophisticated, allowing stage and tissue-specific activation and inactivation of any gene of interest. Together, these advances give reason to believe that the signals guiding chondrogenesis will be increasingly understood in the years to come.

ACKNOWLEDGMENTS

This work was supported by National Institutes of Health grant DK 56246.

REFERENCES

Abe, E., Sun, L., Mechanick, J., Iqbal, J., Yamoah, K., Baliram, R., Arabi, A., Moonga, B.S., Davies, T.F., and Zaidi, M. 2007. Bone loss in thyroid disease: Role of low TSH and high thyroid hormone. *Ann. N.Y. Acad. Sci.* **1116:** 383–391.

Abzhanov, A., Rodda, S.J., McMahon, A.P., and Tabin, C.J. 2007. Regulation of skeletogenic differentiation in cranial dermal bone. *Development* **134:** 3133–3144.

Alvarez, J., Horton, J., Sohn, P., and Serra, R. 2001. The perichondrium plays an important role in mediating the effects of TGF-β1 on endochondral bone formation. *Dev. Dyn.* **221:** 311–321.

Amarilio, R., Viukov, S.V., Sharir, A., Eshkar-Oren, I., Johnson, R.S., and Zelzer, E. 2007. HIF1α regulation of *Sox9* is necessary to maintain differentiation of hypoxic pre-chondrogenic cells during early skeletogenesis. *Development* **134:** 3917–3928.

Astorga, J. and Carlsson, P. 2007. Hedgehog induction of murine vasculogenesis is mediated by Foxf1 and Bmp4. *Development* **134:** 3753–3761.

Baffi, M.O., Slattery, E., Sohn, P., Moses, H.L., Chytil, A., and Serra, R. 2004. Conditional deletion of the TGF-β type II receptor in Col2a expressing cells results in defects in the axial skeleton without alterations in chondrocyte differentiation or embryonic development of long bones. *Dev. Biol.* **276:** 124–142.

Baker, N.E. 2007. Patterning signals and proliferation in *Drosophila* imaginal discs. *Curr. Opin. Genet. Dev.* **17:** 287–293.

Bandyopadhyay, A., Tsuji, K, Cox, K., Harfe, B.D., Rosen, V. and Tabin, C.J. 2006. Genetic analysis of the roles of BMP2, BMP4, and BMP7 in limb patterning and skeletogenesis. *PLoS Genet.* **2:** e216.

Bellus, G.A., Hefferon, T.W., Ortiz de Luna, R.I., Hecht, J.T., Horton, W.A., Machado, M., Kaitila, I., McIntosh, I., and Francomano, C.A. 1995. Achondroplasia is defined by recurrent G380R mutations of FGFR3. *Am. J. Hum. Genet.* **56:** 368–373.

Bitgood, M.J. and McMahon, A.P. 1995. *Hedgehog* and *BMP* genes are coexpressed at many diverse sites of cell–cell interaction in the mouse embryo. *Dev. Biol.* **172:** 126–138.

Bonewald, L.F. and Johnson, M.L. 2008. Osteocytes, mechanosensing and Wnt signaling. *Bone* **42:** 606–615.

Brunet, L.J., McMahon, J.A., McMahon, A.P., and Harland, R.M. 1998. Noggin, cartilage morphogenesis, and joint formation in the mammalian skeleton. *Science* **280:** 1455–1457.

Capdevila, J. and Johnson, R.L. 1998. Endogenous and ectopic expression of *noggin* suggests a conserved mechanism for regulation of BMP function during limb and somite patterning. *Dev. Biol.* **197:** 205–217.

Caspary, T., Garcia-Garcia, M.J., Huangfu, D., Eggenschwiler, J.T., Wyler, M.R., Rakeman, A.S., Alcorn, H.L., and Anderson, K.V. 2002. Mouse *Dispatched homolog1* is required for long-range, but not juxtacrine, Hh signaling. *Curr. Biol.* **12:** 1628–1632.

Chen, L., Adar, R., Yang, X., Monsonego, E.O., Li, C., Hauschka, P.V., Yayon, A., and Dent, C.-X. 1999. Gly369Cys mutation in mouse *FGFR3* causes achondroplasia by affecting both chondrogenesis and osteogenesis. *J. Clin. Invest.* **104:** 1517–1525.

Chung, U.I., Schipani, E., McMahon, A.P., and Kronenberg, H.M. 2001. *Indian hedgehog* couples chondrogenesis to osteogenesis in endochondral bone development (comments). *J. Clin. Invest.* **107:** 295–304.

Colnot, C., de la Fuente, L., Huang, S., Hu, D., Lu, C., St-Jacques, B., and Helms, J.A. 2005. Indian hedgehog synchronizes skeletal angiogenesis and perichondrial maturation with cartilage development. *Development* **132:** 1057–1067.

Day, T.F., Guo, X., Garrett-Beal, L., and Yang, Y. 2005. Wnt/β-catenin signaling in mesenchymal progenitors controls osteoblast and chondrocyte differentiation during vertebrate skeletogenesis. *Dev. Cell* **8:** 739–750.

Deng, C., Wynshaw-Boris, A., Zhou, F., Kuo, A., and Leder, P. 1996. Fibroblast growth factor receptor 3 is a negative regulator of bone growth. *Cell* **84:** 911–921.

Ellies, D.L., Viviano, B., McCarthy, J., Rey, J.P., Itasaki, N., Saunders, S., and Krumlauf, R. 2006. Bone density ligand, Sclerostin, directly interacts with LRP5 but not LRP5[G171V] to modulate Wnt activity. *J. Bone Miner. Res.* **21:** 1738–1749.

Ellsworth, J.L., Berry, J., Bukowski, T., Claus, J., Feldhaus, A., Holderman, S., Holdren, M.S., Lum, K.D., Moore, E.E., Raymond, F., et al. 2002. Fibroblast growth factor-18 is a trophic factor for mature chondrocytes and their progenitors. *Osteoarthritis Cartilage* **10:** 308–320.

Enomoto-Iwamoto, M., Nakamura, T., Aikawa, T., Higuchi, Y., Yuasa, T., Yamaguchi, A., Nohno, T., Noji, S., Matsuya, T., Kurisu, K., et al. 2000. Hedgehog proteins stimulate chondrogenic cell differentiation and cartilage formation. *J. Bone Miner. Res.* **15:** 1659–1668.

Eswarakumar, V.P., Monsonego-Ornan, E., Pines, M., Antonopoulou, I., Morriss-Kay, G.M., and Lonai, P. 2002. The *IIIc* alternative of *Fgfr2* is a positive regulator of bone formation. *Development* **129:** 3783–3793.

Ferguson, C.M., Schwarz, E.M., Reynolds, P.R., Puzas, J.E., Rosier, R.N., and O'Keefe, R.J. 2000. Smad2 and 3 mediate transforming growth factor-β1-induced inhibition of chondrocyte maturation. *Endocrinology* **141:** 4728–4735.

Francis-West, P.H., Parish, J., and Archer, C.W. 1999. BMP/GDF-signaling interactions during synovial joint development. *Cell Tissue Res.* **296:** 111–119.

Fretz, J.A., Zella, L.A., Kim, S., Shevde, N.K., and Pike, J.W. 2007. 1,25-Dihydroxyvitamin D$_3$ induces expression of the Wnt signaling co-regulator LRP5 via regulatory elements located significantly downstream of the gene's transcriptional start site. *J. Steroid Biochem. Mol. Biol.* **103:** 440–445.

Glass II, D.A., Bialek, P., Ahn, J.D., Starbuck, M., Patel, M.S., Clevers, H., Taketo, M.M., Long, F., McMahon, A.P., Lang, R.A., et al. 2005. Canonical Wnt signaling in differentiated osteoblasts controls osteoclast differentiation. *Dev. Cell* **8:** 751–764.

Gong, Y., Slee, R.B., Fukai, N., Rawadi, G., Roman-Roman, S., Reginato, A.M., Wang, H., Cundy, T., Glorieux, F.H., Lev, D., et al. 2001. LDL receptor-related protein 5 (LRP5) affects bone accrual and eye development. *Cell* **107:** 513–523.

Gordon, M.D. and Nusse, R. 2006. Wnt signaling: Multiple pathways, multiple receptors, and multiple transcription factors. *J. Biol. Chem.* **281:** 22429–22433.

Guo, J., Chung, U.I., Kondo, H., Bringhurst, F.R., and Kronenberg, H.M. 2002. The PTH/PTHrP receptor can delay chondrocyte hypertrophy in vivo without activating phospholipase C. *Dev. Cell* **3:** 183–194.

Guo, X., Day, T.F., Jiang, X., Garrett-Beal, L., Topol, L., and Yang, Y. 2004. Wnt/β-catenin signaling is sufficient and necessary for synovial joint formation. *Genes Dev.* **18:** 2404–2417.

Haines, R.W. 1947. The development of joints. *J. Anat.* **81:** 3–55.

Hartmann, C. and Tabin, C.J. 2001. Wnt-14 plays a pivotal role in inducing synovial joint formation in the developing appendicular skeleton. *Cell* **104:** 341–351.

Hill, T.P., Spater, D., Taketo, M.M., Birchmeier, W., and Hartmann, C. 2005. Canonical Wnt/β-catenin signaling prevents osteoblasts from differentiating into chondrocytes. *Dev. Cell* **8:** 727–738.

Hilton, M.J., Tu, X., Cook, J., Hu, H., and Long, F. 2005. Ihh controls cartilage development by antagonizing Gli3, but requires additional effectors to regulate osteoblast and vascular development. *Development* **132:** 4339–4351.

Hilton, M.J., Tu, X., and Long, F. 2007. Tamoxifen-inducible gene deletion reveals a distinct cell type associated with trabecular bone, and direct regulation of PTHrP expression and chondrocyte morphology by Ihh in growth region cartilage. *Dev. Biol.* **308:** 93–105.

Hu, H., Hilton, M.J., Tu, X., Yu, K., Ornitz, D.M., and Long, F. 2005. Sequential roles of Hedgehog and Wnt signaling in osteoblast development. *Development* **132:** 49–60.

Huang, H. and He, X. 2008. Wnt/β-catenin signaling: new (and old) players and new insights. *Curr. Opin. Cell Biol.* **20:** 119–125.

Huang, W., Chung, U.I., Kronenberg, H.M., and de Crombrugghe, B. 2001. The chondrogenic transcription factor Sox9 is a target of signaling by the parathyroid hormone-related peptide in the growth plate of endochondral bones. *Proc. Natl. Acad. Sci.* **98:** 160–165.

Hung, I.H., Yu, K., Lavine, K.J., and Ornitz, D.M. 2007. FGF9 regulates early hypertrophic chondrocyte differentiation and skeletal vascularization in the developing stylopod. *Dev. Biol.***307:** 300–313.

Iwata, T., Li, C.-l., Deng, C.-X., and Francomano, C.A. 2001. Highly activated Ffgr3 with the K644M mutation causes prolonged survival in severe dwarf mice. *Hum. Mol. Genet.* **10:** 1255–1264.

Jemtland, R., Divieti, P., Lee, K., and Segre, G.V. 2003. Hedgehog promotes primary osteoblast differentiation and increases *PTHrP* mRNA expression and iPTHrP secretion. *Bone* **32:** 611–620.

Kahler, R.A. and Westendorf, J.J. 2003. Lymphoid enhancer factor-1 and β-catenin inhibit Runx2-dependent transcriptional activation of the osteocalcin promoter. *J. Biol. Chem.* **278:** 11937–11944.

Kahler, R.A., Galindo, M., Lian, J., Stein, G.S., van Wijnen, A.J., and Westendorf, J.J. 2006. Lymphocyte enhancer-binding factor 1 (Lef1) inhibits terminal differentiation of osteoblasts. *J. Cell. Biochem.* **97:** 969–983.

Karaplis, A.C., Luz, A., Glowacki, J., Bronson, R.T., Tybulewicz, V.L., Kronenberg, H.M., and Mulligan, R.C. 1994. Lethal skeletal dysplasia from targeted disruption of the parathyroid hormone-related peptide gene. *Genes Dev.* **8:** 277–289.

Karp, S.J., Schipani, E., St-Jacques, B., Hunzelman, J., Kronenberg, H., and McMahon, A.P. 2000. *Indian hedgehog* coordinates endochondral bone growth and morphogenesis via *parathyroid hormone related-protein*-dependent and -independent pathways. *Development* **127:** 543–548.

Kato, M., Patel, M.S., Levasseur, R., Lobov, I., Chang, B.H., Glass II, D.A., Hartmann, C., Li, L., Hwang, T.H., Brayton, C.F., et al. 2002. *Cbfa1*-independent decrease in osteoblast proliferation, osteopenia, and persistent embryonic eye vascularization in mice deficient in Lrp5, a Wnt coreceptor. *J. Cell Biol.* **157:** 303–314.

Kim, H.J., Rice, D.P., Kettermen, P.S., and Thesleff, I. 1998. FGF, BMP-, and SHH-mediated signalling pathways in the regulation of cranial suture morphogenesis and calvarial bone development. *Development* **17:** 1241–1251.

Kimura, H., Ng, J.M., and Curran, T. 2008. Transient inhibition of the Hedgehog pathway in young mice causes permanent defects in bone structure. *Cancer Cell* **13:** 249–260.

Kobayashi, T., Chung, U.I., Schipani, E., Starbuck, M., Karsenty, G., Katagiri, T., Goad, D.L., Lanske, B., and Kronenberg, H.M. 2002. PTHrP and Indian hedgehog control differentiation of growth plate chondrocytes at multiple steps. *Development* **129:** 2977–2986.

Kobayashi, T., Lyons, K.M., McMahon, A.P., and Kronenberg, H.M. 2005a. BMP signaling stimulates cellular differentiation at multiple steps during cartilage development. *Proc. Natl. Acad. Sci.* **102:** 18023–18027.

Kobayashi, T., Soegiarto, D.W., Yang, Y., Lanske, B., Schipani, E., McMahon, A.P., and

Kronenberg, H.M. 2005b. Indian hedgehog stimulates periarticular chondrocyte differentiation to regulate growth plate length independently of PTHrP. *J. Clin. Invest.* **115:** 1734–1742.

Komori, T., Yagi, H., Nomura, S., Yamaguchi, A., Sasaki, K., Deguchi, K., Shimizu, Y., Bronson, R.T., Gao, Y.H., Inada, M., et al. 1997. Targeted disruption of Cbfa1 results in a complete lack of bone formation owing to maturational arrest of osteoblasts (comment). *Cell* **89:** 755–764.

Koyama, E., Young, B., Nagayama, M., Shibukawa, Y., Enomoto-Iwamoto, M., Iwamoto, M., Maeda, Y., Lanske, B., Song, B., Serra, R., et al. 2007. Conditional Kif3a ablation causes abnormal hedgehog signaling topography, growth plate dysfunction, and excessive bone and cartilage formation during mouse skeletogenesis. *Development* **134:** 2159–2169.

Kozar, K. and Sicinski, P. 2005. Cell cycle progression without cyclin D-CDK4 and cyclin D-CDK6 complexes. *Cell Cycle* **4:** 388–391.

Koziel, L., Wuelling, M., Schneider, S., and Vortkamp, A. 2005. Gli3 acts as a repressor downstream of Ihh in regulating two distinct steps of chondrocyte differentiation. *Development* **132:** 5249–5260.

Lanske, B., Karaplis, A.C., Lee, K., Luz, A., Vortkamp, A., Pirro, A., Karperien, M., Defize, L.H., Ho, C., Mulligan, R.C., et al. 1996. PTH/PTHrP receptor in early development and Indian hedgehog-regulated bone growth. *Science* **273:** 663–666.

Lee, K., Deeds, J.D., and Segre, G.V. 1995. Expression of parathyroid hormone-related peptide and its receptor messenger ribonucleic acids during fetal development of rats. *Endocrinology* **136:** 453–463.

Lee, K., Lanske, B., Karaplis, A.C., Deeds, J.D., Kohno, H., Nissenson, R.A., Kronenberg, H.M., and Segre, G.V. 1996. Parathyroid hormone-related peptide delays terminal differentiation of chondrocytes during endochondral bone development. *Endocrinology* **137:** 5109–5118.

Lee, B., Thirunavukkararsu, K., Zhou, L., Pastore, L., Baldini, A., Hecht, J., Geoffroy, V., Ducy, P., and Karsenty, G. 1997. Missense mutations abolishing DNA binding of the osteoblast-specific transcription factor OSF2/Cbfa1 in cleidocranial dysplasia. *Nat. Genet.* **16:** 307–310.

Little, R.D., Carulli, J.P., Del Mastro, R.G., Dupuis, J., Osborne, M., Folz, C., Manning, S.P., Swain, P.M., Zhao, S.C., Eustace, B., et al. 2002. A mutation in the LDL receptor-related protein 5 gene results in the autosomal dominant high-bone-mass trait. *Am. J. Hum. Genet.* **70:** 11–19.

Liu, Z., Xu, J., Colvin, J.S., and Ornitz, D.M. 2002. Coordination of chondrogenesis and osteogenesis by fibroblast growth factor 18. *Genes Dev.* **16:** 859–869.

Liu, Z., Lavine, K.J., Hung, I.H., and Ornitz, D.M. 2007. FGF18 is required for early chondrocyte proliferation, hypertrophy and vascular invasion of the growth plate. *Dev. Biol.* **302:** 80–91.

Long, F., Zhang, X.M., Karp, S., Yang, Y., and McMahon, A.P. 2001. Genetic manipulation of hedgehog signaling in the endochondral skeleton reveals a direct role in the regulation of chondrocyte proliferation. *Development* **128:** 5099–5108.

Long, F., Chung, U.I., Ohba, S., McMahon, J., Kronenberg, H.M., and McMahon, A.P. 2004. Ihh signaling is directly required for the osteoblast lineage in the endochondral skeleton. *Development* **131:** 1309–1318.

Long, F., Joeng, K.S., Xuan, S., Efstratiadis, A., and McMahon, A.P. 2006. Independent regulation of skeletal growth by Ihh and IGF signaling. *Dev. Biol.* **298:** 327–333.

MacLean, H.E., Guo, J., Knight, M.C., Zhang, P., Cobrinik, D., and Kronenberg, H.M. 2004. The cyclin-dependent kinase inhibitor p57^{Kip2} mediates proliferative actions of PTHrP in chondrocytes. *J. Clin. Invest.* **113:** 1334–1343.

Maeda, Y., Nakamura, E., Nguyen, M.T., Suva, L.J., Swain, F.L., Razzaque, M.S., Mackem, S., and Lanske, B. 2007. Indian Hedgehog produced by postnatal chondrocytes is essential for maintaining a growth plate and trabecular bone. *Proc. Natl. Acad. Sci.* **104:** 6382–6387.

Maes, C., Carmeliet, P., Moermans, K., Stockmans, I., Smets, N., Collen, D., Bouillon, R., and Carmeliet, G. 2002. Impaired angiogenesis and endochondral bone formation in mice lacking the vascular endothelial growth factor isoforms VEGF$_{164}$ and VEGF$_{188}$. *Mech. Dev.* **111:** 61–73.

Maes, C., Stockmans, I., Moermans, K., Van Looveren, R., Smets, N., Carmeliet, P., Bouillon, R., and Carmeliet, G. 2004. Soluble VEGF isoforms are essential for establishing epiphyseal vascularization and regulating chondrocyte development and survival. *J. Clin. Invest.* **113:** 188–199.

Massagué, J., Seoane, J., and Wotton, D. 2005. Smad transcription factors. *Genes Dev.* **19:** 2783–2810.

McMahon, A.P., Ingham, P.W., and Tabin, C.J. 2003. Developmental roles and clinical significance of hedgehog signaling. *Curr. Top. Dev. Biol.* **53:** 1–114.

Minina, E., Wenzel, H.M., Kreschel, C., Karp, S., Gaffield, W., McMahon, A.P., and Vortkamp, A. 2001. BMP and Ihh/PTHrP signaling interact to coordinate chondrocyte proliferation and differentiation. *Development* **128:** 4523–4534.

Minina, E., Schneider, S., Rosowski, M., Lauster, R., and Vortkamp, A. 2005. Expression of Fgf and Tgfβ signaling related genes during embryonic endochondral ossification. *Gene Expr. Patterns* **6:** 102–109.

Mitrovic, D. 1978. Development of the diarthrodial joints in the rat embryo. *Am. J. Anat.* **151:** 475–485.

Montero, A., Okada, Y., Tomita, M., Ito, M., Tsurukami, H., Nakamura, T., Doetschman, T., Coffin, J.D., and Hurley, M.M. 2000. Disruption of the fibroblast growth factor-2 gene results in decreased bone mass and bone formation. *J. Clin. Invest.* **105:** 1085–1093.

Morvan, F., Boulukos, K., Clément-Lacroix, P., Roman Roman, S., Suc-Royer, I., Vayssière, B., Ammann, P., Martin, P., Pinho, S., Pognonec, P., et al. 2006. Deletion of a single allele of the *Dkk1* gene leads to an increase in bone formation and bone mass. *J. Bone Miner. Res.* **21:** 934–945.

Murakami, S., Kan, M., McKeehan, W.L., and de Crombrugghe, B. 2000. Up-regulation of the chondrogenic *Sox9* gene by fibroblast growth factors is mediated by the mitogen-activated protein kinase pathway. *Proc. Natl. Acad. Sci.* **97:** 1113–1118.

Murakami, S., Balmes, G., McKinney, S., Zhang, Z., Givol, D., and de Crombrugghe, B. 2004. Constitutive activation of MEK1 in chondrocytes causes Stat1-independent achondroplasia-like dwarfism and rescues the *Fgfr3*-deficient mouse phenotype. *Genes Dev.* **18:** 290–305.

Naski, M.C., Colvin, J.S., Coffin, J.D., and Ornitz, D.M. 1998. Repression of hedgehog signaling and BMP4 expression in growth plate cartilage by fibroblast growth factor receptor 3. *Development* **125:** 4977–4988.

Ohbayashi, N., Shibayama, M., Kurotaki, Y., Imanishi, M., Fujimori, T., Itoh, N., and Takada, S. 2002. FGF18 is required for normal cell proliferation and differentiation during osteogenesis and chondrogenesis. *Genes Dev.* **16:** 870–879.

Ornitz, D.M. 2005. FGF signaling in the developing endochondral skeleton. *Cytokine Growth Factor Rev.* **16**: 205–213.

Otto, F., Thornell, A.P., Crompton, T., Denzel, A., Gilmour, K.C., Rosewell, I.R., Stamp, G.W., Beddington, R.S., Mundlos, S., Olsen, B.R., et al. 1997. Cbfa1, a candidate gene for cleidocranial dysplasia syndrome, is essential for osteoblast differentiation and bone development (comment). *Cell* **89**: 765–771.

Peters, K.G., Werner, S., Chen, G., and Williams, L.T. 1992. Two FGF receptor genes are differentially expressed in epithelial and mesenchymal tissues during limb formation and organogenesis in the mouse. *Development* **114**: 233–243.

Philbrick, W.M., Wysolmerski, J.J., Galbraith, S., Holt, E., Orloff, J.J., Yang, K.H., Vasavada, R.C., Weir, E.C., Broadus, A.E., and Stewart, A.F. 1996. Defining the roles of parathyroid hormone-related protein in normal physiology. *Physiol. Rev.* **76**: 127–173.

Pittenger, M.F., Mackay, A.M., Beck, S.C., Jaiswal, R.K., Douglas, R., Mosca, J.D., Moorman, M.A., Simonetti, D.W., Craig, S., and Marshak, D.R. 1999. Multilineage potential of adult human mesenchymal stem cells. *Science* **284**: 143–147.

Pizette, S. and Niswander, L. 2000. BMPs are required at two steps of limb chondrogenesis: Formation of prechondrogenic condensations and their differentiation into chondrocytes. *Dev. Biol.* **219**: 237–249.

Pogue, R. and Lyons, K. 2006. BMP signaling in the cartilage growth plate. *Curr. Top. Dev. Biol.* **76**: 1–48.

Porter, J.A., Young, K.E., and Beachy, P.A. 1996. Cholesterol modification of hedgehog signaling proteins in animal development (comments; erratum in *Science* [1996] **274**: 1597). *Science* **274**: 255–259.

Proetzel, G., Pawlowski, S.A., Wiles, M.V., Yin, M., Boivin, G.P., Howles, P.N., Ding, J., Ferguson, M.W., and Doetschman, T. 1995. Transforming growth factor-β 3 is required for secondary palate fusion. *Nat. Genet.* **11**: 409–414.

Provot, S., Kempf, H., Murtaugh, L.C., Chung, U.I., Kim, D.W., Chyung, J., Kronenberg, H.M., and Lassar, A.B. 2006. Nkx3.2/Bapx1 acts as a negative regulator of chondrocyte maturation. *Development* **133**: 651–662.

Rodda, S.J. and McMahon, A.P. 2006. Distinct roles for Hedgehog and canonical Wnt signaling in specification, differentiation and maintenance of osteoblast progenitors. *Development* **133**: 3231–3244.

Rosen, V. 2006. BMP and BMP inhibitors in bone. *Ann. N.Y. Acad. Sci.* **1068**: 19–25.

Rountree, R.B., Schoor, M., Chen, H., Marks, M.E., Hrley, V., Mishina, Y., and Kingsley, D.M. 2004. BMP receptor signaling is required for postnatal maintenance of articular cartilage. *PLoS Biol.* **2**: e355.

Rousseau, F., Bonaventure, J., Legeai-Mallet, L., Pelet, A., Rozet, J.M., Maroteaux, P., Le Merrer, M., and Munnich, A. 1994. Mutations in the gene encoding fibroblast growth factor receptor-3 in achondroplasia. *Nature* **371**: 252–254.

Sahni, M., Ambrosetti, D.C., Mansukhani, A., Gertner, R., Levy, D., and Basilico, C. 1999. FGF signaling inhibits chondrocyte proliferation and regulates bone development through the STAT-1 pathway. *Genes Dev.* **13**: 1361–1366.

Sahni, M., Raz, R., Coffin, J.D., Levy, D., and Basilico, C. 2001. STAT1 mediates the increased apoptosis and reduced chondrocyte proliferation in mice overexpressing FGF2. *Development* **128**: 2119–2129.

Sakamoto, A., Chen, M., Kobayashi, T., Kronenberg, H.M., and Weinstein, L.S. 2005. Chondrocyte-specific knockout of the G protein GSα leads to epiphyseal and growth plate abnormalities and ectopic chondrocyte formation. *J. Bone Miner. Res.* **20**: 663–671.

Sanford, L.P., Ormsby, I., Gittenberger-de Groot, A.C., Sariola, H., Friedman, R., Boivin, G.P., Cardell, E.L., and Doetschman, T. 1997. TGFβ2 knockout mice have multiple developmental defects that are non-overlapping with other TGFβ knockout phenotypes. *Development* **124**: 2659–2670.

Sawakami, K., Robling, A.G., Ai, M., Pitner, N.D., Liu, D., Warden, S.J., Li, J., Maye, P., Rowe, D.W., Duncan, R.L., et al. 2006. The Wnt co-receptor LRP5 is essential for skeletal mechanotransduction but not for the anabolic bone response to parathyroid hormone treatment. *J. Biol. Chem.* **281**: 23698–23711.

Schipani, E., Lanske, B., Hunzelman, J., Luz, A., Kovacs, C.S., Lee, K., Pirro, A., Kronenberg, H.M., and Juppner, H. 1997. Targeted expression of constitutively active receptors for parathyroid hormone and parathyroid hormone-related peptide delays endochondral bone formation and rescues mice that lack parathyroid hormone-related peptide. *Proc. Natl. Acad. Sci.* **94**: 13689–13694.

Schipani, E., Ryan, H.E., Didrickson, S., Kobayashi, T., Knight, M., and Johnson, R.S. 2001. Hypoxia in cartilage: HIF-1α is essential for chondrocyte growth arrest and survival. *Genes Dev.* **15**: 2865–2876.

Semenov, M.V. and He, X. 2006. LRP5 mutations linked to high bone mass diseases cause reduced LRP5 binding and inhibition by SOST. *J. Biol. Chem.* **281**: 38276–38284.

Seo, H.S. and Serra, R. 2007. Deletion of *Tgfbr2* in Prx1-cre expressing mesenchyme results in defects in development of the long bones and joints. *Dev. Biol.* **310**: 304–316.

Serra, R., Karaplis, A., and Sohn, P. 1999. Parathyroid hormone-related peptide (PTHrP)-dependent and -independent effects of transforming growth factor β (TGF-β) on endochondral bone formation. *J. Cell Biol.* **145**: 783–794.

Settle Jr., S.H., Rountree, R.B., Sinha, A., Thacker, A., Higgins, K., and Kingsley D.M. 2003. Multiple joint and skeletal patterning defects caused by single and double mutations in the mouse *Gdf6* and *Gdf5* genes. *Dev. Biol.* **254**: 116–130.

Shiang, R., Thompson, L.M., Zhu, Y.Z., Church, D.M., Fielder, T.J., Bocian, M., Winokur, S.T., and Wasmuth, J.J. 1994. Mutations in the transmembrane domain of FGFR3 cause the most common genetic form of dwarfism, achondroplasia. *Cell* **78**: 335–342.

Shimo, T., Gentili, C., Iwamoto, M., Wu, C., Koyama, E., and Pacifici, M. 2004. Indian hedgehog and syndecans-3 coregulate chondrocyte proliferation and function during chick limb skeletogenesis. *Dev. Dyn.* **229**: 607–617.

Shimoyama, A., Wada, M., Ikeda, F., Hata, K., Matsubara, T., Nifuji, A., Noda, M., Amano, K., Yamaguchi, A., Nishimura, R., et al. 2007. Ihh/Gli2 signaling promotes osteoblast differentiation by regulating Runx2 expression and function. *Mol. Biol. Cell* **18**: 2411–2418.

Shull, M.M., Ormsby, I., Kier, A.B., Pawlowski, S., Diebold, R.J., Yin, M., Allen, R., Sidman, C., Proetzel, G., Calvin, D., et al. 1992. Targeted disruption of the mouse transforming growth factor-β 1 gene results in multifocal inflammatory disease. *Nature* **359**: 693–699.

Song, B., Haycraft, C.J., Seo, H.S., Yoder, B.K., and Serra, R. 2007. Development of the post-natal growth plate requires intraflagellar transport proteins. *Dev. Biol.* **305**: 202–216.

Später, D., Hill, T.P., Gruber, M., and Hartmann, C. 2006a. Role of canonical Wnt-signalling in joint formation. *Eur. Cells Mater.* **12**: 71–80.

Später, D., Hill, T.P., O'Sullivan R, J., Gruber, M., Conner, D.A., and Hartmann, C. 2006b. Wnt9a signaling is required for joint integrity and regulation of *Ihh* during chondrogenesis. *Development* **133**: 3039–3049.

Stewart, A.F. 2005. Clinical practice. Hypercalcemia associated with cancer. *N. Engl. J. Med.* **352:** 373–379.

St-Jacques, B., Hammerschmidt, M., and McMahon, A.P. 1999. Indian hedgehog signaling regulates proliferation and differentiation of chondrocytes and is essential for bone formation (erratum in *Genes Dev.* [1999] **13:** 2617). *Genes Dev.* **13:** 2072–2086.

Storm, E.E. and Kingsley, D.M. 1996. Joint patterning defects caused by single and double mutations in members of the bone morphogenetic protein (BMP) family. *Development* **122:** 3969–3979.

Su, W.C., Kitagawa, M., Xue, N., Xie, B., Garofalo, S., Cho, J., Deng, C., Horton, W.A., and Fu, X.Y. 1997. Activation of Stat1 by mutant fibroblast growth-factor receptor in thanatophoric dysplasia type II dwarfism. *Nature* **386:** 288–292.

Su, N., Du, X., and Chen, L. 2008. FGF signaling: Its role in bone development and human skeleton diseases. *Front. Biosci.* **13:** 2842–2865.

Suzuki, T., Hasso, S.M., and Fallon, J.F. 2008. Unique MAD1/5/8 activity at the phalanx-forming region determines digit identity. *Proc. Natl. Acad. Sci.* **105:** 4185–4190.

Tang, G.H., Rabie, A.B., and Hägg, U. 2004. Indian hedgehog: A mechanotransduction mediator in condylar cartilage. *J. Dent. Res.* **83:** 434–438.

Tsiairis, C.D. and McMahon, A.P. 2008. Disp1 regulates growth of mammalian long bones through the control of Ihh distribution. *Dev. Biol.* **317:** 480–485.

Tu, X., Joeng, K.S., Nakayama, K.I., Nakayama, K., Rajagopal, J., Carroll, T.J., McMahon, A.P., and Long, F. 2007. Noncanonical Wnt signaling through G protein-linked PKCδ activation promotes bone formation. *Dev. Cell* **12:** 113–127.

Urist, M.R. 1965. Bone: Formation by autoinduction. *Science* **150:** 893–899.

van der Horst, G., Farih-Sips, H., Lowik, C.W., and Karperien, M. 2003. Hedgehog stimulates only osteoblastic differentiation of undifferentiated KS483 cells. *Bone* **33:** 899–910.

Varjosalo, M. and Taipale, J. 2008. Hedgehog: Functions and mechanisms. *Genes Dev.* **22:** 2454–2472.

Vokes, S.A., Yatskievych, T.A., Heimark, R.L., McMahon, J., McMahon, A.P., Antin, P.B., and Krieg, P.A. 2004. Hedgehog signaling is essential for endothelial tube formation during vasculogenesis. *Development* **131:** 4371–4380.

Vortkamp, A., Lee, K., Lanske, B., Segre, G.V., Kronenberg, H.M., and Tabin, C.J. 1996. Regulation of rate of cartilage differentiation by Indian hedgehog and PTH-related protein. *Science* **273:** 613–622.

Wechsler-Reya, R.J. and Scott, M.P. 1999. Control of neuronal precursor proliferation in the cerebellum by Sonic Hedgehog. *Neuron* **22:** 103–114.

Weir, E.C., Philbrick, W.M., Amling, M., Neff, L.A., Baron, R., and Broadus, A.E. 1996. Targeted overexpression of parathyroid hormone-related peptide in chondrocytes causes chondrodysplasia and delayed endochondral bone formation. *Proc. Natl. Acad. Sci.* **93:** 10240–10245.

Wolfman, N.M., Hattersley, G., Cox, K. Celeste, A.J., Nelson, R., Yamaji, N., Dube, J.L., DiBlasio-Smith, E., Novbe, J., Song, J.J., et al. 1997. Ectopic induction of tendon and ligament in rats by growth and differentiation factors 5, 6 and 7, members of the TGF-β gene family. *J. Clin. Invest.* **100:** 321–330.

Wozney, J.M., Rosen, V., Celeste, A.J., Mitsock, L.M., Whitters, M.J., Kriz, R.W., Hewick, R.M., and Wang, E.A. 1988. Novel regulators of bone formation: Molecular clones and activities. *Science* **242:** 1528–1534.

Wu, S. and De Luca, F. 2004. Role of cholesterol in the regulation of growth plate chondrogenesis and longitudinal bone growth. *J. Biol. Chem.* **279:** 4642–4647.

Xiao, L., Naganawa, T., Obugunde, E., Gronowicz, G., Ornitz, D.M., Coffin, J.D., and Hurley, M.M. 2004. Stat1 controls postnatal bone formation by regulating fibroblast growth factor signaling in osteoblasts. *J. Biol. Chem.* **279:** 27743–27752.

Yoon, B.S., Ovchinnikov, D.A., Yoshii, I., Mishina, Y., Behringer, R.R., and Lyons, K.M. 2005. *Bmpr1a* and *Bmpr1b* have overlapping functions and are essential for chondrogenesis in vivo. *Proc. Natl. Acad. Sci.* **102:** 5062–5067.

Yoshida, C.A., Yamamoto, H., Fujita, T., Furuichi, T., Ito, K., Inoue, K.-I., Yamana, K., Zanma, A., Takada, K., Ito, Y., et al. 2004. Runx2 and Runx3 are essential for chondrocyte maturation, and Runx2 regulates limb growth through induction of *Indian hedgehog*. *Genes Dev.* **18:** 952–963.

Yu, K., Xu, J., Liu, Z., Sosic, D., Shao, J., Olson, E.N., Towler, D.A., and Ornitz, D.M. 2003. Conditional inactivation of FGF receptor 2 reveals an essential role for FGF signaling in the regulation of osteoblast function and bone growth. *Development* **130:** 3063–3074.

Yu, L., Liu, H., Yan, M., Yang, J., Long, F., Muneoka, K., and Chen, Y. 2007. *Shox2* is required for chondrocyte proliferation and maturation in proximal limb skeleton. *Dev. Biol.* **306:** 549–559.

Zelzer, E. and Olsen, B.R. 2005. Multiple roles of vascular endothelial growth factor (VEGF) in skeletal development, growth, and repair. *Curr. Top. Dev. Biol.* **65:** 169–187.

Zelzer, E., McLean, W., Ng, Y.S., Fukai, N., Reginato, A.M., Lovejoy, S., D'Amore, P.A., and Olsen, B.R. 2002. Skeletal defects in VEGF$^{120/120}$ mice reveal multiple roles for VEGF in skeletogenesis. *Development* **129:** 1893–1904.

Zelzer, E., Mamluk, R., Ferrara, N., Johnson, R.S., Schipani, E., and Olsen, B.R. 2004. VEGFA is necessary for chondrocyte survival during bone development. *Development* **131:** 2161–2171.

Zhang, C., Cho, K., Huang, Y., Lyons, J.P., Zhou, X., Sinha, K., McCrea, P.D., and de Crombrugghe, B. 2008. Inhibition of Wnt signaling by the osteoblast-specific transcription factor Osterix. *Proc. Natl. Acad. Sci.* **105:** 6936–6941.

Zou, H., Wieser, R., Massagué, J., and Niswander, L. 1997. Distinct roles of type I bone morphogenetic protein receptors in the formation and differentiation of cartilage. *Genes Dev.* **11:** 2191–2203.

7

Transcriptional Control of Osteoblast Differentiation

Gerard Karsenty
Department of Genetics and Development
Columbia University Medical Center
New York, New York 10025

IN CONTRAST WITH CHONDROCYTE DIFFERENTIATION, where all maturational stages are morphologically marked as well as spatially distinguishable within the growth plate, osteoblast differentiation is not marked by phenotypic changes in vivo, and osteoblasts in culture are, and remain throughout their differentiation, similar to fibroblasts. This absence of morphological features implies that one has to rely on gene expression studies to assess osteoblast differentiation. However, here again, the osteoblast has a poorly specific genetic program. Most of the proteins expressed by this cell type are also expressed in other cells, notably in fibroblasts.

Another feature of osteoblast differentiation is that its embryonic layout is more complex than the events taking place once the skeleton is formed. Indeed, the developmental process by which osteoblast precursors first appear in the bone collar, begin to differentiate and then migrate within the core of the forming skeletal element along with invading blood vessels, is not observed anymore once the bones are formed. In the mature skeleton osteoblast, progenitor cells are spread out within the bone marrow and differentiate in situ.

These two particularities explain for the most part why identifying the key transcriptional events required for osteoblast differentiation and function has been slower than for other cell types. However, in the last decade, these limitations have been overcome due to a combination of molecular efforts and genetic studies in mice and humans. This chapter

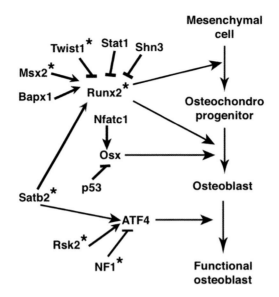

Figure 1. Transcriptional control of osteoblast differentiation and function. *Runx2* expression in mesenchymal progenitor cells is controlled by homeodomain transcription factors such as Msx2, Bpx, and *Hoxa-2*. The activity of Runx2 is first limited by the action of Twist-1, whose expression ceases to allow Runx2-mediated differentiation along the osteoblast lineage. Downstream of Runx2 lie other important transcriptional activators, such as Osx, which is required for proper osteoblast differentiation, and ATF4, which controls osteoblast gene expression and bone matrix deposition. Stars indicate genes that are involved in human pathologies.

summarizes our current knowledge about the transcriptional control of osteoblast differentiation and function (Fig. 1).

CONTROL OF OSTEOBLAST DIFFERENTIATION BY RUNX2

The power of a combined effort between molecular biologists and mouse and human geneticists in identifying key genes regulating osteoblast differentiation is best illustrated by the identification of *Runx2* as the master gene of osteoblast differentiation. Indeed, its crucial role during this process came to light from many different, but complementary, experimental approaches (Ducy et al. 1997; Komori et al. 1997; Lee et al. 1997; Mundlos et al. 1997; Otto et al. 1997).

One approach, molecular in nature, used the promoter of the mouse *Osteocalcin* gene, the most osteoblast-specific gene, as a tool. This analysis first defined two osteoblast-specific *cis*-acting elements, termed OSE1 and OSE2, and then identified *Runx2* as the transcription factor expressed in osteoblasts that binds to OSE2 (Ducy and Karsenty 1995; Ducy et al. 1997). In situ hybridization further revealed that, during mouse development, *Runx2* is first expressed in the lateral plate mesoderm at 10.5 days postcoitum, later marks the osteochondro progenitor cells of the skeletal condensations, and then at birth becomes restricted to osteoblasts and cells of the perichondrium. In addition to the *Osteocalcin* promoter, functional OSE2-like elements were identified in the promoter regions of most other genes that are expressed at relatively high levels in osteoblasts, such as *αI(I) collagen, Osteopontin,* and *Bone sialoprotein* (Ducy et al. 1997; Kern et al. 2001).

The ultimate demonstration that Runx2 acts as a pivotal activator of osteoblast differentiation came from genetic studies in mice and humans. Two groups reported that in *Runx2*-deficient mice, osteoblasts fail to differentiate, leading to a complete absence of endochondral and intramembranous bone formation (Komori et al. 1997; Otto et al. 1997). The critical importance of Runx2 for osteoblast differentiation was further underscored by the finding that mice lacking only one allele of Runx2 display hypoplastic clavicles and delayed closure of the fontanelles, a phenotype characterizing the human disease Cleidocranial dysplasia (CCD). Subsequent genetic analysis of CCD patients revealed disease-causing heterozygous mutations of the *RUNX2* gene, thereby demonstrating the importance of RUNX2 for human osteoblast differentiation (Table 1) (Lee et al. 1997; Mundlos et al. 1997).

In addition to its prominent role in osteoblast differentiation, Runx2 is also involved in the regulation of bone formation beyond development. Indeed, it remains expressed in osteoblasts postnatally, where it contributes to the regulation of genes encoding bone extracellular matrix proteins, such as type I collagen (Ducy et al. 1999; Kern et al. 2001).

Table 1. Association between human diseases and genes regulating cell differentiation in the skeleton

Gene	Type of factor	Human disease(s)
RUNX2	Transcription factor	Cleidocranial dysplasia
MSX2	Transcription factor	Boston-type craniosynostosis
		Enlarged parietal foramina
TWIST1	Transcription factor	Sathre-Chotzen syndrome
SATB2	Nuclear matrix protein	Cleft palate
RSK2	Protein kinase	Coffin-Lowry syndrome
NF1	Ras-GTPase activating factor	Neurofibromatosis type I

REGULATORS OF *RUNX2* EXPRESSION AND FUNCTION

Given the critical functions that *Runx2* exerts during skeletogenesis, it is not surprising that its expression and activity are tightly regulated. In terms of regulators of *Runx2* expression, several lines of evidence indicate that certain homeodomain-containing transcription factors positively regulate *Runx2* expression. For instance, *Msx2*-deficient mice display defective ossificaton of the skull and of bones developing by endochondral ossification, and show a strong decrease of *Runx2* expression (Satokata et al. 2000). Likewise, mice lacking the homeodomain-containing transcription factor Bpx die at birth due to a severe dysplasia of the axial skeleton, and *Runx2* expression in osteochondro progenitor cells of the prospective vertebral column is strongly reduced (Tribioli and Lufkin 1999). In contrast, the homeodomain-containing transcription factor *Hoxa-2* appears to negatively regulate *Runx2* expression. Indeed, *Hoxa-2*-deficient mice show increased *Runx2* expression and ectopic bone formation in the second branchial arch, while transgenic mice expressing *Hoxa-2* in craniofacial bones lack several bones in the craniofacial area (Kanzler et al. 1998).

In addition to these transcriptional regulators, several factors have been identified that affect *Runx2* ability to bind to DNA and/or regulate its *trans*-activation function. *Runx2* was initially identified based on its ability to regulate the expression of *Osteocalcin*, an osteoblast-derived hormone expressed only in fully differentiated osteoblasts (Ducy et al. 1997; Lee et al. 2007). However, during mouse development, *Osteocalcin* expression appears only 4 to 5 days after *Runx2* expression can be detected, suggesting that *Runx2* function is transiently inhibited by another nuclear protein during that time. Conceivably, decreased levels of such a protein should lead to a phenotype opposite of CCD, such as an increase in bone formation in the skull, a condition called craniosynostosis (Wilkie 1997). Among the genes whose inactivation causes craniosynostosis, only one encodes for a nuclear protein: *Twist-1* (see Table 1) (El Ghouzzi et al. 1997; Howard et al. 1997). Mouse genetics and molecular studies verified that a domain present at the carboxyl terminus of Twist-1, called the Twist box, binds Runx2 DNA-binding domain, thereby impairing its activity and delaying osteogenesis (Bialek et al. 2004). Likewise, Stat1, a transcription factor regulated by extracellular signaling molecules, physically interacts with *Runx2* to attenuate its activity. Interestingly, the physical interaction between the two proteins is independent of Stat1 activation by phosphorylation. It has thus been proposed that Stat1 would act by inhibiting the translocation of *Runx2* into the nucleus, since overexpression of Stat1 in osteoblasts leads to cytosolic retention of Runx2, while nuclear translocation of Runx2 is much more prominent in Stat1-deficient osteoblasts (Kim et al. 2003).

Consistent with this hypothesis, Stat1-deficient mice develop a high bone mass phenotype caused by enhanced bone formation (Kim et al. 2003). Schnurri 3 (Shn3) is another protein interacting with Runx2 that acts by decreasing its availability in the nucleus. Shn3 is a zinc finger adapter protein, originally thought to be involved in the VDJ recombination of immunoglobulin genes (Wu et al. 2000). Unexpectedly, Shn3-deficient mice display a severe adult-onset osteosclerotic phenotype due to a cell-autonomous increase of bone matrix deposition (Jones et al. 2006). Interestingly, while *Runx2* expression is not affected by the absence of Shn3, the level of Runx2 protein level is strikingly increased in Shn3-deficient osteoblasts. This discrepancy is caused by a decrease in Runx2 degradation, Shn3 acting as an adapter molecule linking Runx2 to the E3 ubiquitin ligase WWP1 (Jones et al. 2006). Accordingly, RNAi-mediated down-regulation of WWP1 in osteoblasts leads to increased Runx2 protein levels and enhanced extracellular matrix mineralization, thereby virtually mimicking the defects observed in the absence of Shn3 (Jones et al. 2006).

In addition to negative regulators of Runx2 function, there are also interacting factors enhancing its activity. One of them is the nuclear matrix protein SATB2. Human patients carrying a heterozygous chromosomal translocation that inactivates the *SATB2* gene present a cleft palate (see Table 1) (FitzPatrick et al. 2003). The generation of an *SATB2*-deficient mouse model confirmed the importance of this gene in craniofacial development and skeletal patterning but also identified its function in osteoblast differentiation (Dobreva et al. 2006). Part of this latter function is mediated by an increase in the expression of *Hoxa-2*, a gene encoding a negative regulator of *Runx2* expression discussed previously. The binding of SATB2 to an enhancer element of the *Hoxa-2* gene directly represses its expression (Kanzler et al. 1998; Dobreva et al. 2006). Another part is due to a direct action of SATB2 on the promoter of genes expressed in osteoblasts, such as *Bone Sialoprotein* and *Osteocalcin*. In the case of *Bone Sialoprotein*, SATB2 directly binds to an osteoblast-specific element present in the promoter of this gene (Dobreva et al. 2006). In the case of *Osteocalcin*, SATB2 requires a physical interaction with Runx2 to exert its activity (Dobreva et al. 2006). Lastly, SATB2 also interacts with ATF4, another transcription factor involved in osteoblast differentiation and function that will be discussed later in the chapter.

REGULATION OF OSTEOBLAST DIFFERENTIATION BY OSX

Osx is an Sp1-like zinc finger-containing transcription factor expressed in osteoblasts of all skeletal elements (Nakashima et al. 2002). Inactivation of

Osx in mice results in perinatal lethality due to arrested osteoblast differentiation and complete absence of bone formation (Nakashima et al. 2002). Unlike the *Runx2*-deficient mice whose skeleton is entirely non-mineralized, the Osx-deficient mice only lack a mineralized matrix in bones formed by intramembranous ossification. The Osx-deficient bones formed by endochondral ossification contain some mineralized matrix, although it resembled calcified cartilage, not mineralized bone matrix (Nakashima et al. 2002). While *Runx2* is normally expressed in Osx-deficient embryos, Osx is not expressed in *Runx2*-deficient embryos, indicating that Osx is acting downstream of Runx2 in the transcriptional cascade of osteoblast differentiation, and that its expression could be directly regulated by this latter factor (Nakashima et al. 2002). Accordingly, Runx2 binds to a responsive element present in the *Osx* promoter (Nishio et al. 2006). It has also been suggested that p53 acts as a negative regulator of Osx expression since p53$^{-/-}$ osteoblasts have increased Osx levels and p53-deficient mice show an increase in bone formation associated with elevated osteoblast numbers (Wang et al. 2006). However, a direct binding of p53 to the *Osx* promoter could not be demonstrated, suggesting that this regulation could be indirect.

Unlike for *Runx2*, no mutations of the human *Osx* gene have been identified yet that would be associated with decreased bone formation. Additionally, the molecular mechanisms underlying the action of Osx in osteoblasts are less understood than for Runx2. It was recently shown that Osx contributes to the negative effects of NFAT inhibitors on bone mass (Koga et al. 2005). Immunosuppresssion using NFAT inhibitors, such as FK506 or cyclosporin A, often leads to osteopenia (Liu et al. 1991; Rodino and Shane 1998), and in mice, *Nfatc1* deletion or treatment with FK506 also leads to decreased bone mass due to impaired bone formation (Koga et al. 2005). Using chromatin immunoprecipitation assays, Koga et al. (2005) showed that NFATc1 is recruited to the type I collagen promoter via formation of a NFATc1/Osx DNA-binding complex. Further experiments demonstrated that NFATc regulates Osx activity but not its binding to DNA, suggesting that it could activate Osx-mediated *trans*-activation by recruiting other transcriptional coactivators. The identification of these factors will be important to better understand this regulation.

AP1 FAMILY MEMBERS AND SKELETON BIOLOGY

Activator protein 1 (AP1) is a transcriptional complex associating members of the Jun (c-Jun, JunB, or JunD proteins) and Fos (c-Fos, Fra1, Fra2, or Fosb proteins) family of basic leucine-zipper proteins (Karin et al.

1997). AP1-transcription factors have been demonstrated to fulfill various functions in different cell types, including during bone remodeling (Wagner and Eferl 2005). For instance, c-Fos inactivation in mice results in severe osteopetrosis due to an arrest of osteoclast differentiation, while its transgenic overexpression results in osteosarcoma development (Grigoriadis et al. 1993, 1994). Fra1 also controls osteoblast function since mice lacking Fra1 in extraplacental tissues display an osteopenia associated with reduced bone formation, and transgenic mice overexpressing it in osteoblasts develop a severe osteosclerosis (Jochum et al. 2000; Eferl et al. 2004). Similarly, transgenic mice overexpressing Δ*fosB*, a splice variant of FosB, display a severe osteosclerotic phenotype caused by increased osteoblast differentiation and function (Sabatakos et al. 2000). In contrast, inactivation of JunB in extraplacental tissues induces a state of low bone turnover due to cell-autonomous defects of osteoblast and osteoclast diferentiation (Kenner et al. 2004).

There is also a link between AP1-family members and the regulation of bone formation by the sympathetic nervous system (see p. 213) and the circadian clock (Fu et al. 2005). Indeed, mice lacking components of the circadian clock, namely the *Per* or *Cry* genes, display a high bone mass phenotype caused by increased bone formation, as well as an increase in the expression of all AP1 family members (Fu et al. 2005). This increase is especially pronounced in the case of the *c-Fos* gene, leading to a direct activation of *c-Myc* transcription, which in turn indirectly increases the intracellular levels of cyclin D1 and promotes osteoblast proliferation (Fu et al. 2005).

TRANSCRIPTIONAL EFFECTORS OF WNT CANONICAL SIGNALING

The gene encoding LRP5, a transmembrane protein believed to act as a coreceptor for ligands of the Wnt family, was identified in both human and mice as a major determinant of bone mass accrual (Gong et al. 2001; Boyden et al. 2002; Kato et al. 2002; Little et al. 2002). These results sparked an interest in evaluating the role of transcription factors mediating Wnt function in bone cells. In the Wnt canonical signaling cascade, intracellular signaling events lead to nuclear translocation of β-catenin, prompting its association with transcription factors of the Tcf/Lef family and thereby gene transcription (Huelsken and Birchmeier 2001; Mao et al. 2001). However, osteoblast-specific inactivation of β-catenin results in a low bone mass phenotype caused by increased bone resorption, whereas *Lrp5* deficiency induces a postnatal bone formation defect (Glass et al. 2005). The bone resorption activity mediated by β-catenin was caused by a decreased expression of *Osteoprotegerin* (*Opg*), a well-known inhibitor of bone resorption expressed by osteoblasts (Teitelbaum and Ross 2003; Glass et al. 2005). Additional analyses identified Tcf1 as the β-catenin partner regulating *Opg* expression in osteoblasts (Glass et al. 2005). Thus, although these results did not explain how LRP5 regulates bone formation, they demonstrated that two known effectors of Wnt canonical signaling regulate gene expression in osteoblasts.

ATF4, A REGULATOR OF OSTEOBLAST DIFFERENTIATION AND FUNCTIONS

The role of ATF4 in skeletal biology was revealed by human genetics as much as by any other type of approach. Indeed, its identification arose from studies aiming at understanding why mutations in the gene encoding the RSK2 kinase caused the Coffin-Lowry syndrome, an X-linked mental retardation condition associated with skeletal abnormalities (see Table 1) (Trivier et al. 1996). Biochemical analyses demonstrated that ATF4 is strongly phosphorylated by Rsk2, and that this phosporylation is undetectable in osteoblasts derived from Rsk2-deficient mice, which display the same decreased bone mass due to impaired bone formation observed in Coffin-Lowry patients (Yang et al. 2004). The subsequent analysis of an ATF4-deficient mouse model confirmed that this transcription factor plays a crucial role in bone formation (Yang et al. 2004). At the molecular level, ATF4 was found to play two distinct, although complementary, functions. On the one hand, ATF4 is the factor binding to the osteoblast-specific element OSE-1 previously identified in the *Osteocalcin* promoter, and via this binding, it acts as a transcriptional activator (Ducy and Karsenty 1995; Yang et al. 2004). On the other hand, and independent of its ability to regulate transcription, ATF4 is required for efficient amino acid import into osteoblasts, thereby controlling the rate of type I collagen synthesis and bone matrix formation (Yang et al. 2004).

Interestingly, the osteoblast-specific function of ATF4 originates from its selective absence of proteosomal degradation in these cells, and treatment of nonosteoblastic cells with a proteasome inhibitor results in their abnormal accumulation of ATF4 protein and ectopic *Osteocalcin* expression (Yang and Karsenty 2004). As mentioned previously, ATF4 binds to the OSE-1 site of the proximal *Osteocalcin* promoter, which is located besides OSE-2, the Runx2 binding sequence (Ducy and Karsenty 1995; Ducy et al. 1997; Yang et al. 2004). Interaction between the nuclear matrix protein SATB2 and ATF4 was actually shown to enhance the synergistic activity of Runx2 and ATF4 that is required for sufficient *Osteocalcin* expression (Xiao et al. 2005; Dobreva et al. 2006). ATF4 also binds to the promoter of the *Rankl* gene, a secreted factor expressed by osteoblasts that promotes osteoclast differentiation (Teitelbaum and Ross 2003; Elefteriou et al. 2005). Accordingly, ATF4-deficient mice have decreased osteoclast numbers due to reduced *Rankl* expression (Elefteriou et al. 2005). Most importantly, however, this function of ATF4 is involved in the control of bone resorption by the sympathetic nervous system (Elefteriou et al. 2005). Indeed, treatment of normal osteoblasts with isoprotenerol, a sympathomimetic, enhances osteoclastogenesis of cocultured bone marrow macrophages by inducing *Rankl* expression, but this effect is abolished when the osteoblasts lack the β2-adrenergic receptor Adrβ2 or ATF4 (Elefteriou et al. 2005).

The other aspect of ATF4 function is that its role in regulating amino acid import led to its identification as the cause of the skeletal abnormalities of another human disease. A reduced type I collagen synthesis was observed in mice lacking ATF4 but also in Rsk2-deficient mice, providing evidence that diminished ATF4 phosphorylation negatively impacts bone formation (Yang et al. 2004). On the opposite, increased Rsk2-dependent phosphorylation of ATF4 could be associated with the development of the skeletal abnormalities in human patients suffering from neurofibromatosis (Elefteriou et al. 2006). This disease, that is primarily known for tumor development within the nervous system, is caused by inactivating mutations of the *NF1* gene, that encodes a Ras-GTPase activating protein (see Table 1) (Klose et al. 1998). The generation of a mouse model lacking *Nf1* specifically in osteoblasts ($Nf1_{ob-/-}$) led to the demonstration that this gene plays a major physiological role in bone remodeling (Elefteriou et al. 2006). Accordingly, transgenic mice overexpressing *ATF4* in osteoblasts were shown to display a phenotype similar to the $Nf1_{ob-/-}$ mice, and the $Nf1_{ob-/-}$ high bone mass phenotype can be significantly reduced by ATF4 haploinsufficiency (Elefteriou et al. 2006). Importantly, these molecular findings may also have therapeutic implications. Given that the skeletal defects of both the Rsk2-deficient and the $Nf1_{ob-/-}$ mice originated from abnormal ATF4-dependent amino acid import, they could be rescued by dietary manipulations (Elefteriou et al. 2006). More specifically, the increased bone formation observed in $Nf1_{ob-/-}$ mice could be normalized by a low-protein diet, while the defects of osteoblast differentiation and bone formation observed in the Rsk2-deficient mice were corrected by a high-protein diet (Elefteriou et al. 2006). Whether these findings will translate into treatments for Neurofibromatosis and Coffin-Lowry patients is still unknown. If they do, these results will be a major demonstration that understanding the molecular bases of osteoblast differentiation and function can be a powerful tool to design novel treatments for skeletal disorders.

SUMMARY AND CONCLUSIONS

In summary, if we look at the cascade of cell differentiation in the skeleton, it is clear that we now have a fairly detailed picture of its molecular bases thanks to the identification of cell-specific transcriptional regulators, such as Runx2, Osx, ATF4, and others (see Table 1). However, although the molecular networks between these factors are much better understood now than a few years earlier, a large research effort is still necessary to fully appreciate their functions. This includes the definition of specific target genes, their regulation by cofactors, and the identification of extracellular signals modulating their expression and function. From a more biomedical point of view, much is still to be gained in connecting this accumulation of knowledge and the diagnosis or treatment of skeletal disorders.

REFERENCES

Bialek, P., Kern, B., Yang, X., Schrock, M., Sosic, D., Hong, N., Wu, H., Yu, K., Ornitz, D.M., Olson, E.N., et al. 2004. A twist code determines the onset of osteoblast differentiation. *Dev. Cell* **6:** 423–435.

Boyden, L.M., Mao, J., Belsky, J., Mitzner, L., Farhi, A., Mitnick, M.A., Wu, D., Insogna, K., and Lifton, R.P. 2002. High bone density due to a mutation in LDL-receptor-related protein 5. *N. Engl. J. Med.* **346:** 1513–1521.

Dobreva, G., Chahrour, M., Dautzenberg, M., Chirivella, L., Kanzler, B., Farinas, I., Karsenty, G., and Grosschedl, R. 2006. SATB2 is a multifunctional determinant of craniofacial patterning and osteoblast differentiation. *Cell* **125:** 971–986.

Ducy, P. and Karsenty, G. 1995. Two distinct osteoblast-specific *cis*-acting elements control expression of a mouse osteocalcin gene. *Mol. Cell. Biol.* **15:** 1858–1869.

Ducy, P., Zhang, R., Geoffroy, V., Ridall, A.L., and Karsenty, G. 1997. Osf2/Cbfa1: A transcriptional activator of osteoblast differentiation. *Cell* **89:** 747–754.

Ducy, P., Starbuck, M., Priemel, M., Shen, J., Pinero, G., Geoffroy, V., Amling, M., and Karsenty, G. 1999. A Cbfa1-dependent genetic pathway controls bone formation beyond embryonic development. *Genes Dev.* **13:** 1025–1036.

Eferl, R., Hoebertz, A., Schilling, A.F., Rath, M., Karreth, F., Kenner, L., Amling, M., and Wagner, E.F. 2004. The Fos-related antigen Fra-1 is an activator of bone matrix formation. *EMBO J.* **23:** 2789–2799.

El Ghouzzi, V., Le Merrer, M., Perrin-Schmitt, F., Lajeunie, E., Benit, P., Renier, D., Bourgeois, P., Bolcato-Bellemin, A.-L., Munnich, A., and Bonaventure, J. 1997. Mutations of the *TWIST* gene in the Saethre-Chotzen syndrome. *Nat. Genet.* **15:** 42–46.

Elefteriou, F., Ahn, J.D., Takeda, S., Starbuck, M., Yang, X., Liu, X., Kondo, H., Richards,

W.G., Bannon, T.W., Noda, M., et al. 2005. Leptin regulation of bone resorption by the sympathetic nervous system and CART. *Nature* **434:** 514–520.

Elefteriou, F., Benson, M.D., Sowa, H., Starbuck, M., Liu, X., Ron, D., Parada, L.F., and Karsenty, G. 2006. ATF4 mediation of *NF1* functions in osteoblast reveals a nutritional basis for congenital skeletal dysplasiae. *Cell Metab.* **4:** 441–451.

FitzPatrick, D.R., Carr, I.M., McLaren, L., Leek, J.P., Wightman, P., Williamson, K., Gautier, P., McGill, N., Hayward, C., Firth, H., et al. 2003. Identification of *SATB2* as the cleft palate gene on 2q32-q33. *Hum. Mol. Genet.* **12:** 2491–2501.

Fu, L., Patel, M.S., Bradley, A., Wagner, E.F., and Karsenty, G. 2005. The molecular clock mediates leptin-regulated bone formation. *Cell* **122:** 803–815.

Glass II, D.A., Bialek, P., Ahn, J.D., Starbuck, M., Patel, M.S., Clevers, H., Taketo, M.M., Long, F., McMahon, A.P., Lang, R.A., et al. 2005. Canonical Wnt signaling in differentiated osteoblasts controls osteoclast differentiation. *Dev. Cell* **8:** 751–764.

Gong, Y., Slee, R.B., Fukai, N., Rawadi, G., Roman-Roman, S., Reginato, A.M., Wang, H., Cundy, T., Glorieux, F.H., Lev, D. et al. 2001. LDL receptor-related protein 5 (LRP5) affects bone accrual and eye development. *Cell* **107:** 513–523.

Grigoriadis, A.E., Schellander, K., Wang, Z.Q., and Wagner, E.F. 1993. Osteoblasts are target cells for transformation in c-*fos* transgenic mice. *J. Cell Biol.* **122:** 685–701.

Grigoriadis, A.E., Wang, Z.Q., Cecchini, M.G., Hofstetter, W., Felix, R., Fleisch, H.A., and Wagner, E.F. 1994. c-Fos: A key regulator of osteoclast-macrophage lineage determination and bone remodeling. *Science* **266:** 443–448.

Howard, T.D., Paznekas, W.A., Green, E.D., Chiang, L.C., Ma, L., Ortiz De Luna, R.I., Delgado, C.G., Gonzalez-Ramos, M., Kline, A.D., and Jabs, E.W. 1997. Mutations in *TWIST*, a basic helix-loop-helix transcription factor, in Sathre-Chotzen syndrome. *Nat. Genet.* **15:** 36–41.

Huelsken, J. and Birchmeier, W. 2001. New aspects of Wnt signaling pathways in higher vertebrates. *Curr. Opin. Genet. Dev.* **11:** 547–553.

Jochum, W., David, J.P., Elliott, C., Wutz, A., Plenk Jr., H., Matsuo, K., and Wagner, E.F. 2000. Increased bone formation and osteosclerosis in mice overexpressing the transcription factor Fra-1. *Nat. Med.* **6:** 980–984.

Jones, D.C., Wein, M.N., Oukka, M., Hofstaetter, J.G., Glimcher, M.J., and Glimcher, L.H. 2006. Regulation of adult bone mass by the zinc finger adapter protein Schnurri-3. *Science* **312:** 1223–1227.

Kanzler, B., Kuschert, S.J., Liu, Y.H., and Mallo, M. 1998. *Hoxa-2* restricts the chondrogenic domain and inhibits bone formation during development of the branchial area. *Development* **125:** 2587–2597.

Karin, M., Liu, Z., and Zandi, E. 1997. AP-1 function and regulation. *Curr. Opin. Cell Biol.* **9:** 240–246.

Kato, M., Patel, M.S., Levasseur, R., Lobov, I., Chang, B.H., Glass II, D.A., Hartmann, C., Li, L., Hwang, T.H., Brayton, C.F., et al. 2002. *Cbfa1*-independent decrease in osteoblast proliferation, osteopenia, and persistent embryonic eye vascularization in mice deficient in Lrp5, a Wnt coreceptor. *J. Cell Biol.* **157:** 303–314.

Kenner, L., Hoebertz, A., Beil, T., Keon, N., Karreth, F., Eferl, R., Scheuch, H., Szremska, A., Amling, M., Schorpp-Kistner, M., et al. 2004. Mice lacking JunB are osteopenic due to cell-autonomous osteoblast and osteoclast defects. *J. Cell Biol.* **164:** 613–623.

Kern, B., Shen, J., Starbuck, M., and Karsenty, G. 2001. Cbfa1 contributes to the osteoblast-specific expression of *type I collagen* genes. *J. Biol. Chem.* **276:** 7101–7107.

Kim, S., Koga, T., Isobe, M., Kern, B.E., Yokochi, T., Chin, Y.E., Karsenty, G., Taniguchi,

T., and Takayanagi, H. 2003. Stat1 functions as a cytoplasmic attenuator of Runx2 in the transcriptional program of osteoblast differentiation. *Genes Dev.* **17**: 1979–1991.

Klose, A., Ahmadian, M.R., Schuelke, M., Scheffzek, K., Hoffmeyer, S., Gewies, A., Schmitz, F., Kaufmann, D., Peters, H., Wittinghofer, A., et al. 1998. Selective disactivation of neurofibromin GAP activity in neurofibromatosis type 1. *Hum. Mol. Genet.* **7**: 1261–1268.

Koga, T., Matsui, Y., Asagiri, M., Kodama, T., de Crombrugghe, B., Nakashima, K., and Takayanagi, H. 2005. NFAT and Osterix cooperatively regulate bone formation. *Nat. Med.* **11**: 880–885.

Komori, T., Yagi, H., Nomura, S., Yamaguchi, A., Sasaki, K., Deguchi, K., Shimizu, Y., Bronson, R.T., Gao, Y.H., Inada, M., et al. 1997. Targeted disruption of *Cbfa1* results in a complete lack of bone formation owing to maturational arrest of osteoblasts. *Cell* **89**: 755–764.

Lee, B., Thirunavukkarasu, K., Zhou, L., Pastore, L., Baldini, A., Hecht, J., Geoffroy, V., Ducy, P., and Karsenty, G. 1997. Missense mutations abolishing DNA binding of the osteoblast-specific transcription factor *OSF2/CBFA1* in cleidocranial dysplasia. *Nat. Genet.* **16**: 307–310.

Lee, N.K., Sowa, H., Hinoi, E., Ferron, M., Ahn, J.D., Confavreux, C., Dacquin, R., Mee, P.J., McKee, M.D., Jung, D.Y., et al. 2007. Endocrine regulation of energy metabolism by the skeleton. *Cell* **130**: 456–469.

Little, R.D., Carulli, J.P., Del Mastro, R.G., Dupuis, J., Osborne, M., Folz, C., Manning, S.P., Swain, P.M., Zhao, S.C., Eustace, B., et al. 2002. A mutation in the LDL receptor-related protein 5 gene results in the autosomal dominant high-bone-mass trait. *Am. J. Hum. Genet.* **70**: 11–19.

Liu, J., Farmer Jr., J.D., Lane, W.S., Friedman, J., Weissman, I., and Schreiber, S.L. 1991. Calcineurin is a common target of cyclophilin-cyclosporin A and FKBP-FK506 complexes. *Cell* **66**: 807–815.

Mao, J., Wang, J.Y., Bo, L., Pan, W., Farr III, G.H., Flynn, C., Yuan, H., Takada, S., Kimelman, D., Lin, L., et al. 2001. Low-density lipoprotein receptor-related protein-5 binds to axin and regulates the canonical Wnt signaling pathway. *Mol. Cell* **7**: 801–809.

Mundlos, S., Otto, F., Mundlos, C., Mulliken, J.B., Aylsworth, A.S., Albright, S., Lindhout, D., Cole, W.G., Henn, W., Knoll, J.H., et al. 1997. Mutations involving the transcription factor CBFA1 cause cleidocranial dysplasia. *Cell* **89**: 773–779.

Nakashima, K., Zhou, X., Kunkel, G., Zhang, Z., Deng, J.M., Behringer, R.R., and de Crombrugghe, B. 2002. The novel zinc finger-containing transcription factor Osterix is required for osteoblast differentiation and bone formation. *Cell* **108**: 17–29.

Nishio, Y., Dong, Y., Paris, M., O'Keefe, R.J., Schwarz, E.M., and Drissi, H. 2006. Runx2-mediated regulation of the zinc finger Osterix/Sp7 gene. *Gene* **372**: 62–70.

Otto, F., Thornell, A.P., Crompton, T., Denzel, A., Gilmour, K.C., Rosewell, I.R., Stamp, G.W., Beddington, R.S., Mundlos, S., Olsen, B.R., et al. 1997. *Cbfa1*, a candidate gene for cleidocranial dysplasia syndrome, is essential for osteoblast differentiation and bone development. *Cell* **89**: 765–771.

Rodino, M.A. and Shane, E. 1998. Osteoporosis after organ transplantation. *Am. J. Med.* **104**: 459–469.

Sabatakos, G., Sims, N.A., Chen, J., Aoki, K., Kelz, M.B., Amling, M., Bouali, Y., Mukhopadhyay, K., Ford, K., Nestler, E.J., et al. 2000. Overexpression of ΔFosB transcription factor(s) increases bone formation and inhibits adipogenesis. *Nat. Med.* **6**: 985–990.

Satokata, I., Ma, L., Ohshima, H., Bei, M., Woo, I., Nishizawa, K., Maeda, T., Takano, Y., Uchiyama, M., Heaney, S., et al. 2000. *Msx2* deficiency in mice causes pleiotropic defects in bone growth and ectodermal organ formation. *Nat. Genet.* **24:** 391–395.

Teitelbaum, S.L. and Ross, F.P. 2003. Genetic regulation of osteoclast development and function. *Nat. Rev. Genet.* **4:** 638–649.

Tribioli, C. and Lufkin, T. 1999. The murine *Bapx1* homeobox gene plays a critical role in embryonic development of the axial skeleton and spleen. *Development* **126:** 5699–5711.

Trivier, E., De Cesare, D., Jacquot, S., Pannetier, S., Zackai, E., Young, I., Mandel, J.L., Sassone-Corsi, P., and Hanauer, A. 1996. Mutations in the kinase Rsk-2 associated with Coffin-Lowry syndrome. *Nature* **384:** 567–570.

Wagner, E.F. and Eferl, R. 2005. Fos/AP-1 proteins in bone and the immune system. *Immunol. Rev.* **208:** 126–140.

Wang, X., Kua, H.Y., Hu, Y., Guo, K., Zeng, Q., Wu, Q., Ng, H.H., Karsenty, G., de Crombrugghe, B., Yeh, J., et al. 2006. p53 functions as a negative regulator of osteoblastogenesis, osteoblast-dependent osteoclastogenesis, and bone remodeling. *J. Cell Biol.* **172:** 115–125.

Wilkie, A.O.M. 1997. Craniosynostosis: Genes and mechanisms. *Hum. Mol. Genet.* **6:** 1647–1656.

Wu, W., Glinka, A., Delius, H., and Niehrs, C. 2000. Mutual antagonism between *dickkopf1* and *dickkopf2* regulates Wnt/β-catenin signalling. *Curr. Biol.* **10:** 1611–1614.

Xiao, G., Jiang, D., Ge, C., Zhao, Z., Lai, Y., Boules, H., Phimphilai, M., Yang, X., Karsenty, G., and Franceschi, R.T. 2005. Cooperative interactions between activating transcription factor 4 and Runx2/Cbfa1 stimulate osteoblast-specific osteocalcin gene expression. *J. Biol. Chem.* **280:** 30689–30696.

Yang, X. and Karsenty, G. 2004. ATF4, the osteoblast accumulation of which is determined post-translationally, can induce osteoblast-specific gene expression in non-osteoblastic cells. *J. Biol. Chem.* **279:** 47109–47114.

Yang, X., Matsuda, K., Bialek, P., Jacquot, S., Masuoka, H.C., Schinke, T., Li, L., Brancorsini, S., Sassone-Corsi, P., Townes, T.M., et al. 2004. ATF4 is a substrate of RSK2 and an essential regulator of osteoblast biology; implication for Coffin-Lowry syndrome. *Cell* **117:** 387–398.

8

Role of Growth Factors in Bone Development and Differentiation

Sayumi Fujimori, Daniela Kostanova-Poliakova, and Christine Hartmann
Research Institute of Molecular Pathology
Dr. Bohrgasse 7
1030 Vienna, Austria

BONE IS A FORM OF HIGHLY SPECIALIZED mineralized connective tissue that provides strength to the skeletal system of higher vertebrates, while still retaining a certain degree of elasticity. The bone matrix is produced by osteoblasts, a cell-type that develops locally from mesenchymal precursors, and is resorbed by the osteoclast, a cell-type of hematopoietic origin. A few elements, such as the flat bones of the skull and part of the clavicle, are formed by the process of intramembranous ossification, whereby osteoblasts differentiate directly from cells within mesenchymal condensations. In contrast, the majority of skeletal elements are formed by endochondral ossification involving the remodeling of initial cartilaginous templates into bony tissue. The latter process requires controlled maturation of chondrocytes from proliferating and prehypertrophic to hypertrophic chondrocytes, as well as signaling from the prehypertrophic cells to the surrounding cells in the perichondrium, resulting in a regional induction of osteoblast differentiation. Osteoblasts start to differentiate in the periosteum, a region flanking prehypertrophic and hypertrophic chondrocytes. The typical appearance of one end of a juvenile long bone still containing a cartilaginous growth plate is shown in Figure 1. Recent lineage studies suggest that osteoblasts and chondrocytes share a common precursor in the limb. Thus, especially in the limb, the activation and/or inhibition of distinct signaling pathways is necessary in order to coordinate the differentiation of neighboring cells into distinct cell lin-

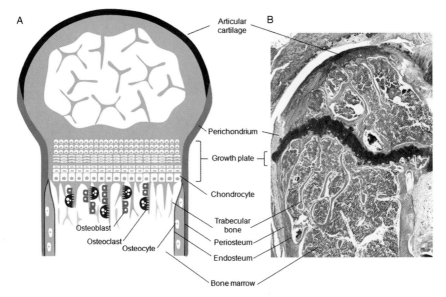

Figure 1. (*A*) Schematic drawing of the distal part of a long bone, indicating the different regions of a mature skeletal element. (*B*) Corresponding histological section through the distal region of the femur of a 12-month-old mouse.

eages and to synchronize their maturation. This chapter focuses on genetic and molecular studies elucidating the role of different locally produced growth factors during embryonic and postnatal bone development, with regard to osteoblast cell-lineage decision and osteoblast proliferation, differentiation, and maturation. For a comprehensive overview of the various mouse models, see Table 1.

GROWTH FACTORS AND THEIR ROLE IN OSTEOBLAST CELL-LINEAGE DECISION DURING DEVELOPMENT

As demonstrated by lineage studies, osteoblasts and chondrocytes in the limb share a common progenitor (Akiyama et al. 2005). Interestingly, recent studies have shown that this is true, in part, for dermal bone as well, since a subset of osteoblasts in dermal bone is derived from cells, which at one time expressed the chondrogenic marker Collagen 2α1 (Abzhanov et al. 2007). Furthermore, various in vivo and in vitro studies revealed that cells from the periosteum have the potential to differentiate into chondrocytes (Fang and Hall 1997; Yoo and Johnstone 1998). Thus, a pathway must exist that controls the lineage switch between pre-

Table 1. Mouse models used to study the role of growth factors in osteogenesis

Gene	Mutation	Phenotype	Reference
WNT signaling			
Wnt5a	Het KO	Osteopenia; decreased trabecular bone mass	Takada et al. (2007a)
Wnt7b	CKO (Dermo1-Cre)	Decrease in ob numbers	Tu et al. (2007)
Wnt10b	Hom KO	Decreased trabecular bone mass; increase in osteocytes	Bennett et al. (2005, 2007)
	Tg (OC-Wnt10b)	Increased bone formation, trabecular bone mass and mineralization; increase in ob numbers	Bennett et al. (2007)
	Tg (FABP4-Wnt10b)	Increased bone mass	Bennett et al. (2005)
Lrp5	Hom KO	Decreased bone mass; reduction of ob proliferation and function	Kato et al. (2002)
Ctnnb1	CKO (Dermo-Cre)	Lack of functional ob	Day et al. (2005); Hu et al. (2005)
	CKO (Prx1-Cre)	Lack of functional ob	Hill et al. (2005)
	CKO (Osx-Cre)	Lack of functional ob	Rodda and McMahon (2006)
	CKO (Col1-Cre)	Osteopenia; increased oc numbers	Glass et al. (2005)
	CKO exon 3 (Ctnnb1 gof allele)	Increased bone mass; decreased oc numbers	Glass et al. (2005)
APC	CKO (OC-Cre)	Increased bone mass; decreased oc numbers	Holmen et al. (2005)
Axin2	Hom KO	Increased ob precursors and ob differentiation	Yu et al. (2005)
Tcf1	Hom KO	Decreased bone mass; no change in ob number or function	Glass et al. (2005)
Sost	Hom KO	Increased bone formation	Li et al. (2008)
	Tg (OC-hSost)	Osteopenia	Winkler et al. (2005)
Sfrp1	Hom KO	Increased bone mass	Bodine et al. (2004); Gaur et al. (2005)
Sfrp3	Hom KO	Increased cortical bone mass	Lories et al. (2007)
Sfrp4	Tg (Col1-Sfrp4)	Decreased bone formation and ob proliferation; osteopenia	Nakanishi et al. (2008)
Dkk1	Het KO	Increased bone mass	Morvan et al. (2006)
	Tg (Col1-Dkk1)	Decreased bone formation and ob proliferation; osteopenia	Li et al. (2006)

(*Continued on following pages.*)

Table 1. (*Continued*)

Gene	Mutation	Phenotype	Reference
Dkk2	Hom KO	Osteopenia; increase in oc numbers; defect in terminal differentiation of ob	Li et al. (2005a)
IHH signaling			
Ihh	Hom KO	Failure of ob development in endochondral bones	St-Jacques et al. (1999)
	Hom KO	Decreased size of dermal flat bone; accelerated ob maturation	Abzhanov et al. (2007)
	Tg (UAS-Ihh; Col2a1-Gal4)	Promotes bone collar expansion towards epiphysis	Long et al. (2004)
Smo	CKO (Col2-Cre)	Failure of ob development in endochondral bones	Long et al. (2004)
Gli2	Hom KO	Reduced ob numbers lining trabecular bone surface	Miao et al. (2004c)
BMP signaling			
Bmp2; Bmp4	CKO (Prx1-Cre)	Defect in ob maturation	Bandyopadhyay et al. (2006)
Bmp3	Hom KO	Increased bone density	Daluiski et al. (2001)
Bmp4	Tg (Col1-Bmp4)	Bone loss; increased oc number	Okamoto et al. (2006)
Bmpr1a	CKO (OC-Cre)	Postnatal defect in ob function	Mishina et al. (2004)
Bmpr1b	Tg (Col1-DNBmpr1b)	Postnatal impairment of bone formation; ob numbers not altered	Zhao et al. (2002)
Smad1	CKO (Col1-Cre)	Postnatal reduced trabecular volume and ob numbers	Chen et al. (2003)
Nog	Tg (Col1-Nog)	Increased bone volume and ob numbers; defect in ob function; decreased oc numbers	Okamoto et al. (2006)
	Tg (hOC-Nog)	Decreased trabecular bone volume and bone formation rate; defect in ob function	Devlin et al. (2003); Wu et al. (2003)
Grem	Tg (OC-hGrem)	Decreased ob proliferation; osteopenia	Gazzerro et al. (2005a)
	CKO (OC-Cre)	Increased ob activity and bone formation	Gazzerro et al. (2007)
Smurf1	Tg (Col1-Smurf1)	Decreased trabecular bone mass; decreased ob differentiation and proliferation	Zhao et al. (2004)

		Phenotype	Reference
	Hom KO	Age-dependent increase of bone mass and ob activity	Yamashita et al. (2005)
Tob1	Hom KO	Increased bone mass and ob numbers	Yoshida et al. (2000)
TGF-β signaling			
Tgfb1	Hom KO	Postnatal decrease in bone mass and mineralization	Geiser et al. (1998); Atti et al. (2002)
Tgfb2	Tg (OC-Tgfb2)	Bone loss; increase in oc activity	Erlebacher and Derynck (1996); Erlebacher et al. (1998)
Tgfbr2	Tg (OC-DNTgfbr2)	Age-dependent increase in trabecular bone mass; imbalance between bone formation and resorption during bone remodeling	Filvaroff et al. (1999)
Smad3	Hom KO	Decreased bone formation rate; increased differentiation and apoptosis of ob; osteopenia	Borton et al. (2001)
Ltbp3	Hom KO	Osteopetrosis; defect in oc function and decreased bone turnover	Dabovic et al. (2002a, b, 2005)
FGF signaling			
Fgf2	Hom KO	Decreased trabecular bone mass, mineralization and bone formation	Montero et al. (2000)
	Tg (PGK-hFgf2)	Postnatal decrease in mineralization; osteopenia	Sobue et al. (2005)
Fgf9	Hom KO	Decreased number of matured ob	Hung et al. (2007)
Fgf18	Hom KO	Decrease in proliferation of mesenchymal osteogenic precursor cells; delayed endochondral ossification and suture closure	Liu et al. (2002); Ohbayashi et al. (2002)
Fgfr1	CKO (Col1-Cre)	Increased trabecular bone mass and mineralization	Jacob et al. (2006)
	KI (Fgfr1^{P250R})	Craniosynostosis	Zhou et al. (2000)
Fgfr2c	KI (Fgfr2c$^{C342Y/+}$)	Craniosynostosis	Eswarakumar et al. (2004)
Fgfr2IIIc	Hom KO	Delay in ossification	Eswarakumar et al. (2002)
Fgfr2	CKO (Dermo1-Cre)	Postnatal dwarfism; decreased bone formation and proliferation of ob progenitors	Yu et al. (2003)

(*Continued on following pages.*)

Table 1. (*Continued*)

Gene	Mutation	Phenotype	Reference
Fgfr3	Hom KO	Overgrowth of skeleton prior to birth	Colvin et al. (1996); Deng et al. (1996)
	Hom KO	Postnatal decrease in trabecular bone mineralization; osteopenia, despite increase in ob numbers	Valverde-Franco et al. (2004)
	KI (Fgfr3^{G369C})	Enhanced expression of osteogenic markers; decreased bone density, due to increased oc activity	Chen et al. (1999)
Stat1	Hom KO	Postnatal increase in bone mineral density	Xiao et al. (2004)
PTH/PTHrP signaling			
Pth	Hom KO	Increase of cortical bone thickness	Miao et al. (2002, 2004b)
Pthrp	Hom KO	Increase in bone mineralization and ob numbers	Karaplis et al. (1994); Miao et al. (2002)
	Het KO	Decreased trabecular bone volume and bone formation rate; cortical bone is not affected	Amizuka et al. (1996); Miao et al. (2005)
	CKO (Col1-Cre)	Decreased trabecular bone volume and bone formation rate	Miao et al. (2005)
PPR	Hom KO	Decrease in trabecular bone mass; increase in cortical bone mass and ob numbers	Lanske et al. (1999)
	Tg(Col1-DAhPPR^{H223R})	Increase in trabecular bone mass and ob proliferation; decreased activity of periosteal ob	Calvi et al. (2001)
IGF signaling			
Igf1	Hom KO	Postnatal growth retardation	Baker et al. (1993); Liu et al. (1993)
	CKO (Alb-Cre)	Postnatal growth retardation	Yakar et al. (2002)
	Tg (Col1-Igf1)	Increased bone formation rate	Jiang et al. (2006)
	Tg (OC-Igf1)	Increased bone formation rate	Zhao et al. (2000)
Igf2	Hom KO	Embryonic growth retardation	Baker et al. (1993)
Igf1r	Hom KO	Embryonic growth retardation	Liu et al. (1993); Baker et al. (1993)

Gene	Model	Phenotype	Reference
Igfbp3	CKO (OC-Cre)	Decrease in ob function and bone formation	Zhang et al. (2002)
	Tg (PGK-Igfbp3)	Decreased bone formation and ob proliferation	Silha et al. (2003)
Igfbp4	Tg (OC-Igfbp4)	Decreased cancellous bone formation and ob numbers	Zhang et al. (2003)
Igfbp5	Tg (OC-Igfbp5)	Transient decrease in trabecular bone volume and in bone mineralization; decreased ob function	Devlin et al. (2002); Atti et al. (2005)
Igfbp5	Tg (ACTB-Igfbp5/+)	Decreased bone formation rate and mineralization	Salih et al. (2005)
EGF signaling			
Egf	Tg (ACTB-hEgfp)	Decreased cortical bone thickness	Chan and Wong (2000)
Btc	Tg (ACTB-Btc)	Growth retardation	Schneider et al. (2005)
Areg	Hom KO	Decreased trabecular bone mass	Qin et al. (2005)
Egfr	Hom KO	Accelerated osteoblastogenesis	Sibilia et al. (2003)
	KI (hEgfr)	Accelerated osteoblastogenesis	Sibilia et al. (2003)

(ACTB) actin beta, (Alb) Albumin, (CKO) conditional knock-out, (DA) dominant active, (DN) dominant negative, (gof) gain-of-function, (Het) heterozygous, (Hom) homozygous, (KI) knock-in, (KO) knock-out, (ob) osteoblast, (oc) osteoclast, (OC) osteocalcin, (PGK) phosphoglycerate kinase, (Tg) transgene.

cursors of osteoblasts and chondrocytes. In recent years, it has been demonstrated by many groups that the canonical Wnt pathway is a critical component of this switch, since mice lacking β-catenin (*Ctnnb1*), the key intracellular component of the pathway, are deficient of functional osteoblasts, and their precursors instead develop into chondrocytes (Day et al. 2005; Hill et al. 2005; Hu et al. 2005; Rodda and McMahon 2006). The relevant Wnt ligands for this process have not yet been identified. Possible candidates based on their expression are Wnt5a, Wnt7b, and Wnt9a (Hu et al. 2005), but their knockout phenotypes revealed no defects in early osteoblast lineage differentiation (Yamaguchi et al. 1999; Rodda and McMahon 2006; Spater et al. 2006; Tu et al. 2007). Another possible candidate is Wnt10b, which can shift the balance between adipocytes and osteoblasts derived from mesenchymal progenitors in vitro and in vivo (Bennett et al. 2005, 2007). However, the in vivo effects are only occurring postnatal.

Genetic epistasis studies and in vitro experiments revealed that canonical Wnt/β-catenin signaling is required downstream of the Indian hedgehog (Ihh) pathway for osteoblast development during endochondral ossification (Hill et al. 2005; Hu et al. 2005; Mak et al. 2006; Rodda and McMahon 2006). The Ihh pathway was considered to be absolutely essential for the initiation of osteoblast differentiation in the perichondrium surrounding the long bones, since no *Runx2*-positive osteoblast precursors are detectable in mutants for *Ihh*, or its signaling receptor *Smoothened* (*Smo*) (St-Jacques et al. 1999; Long et al. 2004). The effect of Ihh on osteoblasts is probably mediated via the downstream transcription factor Gli2, but not Gli3 (Miao et al. 2004c; Hilton et al. 2005; Shimoyama et al. 2007). Recent ex vivo data now suggest that Ihh signaling might not be absolutely necessary for osteoblast differentiation, since *Ihh* mutant skeletal elements eventually ossify when explanted under the kidney capsule (Colnot et al. 2005). Similar observations have been made in *Ihh/Gli3* double mutants (Hilton et al. 2005). In this context, it is interesting to note that Ihh is also expressed in osteoblasts, and it has been suggested that autocrine Ihh signaling might play a role in osteoblast recruitment and differentiation (Murakami et al. 1997; Jemtland et al. 2003). This idea is controversial to the recent proposal that Ihh acts as a negative regulator of osteoblast maturation (Abzhanov et al. 2007), which is based on the size reduction of dermal flat bones due to accelerated osteoblast maturation in $Ihh^{-/-}$ skulls (St-Jacques et al. 1999; Abzhanov et al. 2007). What remains to be shown is where Ihh is expressed and active in vivo during dermal bone formation.

Bone morphogenetic proteins (BMPs), a subfamily of growth factors amongst the transforming growth factor β (TGF-β) superfamily have been the primary suspects for in vivo bone inducers, due to their astonishing ability to induce ectopic bone (Wozney et al. 1988; Reddi 1997). BMPs signal through a type I/II receptor complex, leading to the activation of the intracellular effector molecules Smad1, -5, -8, or MAP kinases (Nohe et al. 2004). In various in vitro experiments, using cell lines as well as primary cells, it has been shown that BMPs can stimulate osteoblastogenesis and induce *Runx2* and *Osterix* expression (Lee et al. 2000, 2003a,b; Ulsamer et al. 2008). Differential roles for BMP receptors type IA and IB in lineage specification are proposed based on experiments using constitutively active and truncated versions of the receptors in different cell lines, but the conclusions reached are contradictory (Chen et al. 1998; Kaps et al. 2004). *Bmpr1a* and *Bmpr1b* single knockout phenotypes support the idea for differential roles within cartilage, but not with regard to osteoblastogenesis (Baur et al. 2000; Yi et al. 2000; Ovchinnikov et al. 2006). The importance of BMP signaling in early cartilage development was confirmed by the double knockout of the two receptors, but the severity of the cartilage defects makes it impossible to address their role in osteoblastogenesis (Yoon et al. 2005). Osteoblast lineage differentiation is also not affected upon conditional deletion of the two BMP ligands, *Bmp2* and *-4*, but those mice are defective in osteoblast maturation (Bandyopadhyay et al. 2006). Thus, to date, there are no compelling in vivo data supporting a role for BMPs or TGF-βs during embryonic osteoblast lineage formation.

OSTEOBLASTS VERSUS ADIPOCYTES: POSTNATAL MESENCHYMAL STEM-CELL LINEAGE DECISION

Wnt10b signaling through the β-catenin pathway stimulates osteoblastogenesis by inhibiting adipocyte lineage differentiation in bone marrow mesenchymal-derived stem cells (MSCs) and preadipocytes (Ross et al. 2000; Bennett et al. 2005; Kang et al. 2007). Similar in vitro activities were shown for Wnt3a (Kawai et al. 2007). However, no increased adipocyte development has been reported in the various *Ctnnb1* knockout models published. For Wnt5a, which can also inhibit adipogenesis in vitro, it has been shown that the effect is mediated by a noncanonical Wnt pathway (Takada et al. 2007a, b). In addition, *Wnt5a* heterozygous mice exhibit an increase in bone marrow adipocytes coincident with osteopenia, meaning they show bone loss (Takada et al. 2007a). Besides Wnt10b and Wnt5a, a member of the fibroblast growth factor (Fgf) family, Fgf1, was identified as a promising factor suppressing adipocyte differentiation of

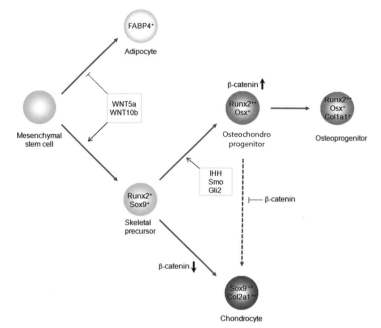

Figure 2. Involvement of growth factor signaling in lineage decisions based on in vivo data: adipocytes versus osteoblasts, and chondrocytes versus osteoblasts. Note: In vivo evidence suggests that in the limb mesenchyme osteoblasts and chondrocytes share a common hypothetical "skeletal precursor." However, it is not clear whether ostoblasts from bone marrow-derived mesenchymal stem cells also differentiate via such a skeletal precursor. Arrows indicate positive effects of growth factors, while blunt-ended lines indicate negative effects. Changes in β-catenin levels are indicated by bold arrows.

human bone-marrow-derived MSCs in a microarray screen (Schilling et al. 2008). However, there are no in vivo data for Fgf1, showing that this activity is of biological relevance. BMPR1A and BMPR1B have also been implicated in regulating the balance between osteoblasts and adipocytes based on in vitro culture experiments using calvarial-derived MSCs (Chen et al. 1998). For a summary of growth factor activities in lineage specification, see Figure 2.

ROLE OF GROWTH FACTORS IN OSTEOBLAST PROLIFERATION AND MATURATION

After osteoblast lineage induction, the precursor population proliferates to expand and the precursors become committed to the lineage and

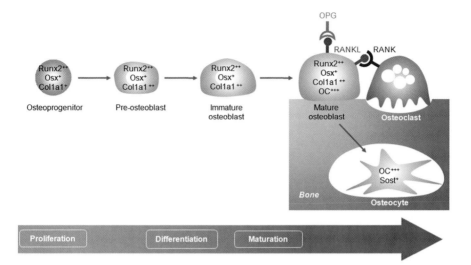

Figure 3. Schematic representation of osteoblast differentiation steps subsequent to entering the lineage. Runx2, Osterix (Osx), Collagen 1a1 (Col1a1), Osteocalcin (OC), and Sclerostin (Sost) are markers for the different differentiation stages of osteoblasts. Osteoprogerin (OPG), secreted by mature osteoblasts, acts as a decoy receptor for RANKL, which is also produced by osteoblasts and prevents the interaction of RANKL with its receptor RANK. The RANK receptor is expressed on osteoclasts and their precursors, thereby inhibiting osteoclastogenesis. (Modified from Aubin 2001.)

undergo maturation (Fig. 3). Proliferation and maturation of osteoblast precursors is regulated by a number of different growth factors, which will be discussed in the following sections.

Wnts

Genetic evidence in humans and mice reveals that Wnt signaling also regulates osteoblast proliferation and maturation. In humans, loss- and gain-of-function mutations in the *Lrp5* Wnt/coreceptor cause osteoporosis-pseudoglioma syndrome and high bone density syndrome, respectively (Gong et al. 2001; Boyden et al. 2002; Little et al. 2002). In mice, loss of Lrp5 results in reduced bone mass due to a reduction in osteoblast proliferation and function (Kato et al. 2002). Bone mass is further reduced upon loss of one allele of *Lrp6* in the $Lrp5^{-/-}$ mutant mice (Holmen et al. 2004). Two of the canonical transcriptional mediators, Tcf1 and Tcf4, are expressed in osteoblasts, and mice lacking Tcf1 have decreased bone mass, without any changes in osteoblast number or function, due to an increase in osteoclasts (Glass et al. 2005). Loss of Wnt10b also results in reduced bone mass, probably due to a defect in osteoblast proliferation or function, while Wnt10b gain-of-function mice, using the *Oc*-promoter or the adipocyte-specific *FABP4*-promoter, show an increase in bone mass due to increased osteoblast numbers (Bennett et al. 2005, 2007). In the *FABP4-Wnt10b* tg mice, this increase is at the expense of fat cells (Bennett et al. 2005). There is also evidence from mouse models in which negative regulators of the Wnt pathway are altered. Mice lacking the Wnt-antagonist sclerostin (Sost) show increased bone formation (Li et al. 2008). In contrast, Sost over-expressing mice are osteopenic, supporting the idea that Sost is a negative regulator of bone formation (Winkler et al. 2003). Sost is expressed by osteocytes (fully matured osteoblasts acting as mechanosensors located within the bone lacunae) and antagonizes Wnt signaling by binding to the Wnt/coreceptors Lrp5/6 (van Bezooijen et al. 2004, 2007; Li et al. 2005b; Semenov et al. 2005; Ellies et al. 2006). There is additional evidence that Wnt signaling has to be tightly controlled, since mice lacking either of the following secreted Wnt antagonists, secreted Frizzled-related protein 1 (Sfrp1), Sfrp3 (Frzb1), or one allele of *dickkopf1* (*Dkk1*), show an increased bone mass phenotype (Bodine et al. 2004; Gaur et al. 2005; Morvan et al. 2006; Lories et al. 2007). In contrast, targeted overexpression of Dkk1 or Sfrp4, Wnt antagonists expressed in osteoblasts, under the *Collagen 1* (*Col1*)-promoter, suppresses osteoblast proliferation, resulting in decreased bone formation and osteopenia (Li et al. 2006; Nakanishi et al. 2008). In light of these findings, it came as a surprise that mice lacking the Wnt-antagonist Dkk2 are also osteopenic. The bone phenotype of *Dkk2*$^{-/-}$ mice is rather complex, affecting terminal differentiation of osteoblasts and matrix mineralization, and leading to an increase in osteoclasts (Li et al. 2005a). Defects in terminal differentiation are potentially associated with elevated levels of β-catenin signaling, since forced activation of β-catenin signaling in osteoblast precursors interferes with their final maturation, but at the same time it enhances formation of Osx-positive precursors (Rodda and McMahon 2006). In addition to endochondral ossification, dermal ossification is regulated by Wnts, as mutants for *Axin2* (*conductin, Axil*), a component of the β-catenin destruction complex and hence a negative regulator of canonical Wnt signaling, and show premature fusion of cranial sutures (craniosynostosis). This is due to enhanced expansion of osteoblast precursors and osteoblast differentiation associated with an increase in Wnt/β-catenin signaling (Yu et al. 2005). Thus, this data clearly suggests that the levels of canonical Wnt/β-catenin signaling have to be tightly controlled during osteoblast proliferation and maturation. There is recent evidence that noncanonical Wnt signaling regulates osteogenesis as well, since conditional *Dermo1-Cre*-mediated deletion of *Wnt7b*, which signals through Protein kinase C delta (PKCδ), results in a decrease in osteoblasts before birth (Tu et al. 2007).

BMPs

As mentioned earlier, BMPs have the amazing capability to induce ectopic bone by recapitulating the entire sequence of events occurring during endogenous endochondral ossification. More direct evidence for an in vivo role was provided by the analysis of conditional double knockout *Bmp2/Bmp4* mice, which showed impaired osteoblast maturation (Bandyopadhyay et al. 2006). Furthermore, exogenous application of recombinant BMP2 can stimulate periosteal bone formation of mouse calvariae without prior cartilage formation (Chen et al. 1997). Surprisingly, osteoblast-specific overexpression of BMP4 under the *Col1*-promoter revealed a severe bone loss during embryonic development and an increase in osteoclasts. However, it is not clear whether this is primarily due to a defect in osteoblast proliferation, maturation, or a BMP4 stimulatory effect on osteoclastogenesis (Okamoto et al. 2006). BMPs are also required postnatally to regulate mature osteoblast function, as demonstrated by *osteocalcin (Oc, Og2)-Cre*-mediated conditional deletion of *Bmpr1a* in osteoblasts, which affects bone matrix mineralization but not osteoblast number (Mishina et al. 2004). Mice in which a dominant-negative form of BMPR1B (dnBMPR1B) is expressed under the *Col1*-promoter show a similar phenotype: a reduced bone mineral density and bone volume and bone formation rate. However, osteoblast numbers are not altered in these mice (Zhao et al. 2002). The in vivo findings are in contrast to the in vitro results, where overexpression of dnBMPR1B blocked osteoblast precursor differentiation (Chen et al. 1998). In light of these findings, it was surprising that *Col1*-driven overexpression of noggin (nog), a BMP antagonist, resulted in a postnatal increase in bone volume associated with a reduction in osteoclasts and an increase in osteoblasts, which were less functional (Okamoto et al. 2006). In contrast, overexpression of nog in more mature osteoblasts, under the *osteocalcin* promoter, resulted in a decrease in trabecular bone volume and bone formation rate, due to impaired osteoblast function without affecting osteoclasts (Devlin et al. 2003; Wu et al. 2003). *Nog* loss-of-function mutants have complex skeletal phenotypes. In some elements, endochondral ossification is accelerated, while it is delayed in others. Whether this is a direct or indirect cartilage-mediated effect remains to be investigated (Tylzanowski et al. 2006). Due to impaired osteoblast proliferation and maturation, overexpression of gremlin (grem), another BMP antagonist in mature osteoblasts (Oc-promoter), results in osteopenia (Gazzerro et al. 2005b). The effect on osteoblast maturation can be recapitulated upon conditional deletion of grem using Oc-Cre, which results in enhanced bone formation due to more active osteoblasts; however, osteoblast numbers are not altered (Gazzerro et al. 2007). Moreover, overexpression of twisted gastrulation (tsg), another potential inhibitor of BMP action in osteoblasts, inhibits osteoblast maturation in vitro (Gazzerro et al. 2005a). There is further indirect evidence for a potential role of BMP signaling in osteogenesis. Mice, which express the ubiquitin-ligase 3, Smurf1 that is involved in inhibiting BMP signaling (Zhu et al. 1999) under the Col1-promoter, show a decrease in trabecular bone volume and bone formation rates due to reduced osteoblast proliferation and differentiation (Zhao et al. 2004). On the contrary, mice deficient in Smurf1 show an age-dependent increase in bone mass due to an increase in osteoblast activity, which are more sensitized to BMPs (Yamashita et al. 2005). There is one potential caveat: Smurf1 is also involved in degradation of the transcription factor Runx2, which plays an essential role in osteoblastogenesis (Zhao et al. 2003). Mice carrying a targeted deletion of the Tob gene, another negative regulator of BMP signaling, also have increased bone mass, but this is due to increased osteoblast numbers. The study further shows that Tob, which encodes an antiproliferative protein expressed in osteoblasts and hypertrophic chondrocytes, acts as a negative regulator of BMP/Smad signaling through association with Smad1 and -5. Thus, endogenous BMP signaling is probably augmented in $Tob^{-/-}$ mice, resulting in increased osteoblast proliferation and accelerated bone formation (Yoshida et al. 2000). There is additional evidence that Smad1 is required for the effects of BMPs on postnatal bone, as osteoblast-specific deletion of Smad1 using Col1-Cre results in reduced trabecular bone volume, osteoblast numbers, and bone formation rates (Chen et al. 2003). Surprisingly, mice deficient in Bmp3, a nonosteogenic BMP that is expressed in osteoblasts and its precursors, show an increase in bone density (Daluiski et al. 2001). Despite the fact that in vitro experiments suggest that BMP3 antagonizes the osteogenic effects of BMP2, the cellular and mechanistic basis for the $BMP3^{-/-}$ phenotype is still unclarified (Daluiski et al. 2001). All of the above data suggest that BMP signaling plays complex roles in osteogenesis, affecting primarily maturation of osteoblasts, but to a certain extent, it is likely it also affects proliferation of precursors.

TGF-βs

There is also evidence that TGF-β signaling (Nohe et al. 2004) regulates osteoblast proliferation and maturation. All three isoforms, TGF-β1, -2, and -3 are expressed in osteoblasts (Pelton et al. 1991a,b). Knockout mice for *Tgf-β* show a postnatal reduction in bone mass and a decrease in mineral content (Geiser et al. 1998). Subsequent studies revealed that primarily the secondary ossification center and cortical bone are affected in these mice (Atti et al. 2002). Mice deficient for *Smad3*, an intracellular TGF-β signaling component, have a reduced bone formation rate and develop osteopenia due to accelerated differentiation of osteoblasts into terminal differentiated osteocytes and increased apoptosis of osteoblasts (Borton et al. 2001). Mice mutant for *Ltbp3*, a member of the latent TGF-β binding protein family, which modulates the availability of TGF-β1, display an osteopetrosis-like phenotype, which is due to defects in osteoclast function, resulting in reduced bone turnover (Dabovic et al. 2002a,b, 2005). Similarly, inhibition of TGF-β-receptor signaling in *Oc*-expressing osteoblasts results in an increase of trabecular bone without altering osteoblast numbers or their activity, but affects osteoclast maturation and activity (Filvaroff et al. 1999). In contrast, osteoblast-specific overexpression of TGF-β2 under the *Oc*-promoter results in progressive bone loss due to increased osteoclastic bone resorption (Erlebacher and Derynck 1996). A follow-up study of the transgenic *Oc-Tgf-β2* mice and mice expressing *Oc-Tgf-β2*, as well as a dominant-negative type II TGF-β receptor in osteoblasts, revealed that TGF-β2 signaling directly increases the steady state of osteoblast differentiation, leading to an increase of terminal differentiated osteocytes, a phenotype which is attenuated indirectly through increased osteoclastic bone resorption (Erlebacher et al. 1998). Hence TGF-βs have divergent effects on bone, by directly controlling osteoblast maturation and survival, and indirectly through regulation of osteoclastogenesis.

Fgfs

The importance of Fgf signaling in regulating osteogenesis was realized early on by the discovery of activating mutations in the Fgf receptors, *Fgfr1*, -2, and -3, in human craniosynostosis syndromes (Webster and Donoghue 1997; Burke et al. 1998; Fragale et al. 1999). From studies with human material and genetic studies in mouse, we now know that Fgf signaling is most likely regulating osteogenesis at the level of proliferation and maturation (Fragale et al. 1999; Marie et al. 2005; Ornitz 2005). Mice lacking Fgf18 show a transient decrease in proliferation of mesenchymal osteogenic calvarial cells, resulting in delayed suture closure. In addition, the mice display a delay in endochondral ossification associated with decreased expression of osteogenic markers (Liu et al. 2002; Ohbayashi et al. 2002). In $Fgf9^{-/-}$ mice, maturated osteoblasts are decreased, while the $Runx2$ positive precursor pool is unaffected (Hung et al. 2007). Mice deficient in $Fgf2$ also show a decrease in bone formation rate in vivo, and decreased proliferation and mineralization in vitro (Montero et al. 2000). The source for Fgf2 during calvarial ossification is most likely the dura mater (the outermost membranous layer surrounding the brain and separating it from the overlying developing bony shelves) (Li et al. 2007). Surprisingly, nontargeted overexpression of Fgf2 results in a postnatal decrease in mineralization, bone formation rate, and osteopenia of membranous and endochondral skeletal elements (Sobue et al. 2005), while local application of Fgf2 stimulates osteoblast precursor proliferation and suture closure (Moursi et al. 2002). Various in vitro studies demonstrated that phenotypic effects vary based on the Fgf used or the concentration thereof. Furthermore, these studies revealed that Fgf signaling can have dual effects on osteoblast proliferation and maturation, dependent on the differentiation stage of precursors. It stimulates proliferation of more differentiated precursors, while inhibiting final maturation and increasing apoptosis in differentiated osteoblasts (Rodan et al. 1989; Tang et al. 1996; Debiais et al. 1998; Mansukhani et al. 2000; Shimoaka et al. 2002; Kalajzic et al. 2003; Fakhry et al. 2005; Dupree et al. 2006). Dual effects were also uncovered in vivo in mice in which the $Fgfr1$ was conditionally deleted in different cell populations. Similar to the in vitro results, the $Fgfr1$ mutant analysis suggests that Fgf signaling represses proliferation of progenitors, while stimulating their differentiation into immature precursors and repressing their final maturation (Jacob et al. 2006). In contrast, in mice carrying the $Fgfr1$-$P250R$ mutation or heterozygous $Fgfr2c$-$C342Y$ mice, which recapitulated the human syndromes, a stimulatory effect of Fgf signaling on early osteoblast precursor proliferation and an increase in the $Runx2$ positive population is observed (Zhou et al. 2000; Eswarakumar et al. 2004). Mice deficient for the $Fgfr2IIIc$ isoform show a delay in ossification, suggesting a positive role for Fgfr2IIIc signaling in bone development (Eswarakumar et al. 2002). This is further supported by conditional removal of both receptor isoforms, $Fgfr2IIIc$ and $Fgfr2IIIb$, in the mesenchyme using $Dermo1$-Cre, which results in postnatal dwarfism and a decreased bone formation rate, associated with a decreased proliferation of osteoblast progenitors (Yu et al. 2003). The interpretation of osteoblastogenesis-related phenotypes in the various conditional *Fgfr* mutants is often complicated by the fact that due to the nature of the Cre lines, cartilage development is also affected. On the contrary, Fgfr3 signaling is supposedly a negative regulator of bone development, since $Fgfr3^{-/-}$ mice show skeletal overgrowth and enhanced ossification prior to birth (Colvin et al. 1996; Deng et al. 1996). However, a recent report revealed that cortical bone of young adult $Fgfr3^{-/-}$ mice is osteopenic and trabecular bone hypomineralized, despite the fact that osteoblast numbers are increased similar to osteoclast numbers (Valverde-Franco et al. 2004). Mice expressing a constitutively active Fgfr3-G369C mutant form have defects in chondrogenesis and osteoblastogenesis. Interestingly, these mice show enhanced expression of osteogenic markers, but have reduced bone density due to increased activity of osteoclasts (Chen et al. 1999). These data suggest that Fgfr3 signaling has diverse effects on different populations of osteoblasts. What complicates the issue is that it is still controversial as to whether Fgfr3 is expressed in osteoblasts (Rice et al. 2000; Funato et al. 2001; Nakajima et al. 2003). Different downstream signaling pathways have been implicated in mediating the Fgf effects in osteogenesis. The Stat1 pathway is playing an important role downstream of Fgfr3 in chondrogenesis and endochondral bone formation during embryogenesis (Su et al. 1997; Lievens and Liboi 2003). However, *Stat1* mutant mice also display a postnatal increase in bone mineral density and mineral content, a phenotype opposite to that observed in $Fgfr3^{-/-}$ mice (Xiao et al. 2004). This can be explained by the observation that loss of Stat1 in osteoblasts results in a decrease in the cell cycle inhibitor p21/WAF and Fgfr3 expression, and an increase of Fgf18 expression and responsiveness to Fgf18. Based on these results, it was suggested that Stat1 might be involved in regulating the balance between positive and negative Fgf signaling (Xiao et al. 2004). PKC is mediating some downstream events of active Fgfr2 signaling and is involved in matrix mineralization (Suzuki et al. 2000; Lemonnier et al. 2001a,b; Kim et al. 2003a). In addition, the ERK pathway has been implicated in mediating some of the Fgf2-stimulated effects (Kim et al. 2003b). More studies will be necessary to decipher the complex activities of Fgf signaling in osteogensis.

Parathyroid Hormone-related Peptide and Parathyroid Hormone

The parathyroid hormone-related peptide (PTHrP) is well known for its important functions in regulating chondrocyte maturation during endochondral skeletal development (Kronenberg 2006). However, PTHrP is also produced by osteoblasts and recent studies have shown that it plays an anabolic role in osteogenesis (Karaplis 2001). Mice that have only one copy of *Pthrp* (*Pthrp$^{+/-}$*) have a reduced trabecular bone volume and bone formation rate associated with an increase in apoptotic osteoblasts (Amizuka et al. 1996; Miao et al. 2005). Similar changes are observed upon osteoblast-specific inactivation of *Pthrp* using the *Col1-Cre* line (Miao et al. 2005). In contrast, mice completely deficient in *Pthrp* show enhanced mineralization of skull bones and increased cortical thickness in the long bones, while trabecular bone is only slightly increased, associated with an increased number of osteoblasts in endosteum and primary spongiosa (Karaplis et al. 1994; Miao et al. 2002). Parathyroid hormone (PTH), exclusively produced by the parathyroid gland, is well known for its role in regulating calcium homeostasis and bone remodeling, but it also has anabolic effects on bone when used intermittently (Liu and Kalu 1990; Dempster et al. 1993; Uzawa et al. 1995; Locklin et al. 2003; Qin et al. 2004; Masi and Brandi 2005). In contrast, continuous treatment with PTH, PTHrP, or hyperparathyroidism results in predominantly catabolic effects (Parisien et al. 1990; Horwitz et al. 2005). The anabolic effects of PTH require the Igf1/Igf1R pathway (Miyakoshi et al. 2001a; Yamaguchi et al. 2005; Wang et al. 2006, 2007; Yakar et al. 2006) and activity of Fgf2, which is up-regulated in response to PTH (Hurley et al. 1999, 2006). Furthermore, recent findings that PTH can increase β-catenin levels suggest that in addition to Igf1 and Fgf2, β-catenin mediates part of the anabolic action (Tobimatsu et al. 2006; Suzuki et al. 2008). Based on in vitro data, PTH acts at the level of β-catenin. Thus, it is not surprising that PTH treatment can improve bone mass in *Lrp5$^{-/-}$* mice (Sawakami et al. 2006; Iwaniec et al. 2007). Part of the in vivo PTH anabolic function might be due to its ability to negatively regulate Sost level, thus further contributing to increased Wnt/β-catenin signaling (Keller and Kneissel 2005; Bellido 2006). The molecular and cellular mechanisms of the anabolic effects of PTH are probably even more diverse (Jilka 2007). Like in the case of the Fgfs, the effects of PTH have been reported to differ dependent on the osteoblastic differentiation state and on the exposure time of osteoblastic cells to PTH (Isogai et al. 1996; Ishizuya et al. 1997). In contrast to *Pthrp$^{-/-}$*, *Pth* knockout embryos show a slightly increased cortical bone thickness. While trabecular bone volume is decreased, osteoblast numbers are decreased in the endosteum and primary spongiosa, but not in the periosteum (Miao et al. 2002). Interestingly, at postnatal stages, *Pth$^{-/-}$* mice display the opposite phenotype, a slight increase in trabecular and a dramatic increase in cortical bone volume, which is due to a compromised bone turnover rate (Miao et al. 2004a). The local effects of PTH and PTHrP on osteoblasts are mediated through signaling, via the PTH/PTHrP receptor 1 (PPR) expressed by osteoblasts (Juppner et al. 1991). $Ppr^{-/-}$ fetuses showed a reduction in trabecular bone formation and an increase in cortical bone associated with a dramatic increase in osteoblast numbers and matrix accumulation in the periosteum (Lanske et al. 1999). Differential effects on osteoblast numbers and their activity in trabecular bone and periosteal cortical bone have also been observed in mice expressing a con-stitutively active form of PPR in the osteoblastic lineage, under the *Col1*-promoter. Here, trabecular osteoblast proliferation was stimulated and apoptosis reduced, while activity of periosteal osteoblasts was reduced (Calvi et al. 2001).

Insulin-like Growth Factors

Insulin-like growth factors (IGFs), Igf1 and Igf2, produced by osteoblasts, are the most abundant growth factors found in bone. Igf signaling exerts profound effects on osteoblasts and skeletal growth in vivo and in vitro (Conover 2000; Clemens and Chernausek 2004). In vivo effects of the IGF system on bone mineralization and bone formation are difficult to assess in humans because of the complexity and the intimate connection to growth hormone (Benbassat et al. 1999, 2003; Bex et al. 2002). The recent development of various mouse models targeting the IGF system helped to shed light on its involvement in osteogenesis. Osteoblast-spe-cific overexpression of Igf1 under the *Oc*-promoter results in an increased bone formation rate in young adult mice, most likely affecting osteoblast activity, since neither osteoblast nor osteoclast numbers are altered (Zhao et al. 2000; Rutter et al. 2005). Slightly broader *Col1*-driven Igf1 overex-pression also has a positive effect on bone formation rate, but in addi-tion it increases bone resorption by altering osteoclast numbers (Jiang et al. 2006). Consistent with a role for Igf1 signaling in osteogenesis, *Igf1* and *Igf1r* deficient mice, and mice in which IGF-serum levels are reduced, display growth retardation, alterations in bone mineral density, and cor-tical bone thickness (Baker et al. 1993; Liu et al. 1993; Yakar et al. 2002; Rosen et al. 2004). Interestingly, mice deficient for *Igf2* are embryonically growth retarded, while $Igf1^{-/-}$ mice show postnatal growth retardation (Baker et al. 1993; Liu et al. 1993). Osteoblast-specific inactivation of *Igf1r* using an *Oc-Cre* line demonstrates a direct requirement of Igf signaling for osteoblast function. Those mice show primarily reduction in trabec-ular bone formation associated with reduced osteoblast numbers, and impaired matrix mineralization, despite an increase in osteoid deposition.

This suggests that *Igf1r*$^{-/-}$ osteoblasts can mature but are defective in mineralizing matrix (Zhang et al. 2002). IGF activity is modulated by at least six high-affinity IGF binding proteins (IGFBP1-6) (Mohan 1993; Govoni et al. 2005). The use of animal models has resulted in somewhat contradictory results with regard to their modulating activity on osteoblasts (Silha and Murphy 2002). Osteoblast-targeted IGFBP4 over-expression under the *Oc*-promoter markedly reduces cancellous bone formation associated with a reduction in osteoblast numbers, most likely by inhibiting Igf1 action on osteoblast proliferation, as previously shown in vitro (Mohan et al. 1995; Zhang et al. 2003). However, opposite effects are observed upon systemic administration. This effect is presumably due to an increase in Igf1 bioavailability in the serum via an IGFBP4 protease-dependent mechanism (Miyakoshi et al. 2001b; Zhang et al. 2003). In vitro and in vivo experiments have shown that IGFBP2 is a potent stimulator of Igf2 function in the presence of extracellular matrix (ECM), stimulating proliferation and maturation of osteoblasts (Conover et al. 2002; Conover and Khosla 2003; Palermo et al. 2004). There are also reports suggesting a stimulatory role for the Igf1/IGFBP3 complex (Ernst and Rodan 1990; Tanaka et al. 1998). However, ubiquitous IGFBP3 over-expression results in reduced bone formation and osteoblast proliferation rate in response to Igf1 and an increase in bone resorption (Silha et al. 2003). Mechanistically, IGFBP3 could either directly inhibit Igf1 function (Schmid et al. 1991) or it might act by inhibiting the IGFBP4 protease, thus possibly increasing IGFBP4 levels (Fowlkes et al. 1995). IGFBP5, which stimulates osteoblast proliferation in vitro (Andress 2001; Miyakoshi et al. 2001c), also stimulates osteoblast proliferation and bone formation in vivo in ovariectomized (OVX) mice (Andress 2001). The stimulatory effect observed in vitro is Igf1 independent and probably mediated through IGFBP5 receptor activation (Mohan et al. 1995; Andress 1998; Miyakoshi et al. 2001c). Conflicting observations are reported in other IGFBP5 overexpression studies, such as decreased and impaired osteoblast function resulting in osteopenia and hypomineralized bone matrix in noncompromised mice (Devlin et al. 2002; Durant et al. 2004; Atti et al. 2005; Salih et al. 2005). The different responses can possibly be explained by the recent finding that IGFBP5 interacts with nuclear Vitamin D receptor (VDR), modulating Vitamin D response of osteoblasts (Schedlich et al. 2007). IGFBP6, with high affinity to Igf2, is regulated by retinoic acid (RA) and mediates the repressive effects of RA on osteoblasts (Zhou et al. 1996; Yan et al. 2001). Overall, the transgenic mouse studies suggest that continuous overexpression of IGFBPs can in some cases lead to effects opposite to those observed in vitro or when systemically administered. *Igfbp* knockout mice have not helped to elucidate the role of IGFBPs in osteoblasts, either due to the fact that their bones were not analyzed or because the IGFBPs might act functionally redundant in ostoeblasts (Wood et al. 1993, 2000; Ning et al. 2006, 2007, 2008; Lofqvist et al. 2007; Murali et al. 2007).

Epidermal Growth Factor and Related Molecules

Broad overexpression of the epidermal growth factor (EGF)-like ligands, EGF, betacellulin (Btc), and transforming growth factor alpha (TGFα) in mice results in stunted growth (Chan and Wong 2000; Schneider et al. 2005). More osteoblasts are observed in these mice in the periosteum and endosteum, suggesting a positive role of EGF on osteoblast proliferation. However, cortical bone thickness was reduced in *Egf*-transgenic animals (Chan and Wong 2000). This finding is consistent with in vitro observations that EGF-like ligands can inhibit osteoblast differentiation or affect osteoblast function (Ibbotson et al. 1986; Chien et al. 2000; Zhu et al. 2007). In vitro studies have further revealed that EGF reduces IGF and IGFBP synthesis (Hembree et al. 1994; Edmondson et al. 1999). The reduced IGFBP3 serum level in EGF-transgenic animals probably contributes to the observed growth retardation. EGFR, the receptor of EGF-like ligands, is expressed in osteoblasts and osteoclasts. In $Egfr^{-/-}$ mice and humanized mice expressing human instead of mouse EGFR, which exhibits a lower activity in bone, osteoblastogenesis is accelerated (Sibilia et al. 2003). These results comply with an inhibitory effect of EGF signaling on osteogenesis. However, there are also contradictory observations: Amphiregulin (AR, Areg), another member of the EGF family expressed in osteoblasts, is up-regulated upon PTH treatment in vitro and in vivo. In vitro studies demonstrated that AR stimulates pre-osteoblast proliferation, however, it also blocks their terminal maturation. Furthermore, $Ar^{-/-}$ mice display significantly less trabecular bone compared to wild-type mice, suggesting a positive role in bone formation (Qin et al. 2005). Undoubtedly, the role of EGF-like ligands in osteogenesis has to be further investigated in the future.

ROLE OF GROWTH FACTORS IN BONE HOMEOSTASIS, FRACTURE REPAIR, AND THERAPEUTIC USE

During postnatal life, bone undergoes constant remodeling, involving bone destruction by osteoclasts and new bone formation by osteoblasts.

Bone remodeling is playing important roles in regulating calcium and phosphate homeostasis and during fracture repair. Canonical Wnt signaling is an intricate player of bone homeostasis, regulating the expression of two molecules, osteoprotegerin (OPG) and Rank ligand (RANKL), which are negative and positive regulators of osteoclastogenesis, respectively (Glass et al. 2005; Holmen et al. 2005; Spencer et al. 2006). Furthermore, recent data suggest that Wnt signaling is one of the essential pathways used by osteoblasts to directly control lineage commitment of mesodermal progenitor cells (Zhou et al. 2008). Studies using secondary fracture repair models (involving formation of cartilage that is remodeled into bone) have revealed a disparate role for β-catenin at different stages of fracture repair (Chen et al. 2007). In addition, PTH-stimulated fracture repair is also in part mediated through β-catenin (Kakar et al. 2007). Thus, therapeutically, stimulation of β-catenin activity is probably beneficial for fracture repair and to increase bone mass in osteoporotic patients. Evidence for the later has been provided in $Lrp5^{-/-}$ mice treated with lithium (Clement-Lacroix et al. 2005). However, reports based on human studies are less clear. Beneficial effects on fracture risk are reported (Vestergaard et al. 2005), while contradictory results are reported regarding osteoporosis/osteopenia (Christiansen et al. 1975; Cohen et al. 1998). Bone loss has also been observed in studies of healthy female rats upon sustained lithium treatment (Lewicki et al. 2006). Thus, more studies are necessary to access the efficacy of lithium treatment in osteoporosis therapy. Neutralizing antibodies for Dkk1 and Sost are another therapeutic approach currently explored in mouse models (Ominsky et al. 2006; Diarra et al. 2007; Padhi et al. 2007; Yaccoby et al. 2007).

Recombinant BMPs are already in use in the clinic as therapeutic biological agents in nonhealing and bridging fracture repair (Garrison et al. 2007; Gautschi et al. 2007; Schmidmaier et al. 2007). Recombinant BMPs have also been successfully used in animal models for osteoporosis (Simic et al. 2006). For rhBMP6, it has been shown that its activity on osteoblasts is mediated in part through the IGF and EGF pathway (Grasser et al. 2007). There is also most likely a negative influence of BMP signaling on osteoclastogenesis, since conditional *Bmpr1a* knockout mice show an age-related increase in osteoclast numbers (Mishina et al. 2004). TGF-βs are also tested in animal models for their potential use in fracture healing (Beck et al. 1991; Richardson et al. 1993; Lee et al. 2006), since they along with other TGF-β superfamily members are endogenously highly expressed during fracture healing (Cho et al. 2002). Similar to other signaling molecules, dose-dependent effects have been observed to affect fracture healing (Zellin et al. 1998). Furthermore, TGF-βs up-regulate OPG levels in culture (Takai et al. 1998; Thirunavukkarasu et al. 2001), thus suggesting an inhibitory role of TGF-βs in osteoclastogenesis. However, there are other reports showing that TGF-βs stimulate osteoclastogenesis in vitro (Sells Galvin et al. 1999; Kaneda et al. 2000). Hence, TGF-βs could potentially influence bone homeostasis as well, although this remains to be shown in vivo.

The catabolic function of PTH is required for regulating calcium blood levels via osteoclastic bone remodeling (Boden and Kaplan 1990). However, PTH is also one of the few FDA-approved drugs for osteoporosis treatment. Short-term high dose treatment of postmenopausal osteoporotic women with PTHrP has been shown to be beneficial (Horwitz et al. 2003). Interestingly, administration of an amino-terminal fragment of PTH (PTH 1-34) has a greater anabolic effect on bone formation in $Pthrp^{+/-}$ when compared to wild-type animals. Why this is the case needs to be deciphered in the future (Miao et al. 2005). Again, the dose used and length of administration has to be tightly controlled since other studies have shown that continuous treatment with PTH and PTHrP can lead to bone loss (Horwitz et al. 2005).

Fgfs also have anabolic effects on bone in different models for osteopenia/osteoporosis (Dunstan et al. 1999; Nagai et al. 1999; Iwaniec et al. 2002; Lane et al. 2003; Power et al. 2004). In vitro studies suggest the use of soluble Fgfr2 receptor molecules as a potential therapeutic agent for craniosynostosis syndromes (Tanimoto et al. 2004). Furthermore, Fgfs have been shown to accelerate fracture healing (Kawaguchi et al. 2001). Thus, modulating Fgf signaling, as well as other growth factors, will probably be therapeutically useful.

SUMMARY AND CONCLUSIONS

All of the signaling pathways discussed play important and complex roles in osteogenic skeletal development and bone maintenance (Fig. 4). What becomes apparent from the functional studies is that their dosage and developmental timing of activity is tightly controlled. This important point has to be taken into consideration for their use in translational approaches, such as tissue engineering and clinical studies to treat diseases such as osteoporosis, or to improve fracture repair. Modulating Wnt activity is clearly one of the new therapeutic avenues with promising perspectives for cell/tissue engineering, fracture healing, and osteoporosis therapy. Combinatorial treatment through stimulation of multiple path-

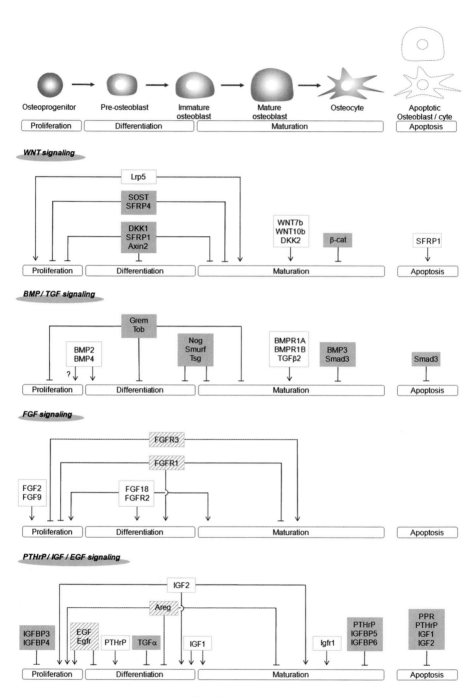

Figure 4. (*See facing page for legend.*)

ways is an additional exploratory therapeutic avenue. This has been looked at in cell culture experiments for a number of signaling molecules/pathways. For example, β-catenin and BMP2 synergize, stimulating osteoblast differentiation and bone formation (Mbalaviele et al. 2005). Similarly, synergistic effects are observed in vitro, combining BMPs with RA (Cowan et al. 2005), or BMPs and PTHrP (Chan et al. 2003). In osteoporosis mouse models, PTH and estrogen show synergistic bone restoring activities (von Stechow et al. 2004). Combining compounds and biological agents affecting different cell types has also proven to be beneficial, as in the case of combining BMPs with biphosphonates to treat neurofibromatosis type-1 (Schindeler et al. 2008). Bisphosphonates have also been used in combination with PTH to treat osteoporosis. However, long-term studies revealed that this impairs the ability of PTH to increase bone mineral density in the patients (Black et al. 2003; Finkelstein et al. 2003). For other pathways, like Notch signaling, so far only in vitro data exist, demonstrating a possible involvement in osteoblastogenesis (Sciaudone et al. 2003; Deregowski et al. 2006). For the future, it will be important to decipher the exact mechanisms of action of the different growth factors and their modes of interaction in order to further exploit them for therapeutic applications.

ACKNOWLEDGMENTS

Studies in the laboratory of Dr. Hartmann are supported by Boehringer Ingelheim and grants from the Austrian Science Fund (FWF), the EU (Cells into Organs), and the Arthritis Research Campaign (ARC). Dr. Kostanova-Poliakova is supported by an FWF grant. We thank Teresa Pizzuto for her help sectioning the adult specimens.

Figure 4. Schematic summary of the activities of WNT, BMP/TGF, FGF, PTHrP, IGF, and EGF signaling on the differentiation stages of osteoblasts. Stimulatory influences are symbolized by an arrow, while inhibitory influences are symbolized by a blunt-ended line. Factors mediating inhibitory activities are boxed gray, factors mediating stimulatory activities are boxed white, and factors having dual activities are indicated by the hatched box.

REFERENCES

Abzhanov, A., Rodda, S.J., McMahon, A.P., and Tabin, C.J. 2007. Regulation of skeletogenic differentiation in cranial dermal bone. *Development* **134:** 3133–3144.

Akiyama, H., Kim, J.E., Nakashima, K., Balmes, G., Iwai, N., Deng, J.M., Zhang, Z., Martin, J.F., Behringer, R.R., Nakamura, T., et al. 2005. Osteo-chondroprogenitor cells are derived from Sox9 expressing precursors. *Proc. Natl. Acad. Sci.* **102:** 14665–14670.

Amizuka, N., Karaplis, A.C., Henderson, J.E., Warshawsky, H., Lipman, M.L., Matsuki, Y., Ejiri, S., Tanaka, M., Izumi, N., Ozawa, H., et al. 1996. Haploinsufficiency of parathyroid hormone-related peptide (PTHrP) results in abnormal postnatal bone development. *Dev. Biol.* **175:** 166–176.

Andress, D.L. 1998. Insulin-like growth factor-binding protein-5 (IGFBP-5) stimulates phosphorylation of the IGFBP-5 receptor. *Am. J. Physiol.* **274:** E744–E750.

Andress, D.L. 2001. IGF-binding protein-5 stimulates osteoblast activity and bone accretion in ovariectomized mice. *Am. J. Physiol. Endocrinol. Metab.* **281:** E283–E288.

Atti, E., Gomez, S., Wahl, S.M., Mendelsohn, R., Paschalis, E., and Boskey, A.L. 2002. Effects of transforming growth factor-β deficiency on bone development: A Fourier transform-infrared imaging analysis. *Bone* **31:** 675–684.

Atti, E., Boskey, A.L., and Canalis, E. 2005. Overexpression of IGF-binding protein 5 alters mineral and matrix properties in mouse femora: An infrared imaging study. *Calcif. Tissue Int.* **76:** 187–193.

Aubin, J.E. 2001. Regulation of osteoblast formation and function. *Rev. Endocr. Metab. Disord.* **2:** 81–94.

Baker, J., Liu, J.P., Robertson, E.J., and Efstratiadis, A. 1993. Role of insulin-like growth factors in embryonic and postnatal growth. *Cell* **75:** 73–82.

Bandyopadhyay, A., Tsuji, K., Cox, K., Harfe, B.D., Rosen, V., and Tabin, C.J. 2006. Genetic analysis of the roles of BMP2, BMP4, and BMP7 in limb patterning and skeletogenesis. *PLoS Genet.* **2:** e216.

Baur, S.T., Mai, J.J., and Dymecki, S.M. 2000. Combinatorial signaling through BMP receptor IB and GDF5: Shaping of the distal mouse limb and the genetics of distal limb diversity. *Development* **127:** 605–619.

Beck, L.S., Deguzman, L., Lee, W.P., Xu, Y., McFatridge, L.A., Gillett, N.A., and Amento, E.P. 1991. Rapid publication. TGF-β 1 induces bone closure of skull defects. *J. Bone Miner. Res.* **6:** 1257–1265.

Bellido, T. 2006. Downregulation of SOST/sclerostin by PTH: A novel mechanism of hormonal control of bone formation mediated by osteocytes. *J. Musculoskelet. Neuronal Interact.* **6:** 358–359.

Benbassat, C.A., Wasserman, M., and Laron, Z. 1999. Changes in bone mineral density after discontinuation and early reinstitution of growth hormone (GH) in patients with childhood-onset GH deficiency. *Growth Horm. IGF Res.* **9:** 290–295.

Benbassat, C.A., Eshed, V., Kamjin, M., and Laron, Z. 2003. Are adult patients with Laron syndrome osteopenic? A comparison between dual-energy X-ray absorptiometry and volumetric bone densities. *J. Clin. Endocrinol. Metab.* **88:** 4586–4589.

Bennett, C.N., Longo, K.A., Wright, W.S., Suva, L.J., Lane, T.F., Hankenson, K.D., and MacDougald, O.A. 2005. Regulation of osteoblastogenesis and bone mass by Wnt10b. *Proc. Natl. Acad. Sci.* **102:** 3324–3329.

Bennett, C.N., Ouyang, H., Ma, Y.L., Zeng, Q., Gerin, I., Sousa, K.M., Lane, T.F., Krishnan, V., Hankenson, K.D., and MacDougald, O.A. 2007. Wnt10b increases postnatal bone formation by enhancing osteoblast differentiation. *J. Bone Miner. Res.* **22:** 1924–1932.

Bex, M., Abs, R., Maiter, D., Beckers, A., Lamberigts, G., and Bouillon, R. 2002. The effects of growth hormone replacement therapy on bone metabolism in adult-onset growth hormone deficiency: A 2-year open randomized controlled multicenter trial. *J. Bone Miner. Res.* **17:** 1081–1094.

Black, D.M., Greenspan, S.L., Ensrud, K.E., Palermo, L., McGowan, J.A., Lang, T.F., Garnero, P., Bouxsein, M.L., Bilezikian, J.P., and Rosen, C.J. 2003. The effects of parathyroid hormone and alendronate alone or in combination in postmenopausal osteoporosis. *N. Engl. J. Med.* **349:** 1207–1215.

Boden, S.D. and Kaplan, F.S. 1990. Calcium homeostasis. *Orthop. Clin. N. Am.* **21:** 31–42.

Bodine, P.V., Zhao, W., Kharode, Y.P., Bex, F.J., Lambert, A.J., Goad, M.B., Gaur, T., Stein, G.S., Lian, J.B., and Komm, B.S. 2004. The Wnt antagonist secreted frizzled-related protein-1 is a negative regulator of trabecular bone formation in adult mice. *Mol. Endocrinol.* **18:** 1222–1237.

Borton, A.J., Frederick, J.P., Datto, M.B., Wang, X.F., and Weinstein, R.S. 2001. The loss of Smad3 results in a lower rate of bone formation and osteopenia through dysregulation of osteoblast differentiation and apoptosis. *J. Bone Miner. Res.* **16:** 1754–1764.

Boyden, L.M., Mao, J., Belsky, J., Mitzner, L., Farhi, A., Mitnick, M.A., Wu, D., Insogna, K., and Lifton, R.P. 2002. High bone density due to a mutation in LDL-receptor-related protein 5. *N. Engl. J. Med.* **346:** 1513–1521.

Burke, D., Wilkes, D., Blundell, T.L., and Malcolm, S. 1998. Fibroblast growth factor receptors: Lessons from the genes. *Trends Biochem. Sci.* **23:** 59–62.

Calvi, L.M., Sims, N.A., Hunzelman, J.L., Knight, M.C., Giovannetti, A., Saxton, J.M., Kronenberg, H.M., Baron, R., and Schipani, E. 2001. Activated parathyroid hormone/parathyroid hormone-related protein receptor in osteoblastic cells differentially affects cortical and trabecular bone. *J. Clin. Invest.* **107:** 277–286.

Chan, S.Y. and Wong, R.W. 2000. Expression of epidermal growth factor in transgenic mice causes growth retardation. *J. Biol. Chem.* **275:** 38693–38698.

Chan, G.K., Miao, D., Deckelbaum, R., Bolivar, I., Karaplis, A., and Goltzman, D. 2003. Parathyroid hormone-related peptide interacts with bone morphogenetic protein 2 to increase osteoblastogenesis and decrease adipogenesis in pluripotent C3H10T 1/2 mesenchymal cells. *Endocrinology* **144:** 5511–5520.

Chen, D., Harris, M.A., Rossini, G., Dunstan, C.R., Dallas, S.L., Feng, J.Q., Mundy, G.R., and Harris, S.E. 1997. Bone morphogenetic protein 2 (BMP-2) enhances BMP-3, BMP-4, and bone cell differentiation marker gene expression during the induction of mineralized bone matrix formation in cultures of fetal rat calvarial osteoblasts. *Calcif. Tissue Int.* **60:** 283–290.

Chen, D., Ji, X., Harris, M.A., Feng, J.Q., Karsenty, G., Celeste, A.J., Rosen, V., Mundy, G.R., and Harris, S.E. 1998. Differential roles for bone morphogenetic protein (BMP) receptor type IB and IA in differentiation and specification of mesenchymal precursor cells to osteoblast and adipocyte lineages. *J. Cell Biol.* **142:** 295–305.

Chen, L., Adar, R., Yang, X., Monsonego, E.O., Li, C., Hauschka, P.V., Yayon, A., and Deng, C.X. 1999. Gly369Cys mutation in mouse FGFR3 causes achondroplasia by affecting both chondrogenesis and osteogenesis. *J. Clin. Invest.* **104:** 1517–1525.

Chen, D., Qiao, M., Story, B., Zhao, M., Jiang, Y., Zhao, J., Feng, J., Xie, Y., Huang, S., Roberts, A., et al. 2003. BMP signaling through the Smad1 pathway is required for normal postnatal bone formation . *J. Bone Miner. Res.* **18:** S6. (Abstr.)

Chen, Y., Whetstone, H.C., Lin, A.C., Nadesan, P., Wei, Q., Poon, R., and Alman, B.A. 2007.

β-Catenin signaling plays a disparate role in different phases of fracture repair: Implications for therapy to improve bone healing. *PLoS Med.* **4:** e249.

Chien, H.H., Lin, W.L., and Cho, M.I. 2000. Down-regulation of osteoblastic cell differentiation by epidermal growth factor receptor. *Calcif. Tissue Int.* **67:** 141–150.

Cho, T.J., Gerstenfeld, L.C., and Einhorn, T.A. 2002. Differential temporal expression of members of the transforming growth factor β superfamily during murine fracture healing. *J. Bone Miner. Res.* **17:** 513–520.

Christiansen, C., Baastrup, P.C., and Transbol, I. 1975. Osteopenia and dysregulation of divalent cations in lithium-treated patients. *Neuropsychobiology* **1:** 344–354.

Clemens, T.L. and Chernausek, S.D. 2004. Genetic strategies for elucidating insulin-like growth factor action in bone. *Growth Horm. IGF Res.* **14:** 195–199.

Clement-Lacroix, P., Ai, M., Morvan, F., Roman-Roman, S., Vayssiere, B., Belleville, C., Estrera, K., Warman, M.L., Baron, R., and Rawadi, G. 2005. Lrp5-independent activation of Wnt signaling by lithium chloride increases bone formation and bone mass in mice. *Proc. Natl. Acad. Sci.* **102:** 17406–17411.

Cohen, O., Rais, T., Lepkifker, E., and Vered, I. 1998. Lithium carbonate therapy is not a risk factor for osteoporosis. *Horm. Metab. Res.* **30:** 594–597.

Colnot, C., de la Fuente, L., Huang, S., Hu, D., Lu, C., St-Jacques, B., and Helms, J.A. 2005. Indian hedgehog synchronizes skeletal angiogenesis and perichondrial maturation with cartilage development. *Development* **132:** 1057–1067.

Colvin, J.S., Bohne, B.A., Harding, G.W., McEwen, D.G., and Ornitz, D.M. 1996. Skeletal overgrowth and deafness in mice lacking fibroblast growth factor receptor 3. *Nat. Genet.* **12:** 390–397.

Conover, C.A. 2000. In vitro studies of insulin-like growth factor I and bone. *Growth Horm. IGF Res.* (suppl. B) **10:** S107–S110.

Conover, C.A. and Khosla, S. 2003. Role of extracellular matrix in insulin-like growth factor (IGF) binding protein-2 regulation of IGF-II action in normal human osteoblasts. *Growth Horm. IGF Res.* **13:** 328–335.

Conover, C.A., Johnstone, E.W., Turner, R.T., Evans, G.L., John Ballard, F.J., Doran, P.M., and Khosla, S. 2002. Subcutaneous administration of insulin-like growth factor (IGF)-II/IGF binding protein-2 complex stimulates bone formation and prevents loss of bone mineral density in a rat model of disuse osteoporosis. *Growth Horm. IGF Res.* **12:** 178–183.

Cowan, C.M., Aalami, O.O., Shi, Y.Y., Chou, Y.F., Mari, C., Thomas, R., Quarto, N., Nacamuli, R.P., Contag, C.H., Wu, B., et al. 2005. Bone morphogenetic protein 2 and retinoic acid accelerate in vivo bone formation, osteoclast recruitment, and bone turnover. *Tissue Eng.* **11:** 645–658.

Dabovic, B., Chen, Y., Colarossi, C., Obata, H., Zambuto, L., Perle, M.A., and Rifkin, D.B. 2002a. Bone abnormalities in latent TGF-β binding protein (Ltbp)-3-null mice indicate a role for Ltbp-3 in modulating TGF-β bioavailability. *J. Cell Biol.* **156:** 227–232.

Dabovic, B., Chen, Y., Colarossi, C., Zambuto, L., Obata, H., and Rifkin, D.B. 2002b. Bone defects in latent TGF-β binding protein (Ltbp)-3 null mice; a role for Ltbp in TGF-β presentation. *J. Endocrinol.* **175:** 129–141.

Dabovic, B., Levasseur, R., Zambuto, L., Chen, Y., Karsenty, G., and Rifkin, D.B. 2005. Osteopetrosis-like phenotype in latent TGF-β binding protein 3 deficient mice. *Bone* **37:** 25–31.

Daluiski, A., Engstrand, T., Bahamonde, M.E., Gamer, L.W., Agius, E., Stevenson, S.L., Cox, K., Rosen, V., and Lyons, K.M. 2001. Bone morphogenetic protein-3 is a negative regulator of bone density. *Nat. Genet.* **27:** 84–88.

Day, T.F., Guo, X., Garrett-Beal, L., and Yang, Y. 2005. Wnt/β-catenin signaling in mesenchymal progenitors controls osteoblast and chondrocyte differentiation during vertebrate skeletogenesis. *Dev. Cell* **8:** 739–750.

Debiais, F., Hott, M., Graulet, A.M., and Marie, P.J. 1998. The effects of fibroblast growth factor-2 on human neonatal calvaria osteoblastic cells are differentiation stage specific. *J. Bone Miner. Res.* **13:** 645–654.

Dempster, D.W., Cosman, F., Parisien, M., Shen, V., and Lindsay, R. 1993. Anabolic actions of parathyroid hormone on bone. *Endocr. Rev.* **14:** 690–709.

Deng, C., Wynshaw-Boris, A., Zhou, F., Kuo, A., and Leder, P. 1996. Fibroblast growth factor receptor 3 is a negative regulator of bone growth. *Cell* **84:** 911–921.

Deregowski, V., Gazzerro, E., Priest, L., Rydziel, S., and Canalis, E. 2006. Notch 1 overexpression inhibits osteoblastogenesis by suppressing Wnt/β-catenin but not bone morphogenetic protein signaling. *J. Biol. Chem.* **281:** 6203–6210.

Devlin, R.D., Du, Z., Buccilli, V., Jorgetti, V., and Canalis, E. 2002. Transgenic mice overexpressing insulin-like growth factor binding protein-5 display transiently decreased osteoblastic function and osteopenia. *Endocrinology* **143:** 3955–3962.

Devlin, R.D., Du, Z., Pereira, R.C., Kimble, R.B., Economides, A.N., Jorgetti, V., and Canalis, E. 2003. Skeletal overexpression of noggin results in osteopenia and reduced bone formation. *Endocrinology* **144:** 1972–1978.

Diarra, D., Stolina, M., Polzer, K., Zwerina, J., Ominsky, M.S., Dwyer, D., Korb, A., Smolen, J., Hoffmann, M., Scheinecker, C., et al. 2007. Dickkopf-1 is a master regulator of joint remodeling. *Nat. Med.* **13:** 156–163.

Dunstan, C.R., Boyce, R., Boyce, B.F., Garrett, I.R., Izbicka, E., Burgess, W.H., and Mundy, G.R. 1999. Systemic administration of acidic fibroblast growth factor (FGF-1) prevents bone loss and increases new bone formation in ovariectomized rats. *J. Bone Miner. Res.* **14:** 953–959.

Dupree, M.A., Pollack, S.R., Levine, E.M., and Laurencin, C.T. 2006. Fibroblast growth factor 2 induced proliferation in osteoblasts and bone marrow stromal cells: A whole cell model. *Biophys. J.* **91:** 3097–3112.

Durant, D., Pereira, R., Stadmeyer, L., and Canalis, E. 2004. Transgenic mice expressing selected insulin-like growth factor-binding protein-5 fragments do not exhibit enhanced bone formation. *Growth Horm. IGF Res.* **14:** 319–327.

Edmondson, S.R., Murashita, M.M., Russo, V.C., Wraight, C.J., and Werther, G.A. 1999. Expression of insulin-like growth factor binding protein-3 (IGFBP-3) in human keratinocytes is regulated by EGF and TGFβ1. *J. Cell. Physiol.* **179:** 201–207.

Ellies, D.L., Viviano, B., McCarthy, J., Rey, J.P., Itasaki, N., Saunders, S., and Krumlauf, R. 2006. Bone density ligand, Sclerostin, directly interacts with LRP5 but not LRP5G171V to modulate Wnt activity. *J. Bone Miner. Res.* **21:** 1738–1749.

Erlebacher, A. and Derynck, R. 1996. Increased expression of TGF-β 2 in osteoblasts results in an osteoporosis-like phenotype. *J. Cell Biol.* **132:** 195–210.

Erlebacher, A., Filvaroff, E.H., Ye, J.Q., and Derynck, R. 1998. Osteoblastic responses to TGF-β during bone remodeling. *Mol. Biol. Cell* **9:** 1903–1918.

Ernst, M. and Rodan, G.A. 1990. Increased activity of insulin-like growth factor (IGF) in osteoblastic cells in the presence of growth hormone (GH): Positive correlation with the presence of the GH-induced IGF-binding protein BP-3. *Endocrinology* **127:** 807–814.

Eswarakumar, V.P., Monsonego-Ornan, E., Pines, M., Antonopoulou, I., Morriss-Kay, G.M., and Lonai, P. 2002. The IIIc alternative of Fgfr2 is a positive regulator of bone formation. *Development* **129:** 3783–3793.

Eswarakumar, V.P., Horowitz, M.C., Locklin, R., Morriss-Kay, G.M., and Lonai, P. 2004. A gain-of-function mutation of Fgfr2c demonstrates the roles of this receptor variant in osteogenesis. *Proc. Natl. Acad. Sci.* **101:** 12555–12560.

Fakhry, A., Ratisoontorn, C., Vedhachalam, C., Salhab, I., Koyama, E., Leboy, P., Pacifici, M., Kirschner, R.E., and Nah, H.D. 2005. Effects of FGF-2/-9 in calvarial bone cell cultures: Differentiation stage-dependent mitogenic effect, inverse regulation of BMP-2 and noggin, and enhancement of osteogenic potential. *Bone* **36:** 254–266.

Fang, J. and Hall, B.K. 1997. Chondrogenic cell differentiation from membrane bone periostea. *Anat. Embryol.* **196:** 349–362.

Filvaroff, E., Erlebacher, A., Ye, J., Gitelman, S.E., Lotz, J., Heillman, M., and Derynck, R. 1999. Inhibition of TGF-β receptor signaling in osteoblasts leads to decreased bone remodeling and increased trabecular bone mass. *Development* **126:** 4267–4279.

Finkelstein, J.S., Hayes, A., Hunzelman, J.L., Wyland, J.J., Lee, H., and Neer, R.M. 2003. The effects of parathyroid hormone, alendronate, or both in men with osteoporosis. *N. Engl. J. Med.* **349:** 1216–1226.

Fowlkes, J.L., Serra, D.M., Rosenberg, C.K., and Thrailkill, K.M. 1995. Insulin-like growth factor (IGF)-binding protein-3 (IGFBP-3) functions as an IGF-reversible inhibitor of IGFBP-4 proteolysis. *J. Biol. Chem.* **270:** 27481–27488.

Fragale, A., Tartaglia, M., Bernardini, S., Di Stasi, A.M., Di Rocco, C., Velardi, F., Teti, A., Battaglia, P.A., and Migliaccio, S. 1999. Decreased proliferation and altered differentiation in osteoblasts from genetically and clinically distinct craniosynostotic disorders. *Am. J. Pathol.* **154:** 1465–1477.

Funato, N., Ohtani, K., Ohyama, K., Kuroda, T., and Nakamura, M. 2001. Common regulation of growth arrest and differentiation of osteoblasts by helix-loop-helix factors. *Mol. Cell. Biol.* **21:** 7416–7428.

Garrison, K.R., Donell, S., Ryder, J., Shemilt, I., Mugford, M., Harvey, I., and Song, F. 2007. Clinical effectiveness and cost-effectiveness of bone morphogenetic proteins in the non-healing of fractures and spinal fusion: A systematic review. *Health Technol. Assess.* **11:** 1–150.

Gaur, T., Lengner, C.J., Hovhannisyan, H., Bhat, R.A., Bodine, P.V., Komm, B.S., Javed, A., van Wijnen, A.J., Stein, J.L., Stein, G.S., et al. 2005. Canonical WNT signaling promotes osteogenesis by directly stimulating Runx2 gene expression. *J. Biol. Chem.* **280:** 33132–33140.

Gautschi, O.P., Frey, S.P., and Zellweger, R. 2007. Bone morphogenetic proteins in clinical applications. *ANZ J. Surg.* **77:** 626–631.

Gazzerro, E., Deregowski, V., Vaira, S., and Canalis, E. 2005a. Overexpression of twisted gastrulation inhibits bone morphogenetic protein action and prevents osteoblast cell differentiation in vitro. *Endocrinology* **146:** 3875–3882.

Gazzerro, E., Pereira, R.C., Jorgetti, V., Olson, S., Economides, A.N., and Canalis, E. 2005b. Skeletal overexpression of gremlin impairs bone formation and causes osteopenia. *Endocrinology* **146:** 655–665.

Gazzerro, E., Smerdel-Ramoya, A., Zanotti, S., Stadmeyer, L., Durant, D., Economides, A.N., and Canalis, E. 2007. Conditional deletion of gremlin causes a transient increase in bone formation and bone mass. *J. Biol. Chem.* **282:** 31549–31557.

Geiser, A.G., Zeng, Q.Q., Sato, M., Helvering, L.M., Hirano, T., and Turner, C.H. 1998. Decreased bone mass and bone elasticity in mice lacking the transforming growth factor-β1 gene. *Bone* **23:** 87–93.

Glass III, D.A., Bialek, P., Ahn, J.D., Starbuck, M., Patel, M.S., Clevers, H., Taketo, M.M., Long, F., McMahon, A.P., Lang, R.A., et al. 2005. Canonical Wnt signaling in differentiated osteoblasts controls osteoclast differentiation. *Dev. Cell* **8:** 751–764.

Gong, Y., Slee, R.B., Fukai, N., Rawadi, G., Roman-Roman, S., Reginato, A.M., Wang, H., Cundy, T., Glorieux, F.H., Lev, D., et al. 2001. LDL receptor-related protein 5 (LRP5) affects bone accrual and eye development. *Cell* **107:** 513–523.

Govoni, K.E., Baylink, D.J., and Mohan, S. 2005. The multi-functional role of insulin-like growth factor binding proteins in bone. *Pediatr. Nephrol.* **20:** 261–268.

Grasser, W.A., Orlic, I., Borovecki, F., Riccardi, K.A., Simic, P., Vukicevic, S., and Paralkar, V.M. 2007. BMP-6 exerts its osteoinductive effect through activation of IGF-I and EGF pathways. *Int. Orthop.* **31:** 759–765.

Hembree, J.R., Agarwal, C., and Eckert, R.L. 1994. Epidermal growth factor suppresses insulin-like growth factor binding protein 3 levels in human papillomavirus type 16-immortalized cervical epithelial cells and thereby potentiates the effects of insulin-like growth factor 1. *Cancer Res.* **54:** 3160–3166.

Hill, T.P., Spater, D., Taketo, M.M., Birchmeier, W., and Hartmann, C. 2005. Canonical Wnt/β-catenin signaling prevents osteoblasts from differentiating into chondrocytes. *Dev. Cell* **8:** 727–738.

Hilton, M.J., Tu, X., Cook, J., Hu, H., and Long, F. 2005. Ihh controls cartilage development by antagonizing Gli3, but requires additional effectors to regulate osteoblast and vascular development. *Development* **132:** 4339–4351.

Holmen, S.L., Giambernardi, T.A., Zylstra, C.R., Buckner-Berghuis, B.D., Resau, J.H., Hess, J.F., Glatt, V., Bouxsein, M.L., Ai, M., Warman, M.L., et al. 2004. Decreased BMD and limb deformities in mice carrying mutations in both Lrp5 and Lrp6. *J. Bone Miner. Res.* **19:** 2033–2040.

Holmen, S.L., Zylstra, C.R., Mukherjee, A., Sigler, R.E., Faugere, M.C., Bouxsein, M.L., Deng, L., Clemens, T.L., and Williams, B.O. 2005. Essential role of β-catenin in postnatal bone acquisition. *J. Biol. Chem.* **280:** 21162–21168.

Horwitz, M.J., Tedesco, M.B., Gundberg, C., Garcia-Ocana, A., and Stewart, A.F. 2003. Short-term, high-dose parathyroid hormone-related protein as a skeletal anabolic agent for the treatment of postmenopausal osteoporosis. *J. Clin. Endocrinol. Metab.* **88:** 569–575.

Horwitz, M.J., Tedesco, M.B., Sereika, S.M., Syed, M.A., Garcia-Ocana, A., Bisello, A., Hollis, B.W., Rosen, C.J., Wysolmerski, J.J., Dann, P., et al. 2005. Continuous PTH and PTHrP infusion causes suppression of bone formation and discordant effects on 1,25(OH)2 vitamin D. *J. Bone Miner. Res.* **20:** 1792–1803.

Hu, H., Hilton, M.J., Tu, X., Yu, K., Ornitz, D.M., and Long, F. 2005. Sequential roles of Hedgehog and Wnt signaling in osteoblast development. *Development* **132:** 49–60.

Hung, I.H., Yu, K., Lavine, K.J., and Ornitz, D.M. 2007. FGF9 regulates early hypertrophic chondrocyte differentiation and skeletal vascularization in the developing stylopod. *Dev. Biol.* **307:** 300–313.

Hurley, M.M., Tetradis, S., Huang, Y.F., Hock, J., Kream, B.E., Raisz, L.G., and Sabbieti, M.G. 1999. Parathyroid hormone regulates the expression of fibroblast growth factor-2 mRNA and fibroblast growth factor receptor mRNA in osteoblastic cells. *J. Bone Miner. Res.* **14:** 776–783.

Hurley, M.M., Okada, Y., Xiao, L., Tanaka, Y., Ito, M., Okimoto, N., Nakamura, T., Rosen, C.J., Doetschman, T., and Coffin, J.D. 2006. Impaired bone anabolic response to parathyroid hormone in Fgf2$^{-/-}$ and Fgf2$^{+/-}$ mice. *Biochem. Biophys. Res. Commun.* **341:** 989–994.

Ibbotson, K.J., Harrod, J., Gowen, M., D'Souza, S., Smith, D.D., Winkler, M.E., Derynck, R., and Mundy, G.R. 1986. Human recombinant transforming growth factor α stimulates bone resorption and inhibits formation in vitro. *Proc. Natl. Acad. Sci.* **83**: 2228–2232.

Ishizuya, T., Yokose, S., Hori, M., Noda, T., Suda, T., Yoshiki, S., and Yamaguchi, A. 1997. Parathyroid hormone exerts disparate effects on osteoblast differentiation depending on exposure time in rat osteoblastic cells. *J. Clin. Invest.* **99**: 2961–2970.

Isogai, Y., Akatsu, T., Ishizuya, T., Yamaguchi, A., Hori, M., Takahashi, N., and Suda, T. 1996. Parathyroid hormone regulates osteoblast differentiation positively or negatively depending on the differentiation stages. *J. Bone Miner. Res.* **11**: 1384–1393.

Iwaniec, U.T., Mosekilde, L., Mitova-Caneva, N.G., Thomsen, J.S., and Wronski, T.J. 2002. Sequential treatment with basic fibroblast growth factor and PTH is more efficacious than treatment with PTH alone for increasing vertebral bone mass and strength in osteopenic ovariectomized rats. *Endocrinology* **143**: 2515–2526.

Iwaniec, U.T., Wronski, T.J., Liu, J., Rivera, M.F., Arzaga, R.R., Hansen, G., and Brommage, R. 2007. PTH stimulates bone formation in mice deficient in Lrp5. *J. Bone Miner. Res.* **22**: 394–402.

Jacob, A.L., Smith, C., Partanen, J., and Ornitz, D.M. 2006. Fibroblast growth factor receptor 1 signaling in the osteo-chondrogenic cell lineage regulates sequential steps of osteoblast maturation. *Dev. Biol.* **296**: 315–328.

Jemtland, R., Divieti, P., Lee, K., and Segre, G.V. 2003. Hedgehog promotes primary osteoblast differentiation and increases PTHrP mRNA expression and iPTHrP secretion. *Bone* **32**: 611–620.

Jiang, J., Lichtler, A.C., Gronowicz, G.A., Adams, D.J., Clark, S.H., Rosen, C.J., and Kream, B.E. 2006. Transgenic mice with osteoblast-targeted insulin-like growth factor-I show increased bone remodeling. *Bone* **39**: 494–504.

Jilka, R.L. 2007. Molecular and cellular mechanisms of the anabolic effect of intermittent PTH. *Bone* **40**: 1434–1446.

Juppner, H., Abou-Samra, A.B., Freeman, M., Kong, X.F., Schipani, E., Richards, J., Kolakowski Jr., L.F., Hock, J., Potts Jr., J.T., and Kronenberg, H.M., et al. 1991. A G protein-linked receptor for parathyroid hormone and parathyroid hormone-related peptide. *Science* **254**: 1024–1026.

Kakar, S., Einhorn, T.A., Vora, S., Miara, L.J., Hon, G., Wigner, N.A., Toben, D., Jacobsen, K.A., Al-Sebaei, M.O., Song, M., et al. 2007. Enhanced chondrogenesis and Wnt signaling in PTH-treated fractures. *J. Bone Miner. Res.* **22**: 1903–1912.

Kalajzic, I., Kalajzic, Z., Hurley, M.M., Lichtler, A.C., and Rowe, D.W. 2003. Stage specific inhibition of osteoblast lineage differentiation by FGF2 and noggin. *J. Cell. Biochem.* **88**: 1168–1176.

Kaneda, T., Nojima, T., Nakagawa, M., Ogasawara, A., Kaneko, H., Sato, T., Mano, H., Kumegawa, M., and Hakeda, Y. 2000. Endogenous production of TGF-β is essential for osteoclastogenesis induced by a combination of receptor activator of NF-κB ligand and macrophage-colony-stimulating factor. *J. Immunol.* **165**: 4254–4263.

Kang, S., Bennett, C.N., Gerin, I., Rapp, L.A., Hankenson, K.D., and MacDougald, O.A. 2007. Wnt signaling stimulates osteoblastogenesis of mesenchymal precursors by suppressing CCAAT/enhancer-binding protein α and peroxisome proliferator-activated receptor γ. *J. Biol. Chem.* **282**: 14515–14524.

Kaps, C., Hoffmann, A., Zilberman, Y., Pelled, G., Haupl, T., Sittinger, M., Burmester, G., Gazit, D., and Gross, G. 2004. Distinct roles of BMP receptors Type IA and IB in osteo-/chondrogenic differentiation in mesenchymal progenitors (C3H10T1/2). *Biofactors* **20**: 71–84.

Karaplis, A.C. 2001. PTHrP: Novel roles in skeletal biology. *Curr. Pharm. Des.* **7:** 655–670.

Karaplis, A.C., Luz, A., Glowacki, J., Bronson, R.T., Tybulewicz, V.L., Kronenberg, H.M., and Mulligan, R.C. 1994. Lethal skeletal dysplasia from targeted disruption of the parathyroid hormone-related peptide gene. *Genes Dev.* **8:** 277–289.

Kato, M., Patel, M.S., Levasseur, R., Lobov, I., Chang, B.H., Glass II, D.A., Hartmann, C., Li, L., Hwang, T.H., Brayton, C.F., et al. 2002. Cbfa1-independent decrease in osteoblast proliferation, osteopenia, and persistent embryonic eye vascularization in mice deficient in Lrp5, a Wnt coreceptor. *J. Cell Biol.* **157:** 303–314.

Kawaguchi, H., Nakamura, K., Tabata, Y., Ikada, Y., Aoyama, I., Anzai, J., Nakamura, T., Hiyama, Y., and Tamura, M. 2001. Acceleration of fracture healing in nonhuman primates by fibroblast growth factor-2. *J. Clin. Endocrinol. Metab.* **86:** 875–880.

Kawai, M., Mushiake, S., Bessho, K., Murakami, M., Namba, N., Kokubu, C., Michigami, T., and Ozono, K. 2007. Wnt/Lrp/β-catenin signaling suppresses adipogenesis by inhibiting mutual activation of PPARγ and C/EBPα. *Biochem. Biophys. Res. Commun.* **363:** 276–282.

Keller, H. and Kneissel, M. 2005. SOST is a target gene for PTH in bone. *Bone* **37:** 148–158.

Kim, H.J., Kim, J.H., Bae, S.C., Choi, J.Y., Kim, H.J., and Ryoo, H.M. 2003a. The protein kinase C pathway plays a central role in the fibroblast growth factor-stimulated expression and transactivation activity of Runx2. *J. Biol. Chem.* **278:** 319–326.

Kim, H.J., Lee, M.H., Park, H.S., Park, M.H., Lee, S.W., Kim, S.Y., Choi, J.Y., Shin, H.I., Kim, H.J., and Ryoo, H.M. 2003b. Erk pathway and activator protein 1 play crucial roles in FGF2-stimulated premature cranial suture closure. *Dev. Dyn.* **227:** 335–346.

Kronenberg, H.M. 2006. PTHrP and skeletal development. *Ann. N.Y. Acad. Sci.* **1068:** 1–13.

Lane, N.E., Kumer, J., Yao, W., Breunig, T., Wronski, T., Modin, G., and Kinney, J.H. 2003. Basic fibroblast growth factor forms new trabeculae that physically connect with pre-existing trabeculae, and this new bone is maintained with an anti-resorptive agent and enhanced with an anabolic agent in an osteopenic rat model. *Osteoporos. Int.* **14:** 374–382.

Lanske, B., Amling, M., Neff, L., Guiducci, J., Baron, R., and Kronenberg, H.M. 1999. Ablation of the PTHrP gene or the PTH/PTHrP receptor gene leads to distinct abnormalities in bone development. *J. Clin. Invest.* **104:** 399–407.

Lee, K.S., Kim, H.J., Li, Q.L., Chi, X.Z., Ueta, C., Komori, T., Wozney, J.M., Kim, E.G., Choi, J.Y., Ryoo, H.M., et al. 2000. Runx2 is a common target of transforming growth factor β1 and bone morphogenetic protein 2, and cooperation between Runx2 and Smad5 induces osteoblast-specific gene expression in the pluripotent mesenchymal precursor cell line C2C12. *Mol. Cell. Biol.* **20:** 8783–8792.

Lee, M.H., Kim, Y.J., Kim, H.J., Park, H.D., Kang, A.R., Kyung, H.M., Sung, J.H., Wozney, J.M., Kim, H.J., and Ryoo, H.M. 2003a. BMP-2-induced Runx2 expression is mediated by Dlx5, and TGF-β 1 opposes the BMP-2-induced osteoblast differentiation by suppression of Dlx5 expression. *J. Biol. Chem.* **278:** 34387–34394.

Lee, M.H., Kwon, T.G., Park, H.S., Wozney, J.M., and Ryoo, H.M. 2003b. BMP-2-induced Osterix expression is mediated by Dlx5 but is independent of Runx2. *Biochem. Biophys. Res. Commun.* **309:** 689–694.

Lee, J.Y., Kim, K.H., Shin, S.Y., Rhyu, I.C., Lee, Y.M., Park, Y.J., Chung, C.P., and Lee, S.J. 2006. Enhanced bone formation by transforming growth factor-β1-releasing collagen/chitosan microgranules. *J. Biomed. Mater. Res. A* **76:** 530–539.

Lemonnier, J., Hay, E., Delannoy, P., Fromigue, O., Lomri, A., Modrowski, D., and Marie, P.J. 2001a. Increased osteoblast apoptosis in apert craniosynostosis: Role of protein kinase C and interleukin-1. *Am. J. Pathol.* **158**: 1833–1842.

Lemonnier, J., Hay, E., Delannoy, P., Lomri, A., Modrowski, D., Caverzasio, J., and Marie, P.J. 2001b. Role of N-cadherin and protein kinase C in osteoblast gene activation induced by the S252W fibroblast growth factor receptor 2 mutation in Apert craniosynostosis. *J. Bone Miner. Res.* **16**: 832–845.

Lewicki, M., Paez, H., and Mandalunis, P.M. 2006. Effect of lithium carbonate on sub-chondral bone in sexually mature Wistar rats. *Exp. Toxicol. Pathol.* **58**: 197–201.

Li, X., Liu, P., Liu, W., Maye, P., Zhang, J., Zhang, Y., Hurley, M., Guo, C., Boskey, A., Sun, L., et al. 2005a. Dkk2 has a role in terminal osteoblast differentiation and mineralized matrix formation. *Nat. Genet.* **37**: 945–952.

Li, X., Zhang, Y., Kang, H., Liu, W., Liu, P., Zhang, J., Harris, S.E., and Wu, D. 2005b. Sclerostin binds to LRP5/6 and antagonizes canonical Wnt signaling. *J. Biol. Chem.* **280**: 19883–19887.

Li, J., Sarosi, I., Cattley, R.C., Pretorius, J., Asuncion, F., Grisanti, M., Morony, S., Adamu, S., Geng, Z., Qiu, W., et al. 2006. Dkk1-mediated inhibition of Wnt signaling in bone results in osteopenia. *Bone* **39**: 754–766.

Li, S., Quarto, N., and Longaker, M.T. 2007. Dura mater-derived FGF-2 mediates mitogenic signaling in calvarial osteoblasts. *Am. J. Physiol. Cell Physiol.* **293**: C1834–C1842.

Li, X., Ominsky, M.S., Niu, Q.T., Sun, N., Daugherty, B., D'Agostin, D., Kurahara, C., Gao, Y., Cao, J., Gong, J., et al. 2008. Targeted deletion of the sclerostin gene in mice results in increased bone formation and bone strength. *J. Bone Miner. Res.* **23**: 860–869.

Lievens, P.M. and Liboi, E. 2003. The thanatophoric dysplasia type II mutation hampers complete maturation of fibroblast growth factor receptor 3 (FGFR3), which activates signal transducer and activator of transcription 1 (STAT1) from the endoplasmic retic-ulum. *J. Biol. Chem.* **278**: 17344–17349.

Little, R.D., Carulli, J.P., Del Mastro, R.G., Dupuis, J., Osborne, M., Folz, C., Manning, S.P., Swain, P.M., Zhao, S.C., Eustace, B., et al. 2002. A mutation in the LDL receptor-related protein 5 gene results in the autosomal dominant high-bone-mass trait. *Am. J. Hum. Genet.* **70**: 11–19.

Liu, C.C. and Kalu, D.N. 1990. Human parathyroid hormone-(1-34) prevents bone loss and augments bone formation in sexually mature ovariectomized rats. *J. Bone Miner. Res.* **5**: 973–982.

Liu, J.P., Baker, J., Perkins, A.S., Robertson, E.J., and Efstratiadis, A. 1993. Mice carrying null mutations of the genes encoding insulin-like growth factor I (Igf-1) and type 1 IGF receptor (Igf1r). *Cell* **75**: 59–72.

Liu, Z., Xu, J., Colvin, J.S., and Ornitz, D.M. 2002. Coordination of chondrogenesis and osteogenesis by fibroblast growth factor 18. *Genes Dev.* **16**: 859–869.

Locklin, R.M., Khosla, S., Turner, R.T., and Riggs, B.L. 2003. Mediators of the biphasic responses of bone to intermittent and continuously administered parathyroid hormone. *J. Cell. Biochem.* **89**: 180–190.

Lofqvist, C., Chen, J., Connor, K.M., Smith, A.C., Aderman, C.M., Liu, N., Pintar, J.E., Ludwig, T., Hellstrom, A., and Smith, L.E. 2007. IGFBP3 suppresses retinopathy through suppression of oxygen-induced vessel loss and promotion of vascular regrowth. *Proc. Natl. Acad. Sci.* **104**: 10589–10594.

Long, F., Chung, U.I., Ohba, S., McMahon, J., Kronenberg, H.M., and McMahon, A.P. 2004. Ihh signaling is directly required for the osteoblast lineage in the endochondral skele-ton. *Development* **131**: 1309–1318.

Lories, R.J., Peeters, J., Bakker, A., Tylzanowski, P., Derese, I., Schrooten, J., Thomas, J.T., and

Luyten, F.P. 2007. Articular cartilage and biomechanical properties of the long bones in Frzb-knockout mice. *Arthritis Rheum.* **56:** 4095–4103.

Mak, K.K., Chen, M.H., Day, T.F., Chuang, P.T., and Yang, Y. 2006. Wnt/β-catenin signaling interacts differentially with Ihh signaling in controlling endochondral bone and synovial joint formation. *Development* **133:** 3695–3707.

Mansukhani, A., Bellosta, P., Sahni, M., and Basilico, C. 2000. Signaling by fibroblast growth factors (FGF) and fibroblast growth factor receptor 2 (FGFR2)-activating mutations blocks mineralization and induces apoptosis in osteoblasts. *J. Cell Biol.* **149:** 1297–1308.

Marie, P.J., Coffin, J.D., and Hurley, M.M. 2005. FGF and FGFR signaling in chondrodysplasias and craniosynostosis. *J. Cell. Biochem.* **96:** 888–896.

Masi, L. and Brandi, M.L. 2005. Molecular, biochemical and cellular biology of PTH anabolic action. *J. Endocrinol. Invest.* (suppl. 8) **28:** 37–40.

Mbalaviele, G., Sheikh, S., Stains, J.P., Salazar, V.S., Cheng, S.L., Chen, D., and Civitelli, R. 2005. β-Catenin and BMP-2 synergize to promote osteoblast differentiation and new bone formation. *J. Cell. Biochem.* **94:** 403–418.

Miao, D., He, B., Karaplis, A.C., and Goltzman, D. 2002. Parathyroid hormone is essential for normal fetal bone formation. *J. Clin. Invest.* **109:** 1173–1182.

Miao, D., He, B., Lanske, B., Bai, X.Y., Tong, X.K., Hendy, G.N., Goltzman, D., and Karaplis, A.C. 2004a. Skeletal abnormalities in Pth-null mice are influenced by dietary calcium. *Endocrinology* **145:** 2046–2053.

Miao, D., Li, J., Xue, Y., Su, H., Karaplis, A.C., and Goltzman, D. 2004b. Parathyroid hormone-related peptide is required for increased trabecular bone volume in parathyroid hormone-null mice. *Endocrinology* **145:** 3554–3562.

Miao, D., Liu, H., Plut, P., Niu, M., Huo, R., Goltzman, D., and Henderson, J.E. 2004c. Impaired endochondral bone development and osteopenia in Gli2-deficient mice. *Exp. Cell Res.* **294:** 210–222.

Miao, D., He, B., Jiang, Y., Kobayashi, T., Soroceanu, M.A., Zhao, J., Su, H., Tong, X., Amizuka, N., Gupta, A., et al. 2005. Osteoblast-derived PTHrP is a potent endogenous bone anabolic agent that modifies the therapeutic efficacy of administered PTH 1-34. *J. Clin. Invest.* **115:** 2402–2411.

Mishina, Y., Starbuck, M.W., Gentile, M.A., Fukuda, T., Kasparcova, V., Seedor, J.G., Hanks, M.C., Amling, M., Pinero, G.J., Harada, S., et al. 2004. Bone morphogenetic protein type IA receptor signaling regulates postnatal osteoblast function and bone remodeling. *J. Biol. Chem.* **279:** 27560–27566.

Miyakoshi, N., Kasukawa, Y., Linkhart, T.A., Baylink, D.J., and Mohan, S. 2001a. Evidence that anabolic effects of PTH on bone require IGF-I in growing mice. *Endocrinology* **142:** 4349–4356.

Miyakoshi, N., Qin, X., Kasukawa, Y., Richman, C., Srivastava, A.K., Baylink, D.J., and Mohan, S. 2001b. Systemic administration of insulin-like growth factor (IGF)-binding protein-4 (IGFBP-4) increases bone formation parameters in mice by increasing IGF bioavailability via an IGFBP-4 protease-dependent mechanism. *Endocrinology* **142:** 2641–2648.

Miyakoshi, N., Richman, C., Kasukawa, Y., Linkhart, T.A., Baylink, D.J., and Mohan, S. 2001c. Evidence that IGF-binding protein-5 functions as a growth factor. *J. Clin. Invest.* **107:** 73–81.

Mohan, S. 1993. Insulin-like growth factor binding proteins in bone cell regulation. *Growth Regul.* **3:** 67–70.

Mohan, S., Nakao, Y., Honda, Y., Landale, E., Leser, U., Dony, C., Lang, K., and Baylink, D.J.

1995. Studies on the mechanisms by which insulin-like growth factor (IGF) binding protein-4 (IGFBP-4) and IGFBP-5 modulate IGF actions in bone cells. *J. Biol. Chem.* **270:** 20424–20431.

Montero, A., Okada, Y., Tomita, M., Ito, M., Tsurukami, H., Nakamura, T., Doetschman, T., Coffin, J.D., and Hurley, M.M. 2000. Disruption of the fibroblast growth factor-2 gene results in decreased bone mass and bone formation. *J. Clin. Invest.* **105:** 1085–1093.

Morvan, F., Boulukos, K., Clement-Lacroix, P., Roman, S., Suc-Royer, I., Vayssiere, B., Ammann, P., Martin, P., Pinho, S., Pognonec, P., et al. 2006. Deletion of a single allele of the Dkk1 gene leads to an increase in bone formation and bone mass. *J. Bone Miner. Res.* **21:** 934–945.

Moursi, A.M., Winnard, P.L., Winnard, A.V., Rubenstrunk, J.M., and Mooney, M.P. 2002. Fibroblast growth factor 2 induces increased calvarial osteoblast proliferation and cranial suture fusion. *Cleft Palate Craniofac. J.* **39:** 487–496.

Murakami, S., Nifuji, A., and Noda, M. 1997. Expression of Indian hedgehog in osteoblasts and its posttranscriptional regulation by transforming growth factor-β. *Endocrinology* **138:** 1972–1978.

Murali, S.G., Liu, X., Nelson, D.W., Hull, A.K., Grahn, M., Clayton, M.K., Pintar, J.E., and Ney, D.M. 2007. Intestinotrophic effects of exogenous IGF-I are not diminished in IGF binding protein-5 knockout mice. *Am. J. Physiol. Regul. Integr. Comp. Physiol.* **292:** R2144–R2150.

Nagai, H., Tsukuda, R., Yamasaki, H., and Mayahara, H. 1999. Systemic injection of FGF-2 stimulates endocortical bone modelling in SAMP6, a murine model of low turnover osteopenia. *J. Vet. Med. Sci.* **61:** 869–875.

Nakajima, A., Shimizu, S., Moriya, H., and Yamazaki, M. 2003. Expression of fibroblast growth factor receptor-3 (FGFR3), signal transducer and activator of transcription-1, and cyclin-dependent kinase inhibitor p21 during endochondral ossification: Differential role of FGFR3 in skeletal development and fracture repair. *Endocrinology* **144:** 4659–4668.

Nakanishi, R., Akiyama, H., Kimura, H., Otsuki, B., Shimizu, M., Tsuboyama, T., and Nakamura, T. 2008. Osteoblast-targeted expression of Sfrp4 in mice results in low bone mass. *J. Bone Miner. Res.* **23:** 271–277.

Ning, Y., Schuller, A.G., Bradshaw, S., Rotwein, P., Ludwig, T., Frystyk, J., and Pintar, J.E. 2006. Diminished growth and enhanced glucose metabolism in triple knockout mice containing mutations of insulin-like growth factor binding protein-3, -4, and -5. *Mol. Endocrinol.* **20:** 2173–2186.

Ning, Y., Hoang, B., Schuller, A.G., Cominski, T.P., Hsu, M.S., Wood, T.L., and Pintar, J.E. 2007. Delayed mammary gland involution in mice with mutation of the insulin-like growth factor binding protein 5 gene. *Endocrinology* **148:** 2138–2147.

Ning, Y., Schuller, A.G., Conover, C.A., and Pintar, J.E. 2008. Insulin-like growth factor (IGF) binding protein-4 is both a positive and negative regulator of IGF activity in vivo. *Mol. Endocrinol.* **22:** 1213–1225.

Nohe, A., Keating, E., Knaus, P., and Petersen, N.O. 2004. Signal transduction of bone morphogenetic protein receptors. *Cell. Signal.* **16:** 291–299.

Ohbayashi, N., Shibayama, M., Kurotaki, Y., Imanishi, M., Fujimori, T., Itoh, N., and Takada, S. 2002. FGF18 is required for normal cell proliferation and differentiation during osteogenesis and chondrogenesis. *Genes Dev.* **16:** 870–879.

Okamoto, M., Murai, J., Yoshikawa, H., and Tsumaki, N. 2006. Bone morphogenetic proteins in bone stimulate osteoclasts and osteoblasts during bone development. *J. Bone*

Miner. Res. **21:** 1022–1033.

Ominsky, M.S., Warmington, K.S., Asuncion, F.J., Tan, H.L., Grisanti, M.S., Geng, Z., Stephens, P., Henry, A., Lawson, A., Lightwood, D., et al. 2006. Sclerostin monoclonal antibody treatment increases bone strength in aged osteopenic ovariectomized rats. *J. Bone Miner. Res.* **21:** S44. (Abstr.)

Ornitz, D.M. 2005. FGF signaling in the developing endochondral skeleton. *Cytokine Growth Factor Rev.* **16:** 205–213.

Ovchinnikov, D.A., Selever, J., Wang, Y., Chen, Y.T., Mishina, Y., Martin, J.F., and Behringer, R.R. 2006. BMP receptor type IA in limb bud mesenchyme regulates distal outgrowth and patterning. *Dev. Biol.* **295:** 103–115.

Padhi, D., Stouch, B., Jang G., Fang, L., Darling, M., Glise, H., Robinson, M., Harris, S., and Posvar, E. 2007. Anti-slerostin antibody increases markers of bone formation in healthy menopausal women. *J. Bone Miner. Res.* (suppl. 1) **22:** S37. (Abstr.)

Palermo, C., Manduca, P., Gazzerro, E., Foppiani, L., Segat, D., and Barreca, A. 2004. Potentiating role of IGFBP-2 on IGF-II-stimulated alkaline phosphatase activity in differentiating osteoblasts. *Am. J. Physiol. Endocrinol. Metab.* **286:** E648–E657.

Parisien, M., Silverberg, S.J., Shane, E., de la Cruz, L., Lindsay, R., Bilezikian, J.P., and Dempster, D.W. 1990. The histomorphometry of bone in primary hyperparathyroidism: Preservation of cancellous bone structure. *J. Clin. Endocrinol. Metab.* **70:** 930–938.

Pelton, R.W., Johnson, M.D., Perkett, E.A., Gold, L.I., and Moses, H.L. 1991a. Expression of transforming growth factor-β1, -β2, and -β3 mRNA and protein in the murine lung. *Am. J. Respir. Cell Mol. Biol.* **5:** 522–530.

Pelton, R.W., Saxena, B., Jones, M., Moses, H.L., and Gold, L.I. 1991b. Immunohistochemical localization of TGFβ1, TGFβ2, and TGFβ3 in the mouse embryo: Expression patterns suggest multiple roles during embryonic development. *J. Cell Biol.* **115:** 1091–1105.

Power, R.A., Iwaniec, U.T., Magee, K.A., Mitova-Caneva, N.G., and Wronski, T.J. 2004. Basic fibroblast growth factor has rapid bone anabolic effects in ovariectomized rats. *Osteoporos. Int.* **15:** 716–723.

Qin, L., Raggatt, L.J., and Partridge, N.C. 2004. Parathyroid hormone: A double-edged sword for bone metabolism. *Trends Endocrinol. Metab.* **15:** 60–65.

Qin, L., Tamasi, J., Raggatt, L., Li, X., Feyen, J.H., Lee, D.C., Dicicco-Bloom, E., and Partridge, N.C. 2005. Amphiregulin is a novel growth factor involved in normal bone development and in the cellular response to parathyroid hormone stimulation. *J. Biol. Chem.* **280:** 3974–3981.

Reddi, A.H. 1997. Bone morphogenetic proteins: An unconventional approach to isolation of first mammalian morphogens. *Cytokine Growth Factor Rev.* **8:** 11–20.

Rice, D.P., Aberg, T., Chan, Y., Tang, Z., Kettunen, P.J., Pakarinen, L., Maxson, R.E., and Thesleff, I. 2000. Integration of FGF and TWIST in calvarial bone and suture development. *Development* **127:** 1845–1855.

Richardson, L., Zioncheck, T.F., Amento, E.P., Deguzman, L., Lee, W.P., Xu, Y., and Beck, L.S. 1993. Characterization of radioiodinated recombinant human TGF-β 1 binding to bone matrix within rabbit skull defects. *J. Bone Miner. Res.* **8:** 1407–1414.

Rodan, S.B., Wesolowski, G., Thomas, K.A., Yoon, K., and Rodan, G.A. 1989. Effects of acidic and basic fibroblast growth factors on osteoblastic cells. *Connect. Tissue Res.* **20:** 283–288.

Rodda, S.J. and McMahon, A.P. 2006. Distinct roles for Hedgehog and canonical Wnt signaling in specification, differentiation and maintenance of osteoblast progenitors. *Development* **133:** 3231–3244.

Rosen, C.J., Ackert-Bicknell, C.L., Adamo, M.L., Shultz, K.L., Rubin, J., Donahue, L.R.,

Horton, L.G., Delahunty, K.M., Beamer, W.G., Sipos, J., et al. 2004. Congenic mice with low serum IGF-I have increased body fat, reduced bone mineral density, and an altered osteoblast differentiation program. *Bone* **35:** 1046–1058.

Ross, S.E., Hemati, N., Longo, K.A., Bennett, C.N., Lucas, P.C., Erickson, R.L., and MacDougald, O.A. 2000. Inhibition of adipogenesis by Wnt signaling. *Science* **289:** 950–953.

Rutter, M.M., Markoff, E., Clayton, L., Akeno, N., Zhao, G., Clemens, T.L., and Chernausek, S.D. 2005. Osteoblast-specific expression of insulin-like growth factor-1 in bone of transgenic mice induces insulin-like growth factor binding protein-5. *Bone* **36:** 224–231.

Salih, D.A., Mohan, S., Kasukawa, Y., Tripathi, G., Lovett, F.A., Anderson, N.F., Carter, E.J., Wergedal, J.E., Baylink, D.J., and Pell, J.M. 2005. Insulin-like growth factor-binding protein-5 induces a gender-related decrease in bone mineral density in transgenic mice. *Endocrinology* **146:** 931–940.

Sawakami, K., Robling, A.G., Ai, M., Pitner, N.D., Liu, D., Warden, S.J., Li, J., Maye, P., Rowe, D.W., Duncan, R.L., et al. 2006. The Wnt co-receptor LRP5 is essential for skeletal mechanotransduction but not for the anabolic bone response to parathyroid hormone treatment. *J. Biol. Chem.* **281:** 23698–23711.

Schedlich, L.J., Muthukaruppan, A., O'Han, M.K., and Baxter, R.C. 2007. Insulin-like growth factor binding protein-5 interacts with the vitamin D receptor and modulates the vitamin D response in osteoblasts. *Mol. Endocrinol.* **21:** 2378–2390.

Schilling, T., Kuffner, R., Klein-Hitpass, L., Zimmer, R., Jakob, F., and Schutze, N. 2008. Microarray analyses of transdifferentiated mesenchymal stem cells. *J. Cell. Biochem.* **103:** 413–433.

Schindeler, A., Ramachandran, M., Godfrey, C., Morse, A., McDonald, M., Mikulec, K., and Little, D.G. 2008. Modeling bone morphogenetic protein and bisphosphonate combination therapy in wild-type and Nf1 haploinsufficient mice. *J. Orthop. Res.* **26:** 65–74.

Schmid, C., Rutishauser, J., Schläpfer, I., Froesch, E.R., and Zapf, J. 1991. Intact but not truncated insulin-like growth factor binding protein-3 (IGFBP-3) blocks IGF I-induced stimulation of osteoblasts: Control of IGF signalling to bone cells by IGFBP-3-specific proteolysis? *Biochem. Biophys. Res. Commun.* **179:** 579–585.

Schmidmaier, G., Schwabe, P., Wildemann, B., and Haas, N.P. 2007. Use of bone morphogenetic proteins for treatment of non-unions and future perspectives. *Injury* (suppl. 4) **38:** S35–S41.

Schneider, M.R., Dahlhoff, M., Herbach, N., Renner-Mueller, I., Dalke, C., Puk, O., Graw, J., Wanke, R., and Wolf, E. 2005. Betacellulin overexpression in transgenic mice causes disproportionate growth, pulmonary hemorrhage syndrome, and complex eye pathology. *Endocrinology* **146:** 5237–5246.

Sciaudone, M., Gazzerro, E., Priest, L., Delany, A.M., and Canalis, E. 2003. Notch 1 impairs osteoblastic cell differentiation. *Endocrinology* **144:** 5631–5639.

Sells Galvin, R.J., Gatlin, C.L., Horn, J.W., and Fuson, T.R. 1999. TGF-β enhances osteoclast differentiation in hematopoietic cell cultures stimulated with RANKL and M-CSF. *Biochem. Biophys. Res. Commun.* **265:** 233–239.

Semenov, M., Tamai, K., and He, X. 2005. SOST is a ligand for LRP5/LRP6 and a Wnt signaling inhibitor. *J. Biol. Chem.* **280:** 26770–26775.

Shimoaka, T., Ogasawara, T., Yonamine, A., Chikazu, D., Kawano, H., Nakamura, K., Itoh, N., and Kawaguchi, H. 2002. Regulation of osteoblast, chondrocyte, and osteoclast functions by fibroblast growth factor (FGF)-18 in comparison with FGF-2 and FGF-10. *J. Biol. Chem.* **277:** 7493–7500.

Shimoyama, A., Wada, M., Ikeda, F., Hata, K., Matsubara, T., Nifuji, A., Noda, M., Amano, K., Yamaguchi, A., Nishimura, R., et al. 2007. Ihh/Gli2 signaling promotes osteoblast differentiation by regulating Runx2 expression and function. *Mol. Biol. Cell* **18:** 2411–2418.

Sibilia, M., Wagner, B., Hoebertz, A., Elliott, C., Marino, S., Jochum, W., and Wagner, E.F. 2003. Mice humanised for the EGF receptor display hypomorphic phenotypes in skin, bone and heart. *Development* **130:** 4515–4525.

Silha, J.V. and Murphy, L.J. 2002. Insights from insulin-like growth factor binding protein transgenic mice. *Endocrinology* **143:** 3711–3714.

Silha, J.V., Mishra, S., Rosen, C.J., Beamer, W.G., Turner, R.T., Powell, D.R., and Murphy, L.J. 2003. Perturbations in bone formation and resorption in insulin-like growth factor binding protein-3 transgenic mice. *J. Bone Miner. Res.* **18:** 1834–1841.

Simic, P., Culej, J.B., Orlic, I., Grgurevic, L., Draca, N., Spaventi, R., and Vukicevic, S. 2006. Systemically administered bone morphogenetic protein-6 restores bone in aged ovariectomized rats by increasing bone formation and suppressing bone resorption. *J. Biol. Chem.* **281:** 25509–25521.

Sobue, T., Naganawa, T., Xiao, L., Okada, Y., Tanaka, Y., Ito, M., Okimoto, N., Nakamura, T., Coffin, J.D., and Hurley, M.M. 2005. Over-expression of fibroblast growth factor-2 causes defective bone mineralization and osteopenia in transgenic mice. *J. Cell. Biochem.* **95:** 83–94.

Spater, D., Hill, T.P., O'Sullivan, R.J., Gruber, M., Conner, D.A., and Hartmann, C. 2006. Wnt9a signaling is required for joint integrity and regulation of *Ihh* during chondrogenesis. *Development* **133:** 3039–3049.

Spencer, G.J., Utting, J.C., Etheridge, S.L., Arnett, T.R., and Genever, P.G. 2006. Wnt signalling in osteoblasts regulates expression of the receptor activator of NFκB ligand and inhibits osteoclastogenesis in vitro. *J. Cell Sci.* **119:** 1283–1296.

St-Jacques, B., Hammerschmidt, M., and McMahon, A.P. 1999. Indian hedgehog signaling regulates proliferation and differentiation of chondrocytes and is essential for bone formation. *Genes Dev.* **13:** 2072–2086.

Su, W.C., Kitagawa, M., Xue, N., Xie, B., Garofalo, S., Cho, J., Deng, C., Horton, W.A., and Fu, X.Y. 1997. Activation of Stat1 by mutant fibroblast growth-factor receptor in thanatophoric dysplasia type II dwarfism. *Nature* **386:** 288–292.

Suzuki, A., Palmer, G., Bonjour, J.P., and Caverzasio, J. 2000. Stimulation of sodium-dependent phosphate transport and signaling mechanisms induced by basic fibroblast growth factor in MC3T3-E1 osteoblast-like cells. *J. Bone Miner. Res.* **15:** 95–102.

Suzuki, A., Ozono, K., Kubota, T., Kondou, H., Tachikawa, K., and Michigami, T. 2008. PTH/cAMP/PKA signaling facilitates canonical Wnt signaling via inactivation of glycogen synthase kinase-3β in osteoblastic Saos-2 cells. *J. Cell. Biochem.* **104:** 304–317.

Takada, I., Mihara, M., Suzawa, M., Ohtake, F., Kobayashi, S., Igarashi, M., Youn, M.Y., Takeyama, K., Nakamura, T., Mezaki, Y., et al. 2007a. A histone lysine methyltransferase activated by non-canonical Wnt signalling suppresses PPAR-γ transactivation. *Nat. Cell Biol.* **9:** 1273–1285.

Takada, I., Suzawa, M., Matsumoto, K., and Kato, S. 2007b. Suppression of PPAR transactivation switches cell fate of bone marrow stem cells from adipocytes into osteoblasts. *Ann. N.Y. Acad. Sci.* **1116:** 182–195.

Takai, H., Kanematsu, M., Yano, K., Tsuda, E., Higashio, K., Ikeda, K., Watanabe, K., and Yamada, Y. 1998. Transforming growth factor-β stimulates the production of osteoprotegerin/osteoclastogenesis inhibitory factor by bone marrow stromal cells. *J. Biol. Chem.* **273:** 27091–27096.

Tanaka, H., Moriwake, T., Matsuoka, Y., Nakamura, T., and Seino, Y. 1998. Potential role of rhIGF-I/IGFBP-3 in maintaining skeletal mass in space. *Bone* (suppl. 5) **22:** 145S–147S.

Tang, K.T., Capparelli, C., Stein, J.L., Stein, G.S., Lian, J.B., Huber, A.C., Braverman, L.E., and DeVito, W.J. 1996. Acidic fibroblast growth factor inhibits osteoblast differentiation in vitro: Altered expression of collagenase, cell growth-related, and mineralization-associated genes. *J. Cell. Biochem.* **61:** 152–166.

Tanimoto, Y., Yokozeki, M., Hiura, K., Matsumoto, K., Nakanishi, H., Matsumoto, T., Marie, P.J., and Moriyama, K. 2004. A soluble form of fibroblast growth factor receptor 2 (FGFR2) with S252W mutation acts as an efficient inhibitor for the enhanced osteoblastic differentiation caused by FGFR2 activation in Apert syndrome. *J. Biol. Chem.* **279:** 45926–45934.

Thirunavukkarasu, K., Miles, R.R., Halladay, D.L., Yang, X., Galvin, R.J., Chandrasekhar, S., Martin, T.J., and Onyia, J.E. 2001. Stimulation of osteoprotegerin (OPG) gene expression by transforming growth factor-β (TGF-β). Mapping of the OPG promoter region that mediates TGF-β effects. *J. Biol. Chem.* **276:** 36241–36250.

Tobimatsu, T., Kaji, H., Sowa, H., Naito, J., Canaff, L., Hendy, G.N., Sugimoto, T., and Chihara, K. 2006. Parathyroid hormone increases β-catenin levels through Smad3 in mouse osteoblastic cells. *Endocrinology* **147:** 2583–2590.

Tu, X., Joeng, K.S., Nakayama, K.I., Nakayama, K., Rajagopal, J., Carroll, T.J., McMahon, A.P., and Long, F. 2007. Noncanonical Wnt signaling through G protein-linked PKCδ activation promotes bone formation. *Dev. Cell* **12:** 113–127.

Tylzanowski, P., Mebis, L., and Luyten, F.P. 2006. The Noggin null mouse phenotype is strain dependent and haploinsufficiency leads to skeletal defects. *Dev. Dyn.* **235:** 1599–1607.

Ulsamer, A., Ortuño, M.J., Ruiz, S., Susperregui, A.R., Osses, N., Rosa, J.L., and Ventura, F. 2008. BMP-2 induces Osterix expression through upregulation of DLX5 and its phosphorylation by p38. *J. Biol. Chem.* **283:** 3816–3826.

Uzawa, T., Hori, M., Ejiri, S., and Ozawa, H. 1995. Comparison of the effects of intermittent and continuous administration of human parathyroid hormone(1-34) on rat bone. *Bone* **16:** 477–484.

Valverde-Franco, G., Liu, H., Davidson, D., Chai, S., Valderrama-Carvajal, H., Goltzman, D., Ornitz, D.M., and Henderson, J.E. 2004. Defective bone mineralization and osteopenia in young adult FGFR3$^{-/-}$ mice. *Hum. Mol. Genet.* **13:** 271–284.

van Bezooijen, R.L., Roelen, B.A., Visser, A., van der Wee-Pals, L., de Wilt, E., Karperien, M., Hamersma, H., Papapoulos, S.E., ten Dijke, P., and Lowik, C.W. 2004. Sclerostin is an osteocyte-expressed negative regulator of bone formation, but not a classical BMP antagonist. *J. Exp. Med.* **199:** 805–814.

van Bezooijen, R.L., Svensson, J.P., Eefting, D., Visser, A., van der Horst, G., Karperien, M., Quax, P.H., Vrieling, H., Papapoulos, S.E., ten Dijke, P., et al. 2007. Wnt but not BMP signaling is involved in the inhibitory action of sclerostin on BMP-stimulated bone formation. *J. Bone Miner. Res.* **22:** 19–28.

Vestergaard, P., Rejnmark, L., and Mosekilde, L. 2005. Reduced relative risk of fractures among users of lithium. *Calcif. Tissue Int.* **77:** 1–8.

von Stechow, D., Zurakowski, D., Pettit, A.R., Muller, R., Gronowicz, G., Chorev, M., Otu, H., Libermann, T., and Alexander, J.M. 2004. Differential transcriptional effects of PTH and estrogen during anabolic bone formation. *J. Cell. Biochem.* **93:** 476–490.

Wang, Y., Sakata, T., Elalieh, H.Z., Munson, S.J., Burghardt, A., Majumdar, S., Halloran, B.P., and Bikle, D.D. 2006. Gender differences in the response of CD-1 mouse bone to parathyroid hormone: Potential role of IGF-I. *J. Endocrinol.* **189:** 279–287.

Wang, Y., Nishida, S., Boudignon, B.M., Burghardt, A., Elalieh, H.Z., Hamilton, M.M., Majumdar, S., Halloran, B.P., Clemens, T.L., and Bikle, D.D. 2007. IGF-I receptor is required for the anabolic actions of parathyroid hormone on bone. *J. Bone Miner. Res.* **22:** 1329–1337.

Webster, M.K. and Donoghue, D.J. 1997. FGFR activation in skeletal disorders: Too much of a good thing. *Trends Genet.* **13:** 178–182.

Winkler, D.G., Sutherland, M.K., Geoghegan, J.C., Yu, C., Hayes, T., Skonier, J.E., Shpektor, D., Jonas, M., Kovacevich, B.R., Staehling-Hampton, K., et al. 2003. Osteocyte control of bone formation via sclerostin, a novel BMP antagonist. *EMBO J.* **22:** 6267–6276.

Winkler, D.G., Sutherland, M.S., Ojala, E., Turcott, E., Geoghegan, J.C., Shpektor, D., Skonier, J.E., Yu, C., and Latham, J.A. 2005. Sclerostin inhibition of Wnt-3a-induced C3H10T1/2 cell differentiation is indirect and mediated by bone morphogenetic proteins. *J. Biol. Chem.* **280:** 2498–2502.

Wood, T.L., Rogler, L., Streck, R.D., Cerro, J., Green, B., Grewal, A., and Pintar, J.E. 1993. Targeted disruption of IGFBP-2 gene. *Growth Regul.* **3:** 5–8.

Wood, T.L., Rogler, L.E., Czick, M.E., Schuller, A.G., and Pintar, J.E. 2000. Selective alterations in organ sizes in mice with a targeted disruption of the insulin-like growth factor binding protein-2 gene. *Mol. Endocrinol.* **14:** 1472–1482.

Wozney, J.M., Rosen, V., Celeste, A.J., Mitsock, L.M., Whitters, M.J., Kriz, R.W., Hewick, R.M., and Wang, E.A. 1988. Novel regulators of bone formation: Molecular clones and activities. *Science* **242:** 1528–1534.

Wu, X.B., Li, Y., Schneider, A., Yu, W., Rajendren, G., Iqbal, J., Yamamoto, M., Alam, M., Brunet, L.J., Blair, H.C., et al. 2003. Impaired osteoblastic differentiation, reduced bone formation, and severe osteoporosis in noggin-overexpressing mice. *J. Clin. Invest.* **112:** 924–934.

Xiao, L., Naganawa, T., Obugunde, E., Gronowicz, G., Ornitz, D.M., Coffin, J.D., and Hurley, M.M. 2004. Stat1 controls postnatal bone formation by regulating fibroblast growth factor signaling in osteoblasts. *J. Biol. Chem.* **279:** 27743–27752.

Yaccoby, S., Ling, W., Zhan, F., Walker, R., Barlogie, B., and Shaughnessy Jr., J.D. 2007. Antibody-based inhibition of DKK1 suppresses tumor-induced bone resorption and multiple myeloma growth in vivo. *Blood* **109:** 2106–2111.

Yakar, S., Rosen, C.J., Beamer, W.G., Ackert-Bicknell, C.L., Wu, Y., Liu, J.L., Ooi, G.T., Setser, J., Frystyk, J., Boisclair, Y.R., et al. 2002. Circulating levels of IGF-1 directly regulate bone growth and density. *J. Clin. Invest.* **110:** 771–781.

Yakar, S., Bouxsein, M.L., Canalis, E., Sun, H., Glatt, V., Gundberg, C., Cohen, P., Hwang, D., Boisclair, Y., Leroith, D., et al. 2006. The ternary IGF complex influences postnatal bone acquisition and the skeletal response to intermittent parathyroid hormone. *J. Endocrinol.* **189:** 289–299.

Yamaguchi, T.P., Bradley, A., McMahon, A.P., and Jones, S. 1999. A Wnt5a pathway underlies outgrowth of multiple structures in the vertebrate embryo. *Development* **126:** 1211–1223.

Yamaguchi, M., Ogata, N., Shinoda, Y., Akune, T., Kamekura, S., Terauchi, Y., Kadowaki, T., Hoshi, K., Chung, U.I., Nakamura, K., et al. 2005. Insulin receptor substrate-1 is required for bone anabolic function of parathyroid hormone in mice. *Endocrinology* **146:** 2620–2628.

Yamashita, M., Ying, S.X., Zhang, G.M., Li, C., Cheng, S.Y., Deng, C.X., and Zhang, Y.E. 2005. Ubiquitin ligase Smurf1 controls osteoblast activity and bone homeostasis by targeting MEKK2 for degradation. *Cell* **121:** 101–113.

Yan, T., Wergedal, J., Zhou, Y., Mohan, S., Baylink, D.J., and Strong, D.D. 2001. Inhibition of human osteoblast marker gene expression by retinoids is mediated in part by insulin-like growth factor binding protein-6. *Growth Horm. IGF Res.* **11:** 368–377.

Yi, S.E., Daluiski, A., Pederson, R., Rosen, V., and Lyons, K.M. 2000. The type I BMP receptor BMPRIB is required for chondrogenesis in the mouse limb. *Development* **127:** 621–630.

Yoo, J.U. and Johnstone, B. 1998. The role of osteochondral progenitor cells in fracture repair. *Clin. Orthop. Relat. Res.* (suppl.) **355:** S73–S81.

Yoon, B.S., Ovchinnikov, D.A., Yoshii, I., Mishina, Y., Behringer, R.R., and Lyons, K.M. 2005. Bmpr1a and Bmpr1b have overlapping functions and are essential for chondrogenesis in vivo. *Proc. Natl. Acad. Sci.* **102:** 5062–5067.

Yoshida, Y., Tanaka, S., Umemori, H., Minowa, O., Usui, M., Ikematsu, N., Hosoda, E., Imamura, T., Kuno, J., Yamashita, T., et al. 2000. Negative regulation of BMP/Smad signaling by Tob in osteoblasts. *Cell* **103:** 1085–1097.

Yu, K., Xu, J., Liu, Z., Sosic, D., Shao, J., Olson, E.N., Towler, D.A., and Ornitz, D.M. 2003. Conditional inactivation of FGF receptor 2 reveals an essential role for FGF signaling in the regulation of osteoblast function and bone growth. *Development* **130:** 3063–3074.

Yu, H.M., Jerchow, B., Sheu, T.J., Liu, B., Costantini, F., Puzas, J.E., Birchmeier, W., and Hsu, W. 2005. The role of Axin2 in calvarial morphogenesis and craniosynostosis. *Development* **132:** 1995–2005.

Zellin, G., Beck, S., Hardwick, R., and Linde, A. 1998. Opposite effects of recombinant human transforming growth factor-β1 on bone regeneration in vivo: Effects of exclusion of periosteal cells by microporous membrane. *Bone* **22:** 613–620.

Zhang, M., Xuan, S., Bouxsein, M.L., von Stechow, D., Akeno, N., Faugere, M.C., Malluche, H., Zhao, G., Rosen, C.J., Efstratiadis, A., et al. 2002. Osteoblast-specific knockout of the insulin-like growth factor (IGF) receptor gene reveals an essential role of IGF signaling in bone matrix mineralization. *J. Biol. Chem.* **277:** 44005–44012.

Zhang, M., Faugere, M.C., Malluche, H., Rosen, C.J., Chernausek, S.D., and Clemens, T.L. 2003. Paracrine overexpression of IGFBP-4 in osteoblasts of transgenic mice decreases bone turnover and causes global growth retardation. *J. Bone Miner. Res.* **18:** 836–843.

Zhao, G., Monier-Faugere, M.C., Langub, M.C., Geng, Z., Nakayama, T., Pike, J.W., Chernausek, S.D., Rosen, C.J., Donahue, L.R., Malluche, H.H., et al. 2000. Targeted overexpression of insulin-like growth factor I to osteoblasts of transgenic mice: Increased trabecular bone volume without increased osteoblast proliferation. *Endocrinology* **141:** 2674–2682.

Zhao, M., Harris, S.E., Horn, D., Geng, Z., Nishimura, R., Mundy, G.R., and Chen, D. 2002. Bone morphogenetic protein receptor signaling is necessary for normal murine postnatal bone formation. *J. Cell Biol.* **157:** 1049–1060.

Zhao, M., Qiao, M., Oyajobi, B.O., Mundy, G.R., and Chen, D. 2003. E3 ubiquitin ligase Smurf1 mediates core-binding factor α1/Runx2 degradation and plays a specific role in osteoblast differentiation. *J. Biol. Chem.* **278:** 27939–27944.

Zhao, M., Qiao, M., Harris, S.E., Oyajobi, B.O., Mundy, G.R., and Chen, D. 2004. Smurf1 inhibits osteoblast differentiation and bone formation in vitro and in vivo. *J. Biol. Chem.* **279:** 12854–12859.

Zhou, Y., Mohan, S., Linkhart, T.A., Baylink, D.J., and Strong, D.D. 1996. Retinoic acid regulates insulin-like growth factor-binding protein expression in human osteoblast cells. *Endocrinology* **137:** 975–983.

Zhou, Y.X., Xu, X., Chen, L., Li, C., Brodie, S.G., and Deng, C.X. 2000. A Pro250Arg substitution in mouse Fgfr1 causes increased expression of Cbfa1 and premature fusion of calvarial sutures. *Hum. Mol. Genet.* **9:** 2001–2008.

Zhou, H., Mak, W., Zheng, Y., Dunstan, C.R., and Seibel, M.J. 2008. Osteoblasts directly control lineage commitment of mesenchymal progenitor cells through Wnt signaling. *J. Biol. Chem.* **283:** 1936–1945.

Zhu, H., Kavsak, P., Abdollah, S., Wrana, J.L., and Thomsen, G.H. 1999. A SMAD ubiquitin ligase targets the BMP pathway and affects embryonic pattern formation. *Nature* **400:** 687–693.

Zhu, J., Jia, X., Xiao, G., Kang, Y., Partridge, N.C., and Qin, L. 2007. EGF-like ligands stimulate osteoclastogenesis by regulating expression of osteoclast regulatory factors by osteoblasts: Implications for osteolytic bone metastases. *J. Biol. Chem.* **282:** 26656–26664.

9

Cytokine and Growth Factor Regulation of Osteoclastogenesis

Hiroshi Takayanagi

Department of Cell Signaling, Graduate School
Tokyo Medical and Dental University
Yushima 1-5-45, Bunkyo-ku, Tokyo 113-8549, Japan

Osteoclast differentiation is an important biological process that determines the level of bone resorption in vivo. Numerous cytokines and growth factors are involved in the regulation of this process by directly acting on osteoclast precursor cells or through their effect on osteoclastogenesis-supporting mesenchymal cells, such as osteoblasts. Receptor activator of NF-κB ligand (RANKL) is an essential cytokine that promotes osteoclastogenesis and, in most cases, the effects of other factors can be explained in the context of cross talk with RANKL signaling. This chapter describes recent advances in the understanding of the intracellular signaling mechanism of RANKL and its interaction with other signaling events during osteoclastogenesis, which may provide a molecular basis for therapeutic intervention in pathological bone resorption.

THE ESSENTIAL ROLE OF THE RANKL-RANK SYSTEM IN OSTEOCLASTOGENESIS

Osteoclast differentiation is a tightly regulated process because the balance between osteoclasts and osteoblasts is critical for bone homeostasis. Osteoclasts differentiate from hematopoietic cells of monocyte/macrophage lineage, but osteoclastogenesis-supporting cells of mesenchymal origin are required for the differentiation commitment. Thus, osteoclasts have traditionally been formed in a coculture of hematopoietic cells derived from bone marrow and calvarial osteoblasts, which were thought to express an unknown osteoclast differentiation factor. Before this factor was identified, information about the molecules involved in osteoclast differentiation had been obtained from the analysis of osteopetrotic mice (Fig. 1) (Asagiri and Takayanagi 2006).

RANKL, a type II membrane protein of the TNF superfamily, was identified as the long sought-after osteoclast differentiation factor expressed by osteoblasts, but interestingly, the same molecule was also cloned in T cells (Anderson et al. 1997; Wong et al. 1997; Lacey et al. 1998; Yasuda et al. 1998). The binding of RANKL to its receptor RANK is negatively regulated by the decoy receptor osteoprotegerin (OPG) (Simonet et al. 1997; Tsuda et al. 1997). Genetically modified mice have underscored the essential role of the RANKL system in osteoclastogenesis (Kong et al. 1999; Theill et al. 2002).

RANKL is mainly expressed by osteoclastogenesis-supporting cells, including osteoblasts and bone marrow stromal cells. The expression of RANKL is regulated by various regulators, such as 1,25-$(OH)_2$ vitamin D_3, prostaglandin E_2 (PGE_2), and parathyroid hormone (PTH), and is a crucial determinant of the level of bone resorption in vivo (Suda et al.

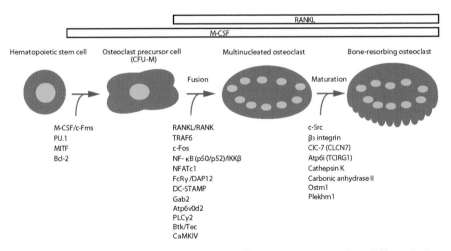

Figure 1. Molecules essential for osteoclast differentiation. Osteoclast differentiation is dependent on two major cytokines, M-CSF and RANKL, at the indicated periods. Hematopoietic stem cells undergo differentiation into macrophage colony-forming units (CFU-M), which are the common precursor cells of macrophages and osteoclasts. The differentiation step from CFU-M to multinucleated osteoclasts is characterized by cell–cell fusion, which is mainly induced by RANKL. Mature osteoclasts acquire bone resorbing activity, which is also dependent on M-CSF and RANKL. Molecules are depicted at the stage in which they are mainly involved.

1999). In humans, a mutation in RANKL causes osteoclast-poor osteopetrosis (Sobacchi et al. 2007), whereas mutations in RANK and OPG cause familial expansile osteolysis and juvenile Paget's disease of bone, respectively (Whyte and Mumm 2004).

RANKL is also released in a soluble form, although the in vivo function of the soluble form remains to be clarified (Theill et al. 2002; Hikita et al. 2006). RANKL functions as a crucial regulator of secondary lymphoid organ development (Kong et al. 1999) and mammary gland formation (Fata et al. 2000). Recent reports suggest that RANKL also regulates regulatory T-cell differentiation in the skin (Loser et al. 2006) and tumor cell chemotaxis (Jones et al. 2006). Considering the potent multifunctionality of RANKL, it is reasonable that the effects of this cytokine would be held under tight control by restricting its predominant effect to the membrane-bound form.

INTRACELLULAR SIGNAL TRANSDUCTION OF RANKL

The interaction of RANKL with RANK results in the trimerization of RANK and recruitment of adaptor molecules such as the TNF receptor-associated factor (TRAF) family of proteins (Wong et al. 1998), among which TRAF6 has been revealed to be the major adaptor molecule (Lomaga et al. 1999; Naito et al. 1999). TRAF6 also trimerizes upon RANK stimulation and activates NF-κB and the mitogen-activated kinases (MAPKs), including Jun amino-terminal kinase (JNK) and p38 (Kobayashi et al. 2001). The essential role of NF-κB in osteoclastogenesis has been demonstrated genetically (Franzoso et al. 1997; Iotsova et al. 1997), but the in vivo function of each MAPK remains to be specifically elucidated. The Lysine 63-linked polyubiquitination mediated by the RING (really interesting new gene)-finger motif of TRAF6 has been shown to be important for NF-κB activation in other cell types (Chen 2005), but deletion analysis has in fact indicated that the RING-finger domain of TRAF6 is dispensable for the formation of osteoclasts (Kobayashi et al. 2001). Therefore, the role of the ubiquitin-ligase activity of TRAF6 in osteoclastogenesis needs to be tested in vivo.

Receptors that activate TRAF6 include CD40, toll-like receptors (TLRs), and the interleukin-1 (IL-1) receptor, but RANK alone is able to stimulate osteoclastogenesis with marked potency. It has been suggested that additional RANK-specific adaptor molecule(s) exist that enhance TRAF6 signaling (Gohda et al. 2005). For example, the molecular scaffold Grab2-associated binding protein 2 (Gab2) has been shown to be associated with RANK and to have an important role in its signal transduction

(Wada et al. 2005). Signaling cross talk between RANKL and M-CSF, another essential cytokine for osteoclast differentiation, might also be crucially important for osteoclastogenesis (Ross and Teitelbaum 2005).

TRANSCRIPTIONAL MACHINERY ACTIVATED BY RANKL

RANK activates the transcription-factor complex, activator protein 1 (AP1), through induction of its component c-Fos (Wagner and Eferl 2005). The induction mechanism of c-Fos is not completely understood, but recent reports suggest that it is dependent on the activation of the Ca^{2+}/calmodulin-dependent kinase (CaMK) IV and the cAMP response element-binding protein (CREB) (Sato et al. 2006a), as well as the activation of NF-κB (Yamashita et al. 2007). RANKL specifically and strongly induces nuclear factor of activated T cells cytoplasmic 1 (NFATc1), the master regulator of osteoclast differentiation (Takayanagi et al. 2002a; Asagiri et al. 2005), and this induction is dependent on both the TRAF6 and c-Fos pathways. Activation of NFAT is mediated by a specific phosphatase, calcineurin, which is activated by calcium–calmodulin signaling. The NFATc1 promoter contains NFAT binding sites and NFATc1 specifically autoregulates its own promoter during osteoclastogenesis, thus affecting the robust induction of NFATc1 (Asagiri et al. 2005). NFATc1 regulates a number of osteoclast-specific genes such as *cathepsin K, tartrate-resistant acid phosphatase (TRAP), calcitonin receptor, Osteoclast-associated receptor (OSCAR)*, and β3 *integrin*, in cooperation with other transcription factors such as AP1, PU.1, MITF, and CREB (Takayanagi 2007).

CROSS TALK WITH OTHER CYTOKINE SYSTEMS

Osteoclasts are derived from the monocyte/macrophage lineage, and the osteoclast precursor cells express various cytokine receptors. The inflammatory cytokines that are mainly produced by macrophages, such as IL-1, TNF-α, and IL-6, promote osteoclastogenesis, and are also called "osteolytic cytokines" based on their bone resorptive effects in vivo. IL-1 stimulates TRAF6 (and therefore activates NF-κB and MAPKs) and synergizes with RANKL to induce mature osteoclasts to carry out bone-resorbing activity, but interestingly, IL-1 alone cannot induce differentiation, indicating that TRAF6 activation by itself is not sufficient. IL-1 also indirectly facilitates osteoclastogenesis by acting on osteoblasts through the induction of PGE_2, and induces the expression of RANKL (Suda et al. 1999). TNF-α stimulates the activation of NF-κB mainly through TRAF2. Although TNF-α alone cannot induce osteoclastogenesis in vivo, and TNF-α overexpression cannot rescue the deficiency of RANKL (Lam et al. 2000; Li et al. 2004), it is reported that TNF-α plus transforming growth factor-β (TGF-β) induces in vitro osteoclastogenesis, even in the absence of RANK or TRAF6 (Kim et al. 2005b). These results indicate that TNF-α has a pivotal role in the pathological activation of osteoclasts associated with inflammation.

Type I interferons (IFN-α and IFN-β) are essential for the host defense against pathogens such as viruses. RANKL induces the *IFN-β* gene in osteoclast precursor cells, and IFN-β functions as a negative-feedback regulator that inhibits the differentiation of osteoclasts by interfering with the RANKL-induced expression of c-Fos. The importance of type I IFNs in bone homeostasis has been underscored by the observation that mice deficient in the type I IFN-receptor component IFNAR1 spontaneously develop marked osteopenia accompanied by enhanced osteoclastogenesis (Takayanagi et al. 2002b).

T-CELL CYTOKINES AND OSTEOCLASTOGENESIS

As T-cell activation is often associated with abnormal bone resorption, the regulation of bone metabolism by T-cell cytokines is of great clinical significance. T cells express RANKL, but major T-cell cytokines such as IFN-γ, IL-4, and IL-10 are all inhibitory to in vitro osteoclastogenesis, and IL-12 and IL-18 also negatively affect osteoclastogenesis (Takayanagi 2007). The T-cell cytokines that positively regulate osteoclastogenesis include RANKL, TNF-α, IL-6, and IL-17. If RANKL is expressed in activated T cells, the T cells might have the capacity to induce osteoclast differentiation by directly acting on osteoclast precursor cells, but the IFN-γ produced by T cells potently suppresses RANKL signaling through downregulation of TRAF6 (Takayanagi et al. 2000). Thus, the effects of T cells on osteoclastogenesis appear to be dependent on the balance between positive and negative factors expressed by the T cells. As the effector CD4$^+$ T helper (T$_H$)-cell subsets T$_H$1 and T$_H$2 produce IFN-γ and IL-4, both are anti-osteoclastogenic, so it is difficult to directly link these T$_H$ subsets to osteoclastogenesis.

It has been recently shown that an IL-17-producing T-cell subset (T$_H$17) comprises an osteoclastogenic T$_H$-cell subset (Sato et al. 2006b). IL-17 exerts positive effects on osteoclastogenesis by stimulating RANKL expression on mesenchymal cells (Kotake et al. 1999) and inducing local inflammation that leads to the production of inflammatory cytokines such as TNF-α, IL-1, and IL-6. Thus, the infiltration of T$_H$17 cells into inflammatory lesion links the abnormal T-cell response to bone damage, and this subset provides an auspicious target for future therapy of the bone destruction associated with autoimmune inflammation.

SURVIVAL FACTORS FOR OSTEOCLASTS

Osteoclast differentiation is not dependent on RANKL alone. M-CSF was revealed to be essential for osteoclastogenesis in an analysis of *op/op* mice with osteopetrosis. M-CSF binds to its receptor c-Fms, which possesses tyrosine kinase activity. M-CSF is crucial for the proliferation and survival of osteoclast and macrophage precursor cells, mainly by activating ERK through Grb2, and Akt through phosphatidylinositol 3-kinase (Ross and Teitelbaum 2005). M-CSF also stimulates the expression of RANK in the monocyte/macrophage precursor cells, thereby rendering them to efficiently respond to RANKL. PU.1, a member of the ETS family of transcription factors, regulates the development of both macrophages and osteoclasts by controlling the expression of c-Fms (Tondravi et al. 1997). MITF, activated by M-CSF, induces Bcl-2, which has a pivotal role in cell survival (McGill et al. 2002). Overexpression of Bcl-2 rescues the osteopetrosis in *op/op* mice, suggesting a major role of M-CSF as a survival factor of osteoclast precursor cells (Lagasse and Weissman 1997).

COSTIMULATORY SIGNALS FOR RANK

Osteoclasts are formed when bone marrow cells are stimulated with RANKL and M-CSF in a culture system, and therefore it has been thought that the RANK and c-Fms together transmit the signals sufficient for osteoclastogenesis (Boyle et al. 2003). However, a novel type of receptor has been revealed to be highly expressed in osteoclasts. OSCAR was shown to be an immunoglobulin-like receptor involved in the cell–cell interaction between osteoblasts and osteoclasts (Kim et al. 2002). Subsequent studies showed that OSCAR associates with an adaptor molecule, Fc receptor common γ subunit (FcRγ) (Koga et al. 2004), which harbors an immunoreceptor tyrosine-based activation motif (ITAM) critical for the activation of calcium signaling in immune cells. Another ITAM-harboring adaptor, DNAX activating protein 12 (DAP12), has been reported to be involved in the formation and function of osteoclasts (Kaifu et al. 2003). Importantly, mice doubly deficient in FcRγ and DAP12 exhibit severe osteopetrosis owing to a differentiation blockade of osteoclasts, demonstrating that the immunoglobulin-like receptors associated with FcRγ and DAP12 are essential for osteoclastogenesis (Koga et al. 2004; Mocsai et al. 2004). These receptors include OSCAR, triggering receptor expressed in myeloid cells (TREM)-2, signal-regulatory protein (SIRP) β1, and paired immunoglobulin-like receptor (PIR)-A, although the ligand and exact function of each of these receptors remain to be determined. The importance of the ITAM-harboring adaptors and the receptors associated with them in bone metabolism is also supported by reports that mutations in the *DAP12* and *TREM-2* genes cause severe skeletal and behavior abnormalities known as Nasu–Hakola disease (Paloneva et al. 2000, 2002).

ITAM-mediated signals cooperate with RANK to stimulate calcium signaling through ITAM phosphorylation and the resulting activation of Syk and PLCγ. Therefore, these signals should properly be called costim- ulatory signals for RANK (Takayanagi 2005). Initially characterized in natural killer and myeloid cells, the immunoglobulin-like receptors asso- ciated with FcRγ or DAP12 are thus identified as previously unexpected but nevertheless essential partners of RANK during osteoclastogenesis. It is not fully understood how RANK can specifically induce osteoclastoge- nesis in cooperation with ITAM signaling, but it is partially explained by the observation that phosphorylation of ITAM is up-regulated by RANKL (Koga et al. 2004). In addition, RANKL stimulation results in an increased expression of immunoreceptors such as OSCAR, thereby aug- menting the ITAM signal (Kim et al. 2005a,c).

It is also conceivable that RANK activates an as yet unknown path- way that specifically synergizes with or up-regulates ITAM signaling. Recent studies have shown that RGS10 is activated by RANK and medi- ates calcium oscillation (Yang and Li 2007). Furthermore, we have shown that Tec family tyrosine kinases, such as Btk and Tec, are activated by RANK and are involved in the phosphorylation of PLCγ (Shinohara et al. 2008). An osteopetrotic phenotype in $Tec^{-/-}Btk^{-/-}$ mice revealed that these two kinases play an essential role in the regulation of osteoclast dif- ferentiation. Tec and Btk are known to have a key role in proximal BCR signaling, but this study established a further, crucial role in linking the RANK and ITAM signals. This study also identified an osteoclastogenic signaling complex composed of Tec kinases and scaffold proteins, which introduces a new paradigm for the signal transduction mechanism of osteoclast differentiation (Fig. 2): ITAM phosphorylation results in the recruitment of Syk, which phosphorylates adaptor proteins such as BLNK and SLP-76, which in turn function as scaffolds to recruit Tec kinases and PLCγ to the osteoclast signaling complex, so as to induce the maximal activation of calcium influx.

The role of inhibitory-type receptors containing the ITIM in osteo- clast differentiation remains elusive, but osteoclastogenesis is enhanced in mice lacking phosphatases, such as Src homology (SH)-2-containing pro- tein tyrosine phosphatase (SHP)-1 (Aoki et al. 1999; Umeda et al. 1999)

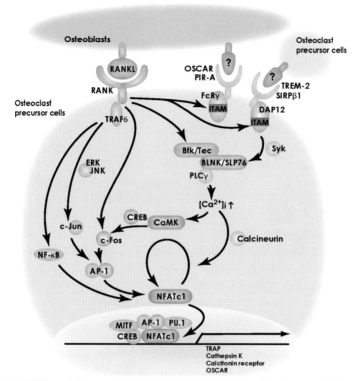

Figure 2. RANK and immunoreceptor tyrosine-based activation motif (ITAM) signals in osteoclastogenesis. RANK activates the NF-κB and c-Fos pathways that stimulate the induction of NFATc1. Immunoglobulin-like receptors associate with Fc receptor common γ subunit (FcRγ) and DNAX-activating protein 12 (DAP12), both of which contain the ITAM motif. RANK and ITAM signaling cooperate to phosphorylate phospholipase Cγ (PLCγ) and activate calcium signaling, which is critical for the activation of NFATc1. Tyrosine kinases Btk and Tec are activated by RANK and are important for the phosphorylation of PLCγ, thus linking the two pathways.

or SH-2-containing inositol 5′-phosphatase (SHIP)-1 (Takeshita et al. 2002), which counterbalance the ITAM signal in the immune system.

SUMMARY AND CONCLUSIONS

Osteopetrotic mice have provided important information on the molecules involved in osteoclastogenesis, but the discovery of RANKL and RANK has greatly enriched our understanding of the mechanisms of osteoclast differentiation. Clearly, commitment to the osteoclast lineage is dependent on the cooperation of the three signaling pathways activated by RANK, c-Fms, and ITAM. Despite advancements in understanding, however, it is still not clear why only RANK induces osteoclastogenesis among the receptors that commonly activate TRAF6. The signaling events activated by RANK are integrated by the transcription factor NFATc1, but nothing is known about the downstream molecule(s) that orchestrate the late phase of the differentiation process. Therefore, a great deal of work remains, but this work will surely be profitable, since the utility of targeting RANK signaling therapeutically has already been shown by the efficacy of the antiRANKL antibody, which has been demonstrated in several human trials against bone diseases.

ACKNOWLEDGMENTS

I would like to thank all the members of my laboratory for the studies on which this review article is based. This work was supported in part by a Grant-in-Aid for Creative Scientific Research from the Japan Society for the Promotion of Science (JSPS); Global Center of Excellence (GCOE) Program, grants for the Genome Network Project from the Ministry of Education, Culture, Sports, Science, and Technology of Japan (MEXT); Grant-in-Aid of The Japan Medical Association; grants from Tokyo Biochemical Research Foundation; Yokoyama Foundation for Clinical Pharmacology; and Takeda Science Foundation.

REFERENCES

Anderson, D.M., Maraskovsky, E., Billingsley, W.L., Dougall, W.C., Tometsko, M.E., Roux, E.R., Teepe, M.C., DuBose, R.F., Cosman, D., and Galibert, L. 1997. A homologue of the TNF receptor and its ligand enhance T-cell growth and dendritic-cell function. *Nature* **390:** 175–179.

Aoki, K., Didomenico, E., Sims, N.A., Mukhopadhyay, K., Neff, L., Houghton, A., Amling, M., Levy, J.B., Horne, W.C., and Baron, R. 1999. The tyrosine phosphatase SHP-1 is a negative regulator of osteoclastogenesis and osteoclast resorbing activity: Increased resorption and osteopenia in me^v/me^v mutant mice. *Bone* **25:** 261–267.

Asagiri, M. and Takayanagi, H. 2006. The molecular understanding of osteoclast differentiation. *Bone* **40:** 251–264.

Asagiri, M., Sato, K., Usami, T., Ochi, S., Nishina, H., Yoshida, H., Morita, I., Wagner, E.F., Mak, T.W., Serfling, E., et al. 2005. Autoamplification of NFATc1 expression determines its essential role in bone homeostasis. *J. Exp. Med.* **202:** 1261–1269.

Boyle, W.J., Simonet, W.S., and Lacey, D.L. 2003. Osteoclast differentiation and activation. *Nature* **423:** 337–342.

Chen, Z. J. 2005. Ubiquitin signalling in the NF-κB pathway. *Nat. Cell Biol.* **7:** 758–765.

Fata, J.E., Kong, Y.Y., Li, J., Sasaki, T., Irie-Sasaki, J., Moorehead, R.A., Elliott, R., Scully, S., Voura, E.B., Lacey, D.L., et al. 2000. The osteoclast differentiation factor osteoprotegerin-ligand is essential for mammary gland development. *Cell* **103:** 41–50.

Franzoso, G., Carlson, L., Xing, L., Poljak, L., Shores, E.W., Brown, K.D., Leonardi, A., Tran, T., Boyce, B.F., and Siebenlist, U. 1997. Requirement for NF-κB in osteoclast and B-cell development. *Genes Dev.* **11:** 3482–3496.

Gohda, J., Akiyama, T., Koga, T., Takayanagi, H., Tanaka, S., and Inoue, J.I. 2005. RANK-mediated amplification of TRAF6 signaling leads to NFATc1 induction during osteo-clastogenesis. *EMBO J.* **24:** 790–799.

Hikita, A., Yana, I., Wakeyama, H., Nakamura, M., Kadono, Y., Oshima, Y., Nakamura, K., Seiki, M., and Tanaka, S. 2006. Negative regulation of osteoclastogenesis by ectodomain shedding of receptor activator of NF-κB ligand. *J. Biol. Chem.* **281:** 36846–36855.

Iotsova, V., Caamano, J., Loy, J., Yang, Y., Lewin, A., and Bravo, R. 1997. Osteopetrosis in mice lacking NF-κB1 and NF-κB2. *Nat. Med.* **3:** 1285–1289.

Jones, D.H., Nakashima, T., Sanchez, O.H., Kozieradzki, I., Komarova, S.V., Sarosi, I., Morony, S., Rubin, E., Sarao, R., Hojilla, C.V., et al. 2006. Regulation of cancer cell migration and bone metastasis by RANKL. *Nature* **440:** 692–696.

Kaifu, T., Nakahara, J., Inui, M., Mishima, K., Momiyama, T., Kaji, M., Sugahara, A., Koito, H., Ujike-Asai, A., Nakamura, A., et al. 2003. Osteopetrosis and thalamic hypomyeli-nosis with synaptic degeneration in DAP12-deficient mice. *J. Clin. Invest.* **111:** 323–332.

Kim, N., Takami, M., Rho, J., Josien, R., and Choi, Y. 2002. A novel member of the leuko-cyte receptor complex regulates osteoclast differentiation. *J. Exp. Med.* **195:** 201–209.

Kim, K., Kim, J.H., Lee, J., Jin, H.M., Lee, S.H., Fisher, D.E., Kook, H., Kim, K.K., Choi, Y., and Kim, N. 2005a. Nuclear factor of activated T cells c1 induces osteoclast-asso-ciated receptor gene expression during tumor necrosis factor-related activation-induced cytokine-mediated osteoclastogenesis. *J. Biol. Chem.* **280:** 35209–35216.

Kim, N., Kadono, Y., Takami, M., Lee, J., Lee, S.H., Okada, F., Kim, J.H., Kobayashi, T., Odgren, P.R., Nakano, H., et al. 2005b. Osteoclast differentiation independent of the TRANCE-RANK-TRAF6 axis. *J. Exp. Med.* **202:** 589–595.

Kim, Y., Sato, K., Asagiri, M., Morita, I., Soma, K., and Takayanagi, H. 2005c. Contribution of nuclear factor of activated T cells c1 to the transcriptional control of immunore-ceptor osteoclast-associated receptor but not triggering receptor expressed by myeloid cells-2 during osteoclastogenesis. *J. Biol. Chem.* **280:** 32905–32913.

Kobayashi, N., Kadono, Y., Naito, A., Matsumoto, K., Yamamoto, T., Tanaka, S., and Inoue, J. 2001. Segregation of TRAF6-mediated signaling pathways clarifies its role in osteo-clastogenesis. *EMBO J.* **20:** 1271–1280.

Koga, T., Inui, M., Inoue, K., Kim, S., Suematsu, A., Kobayashi, E., Iwata, T., Ohnishi, H., Matozaki, T., Kodama, T., et al. 2004. Costimulatory signals mediated by the ITAM motif cooperate with RANKL for bone homeostasis. *Nature* **428:** 758–763.

Kong, Y.Y., Yoshida, H., Sarosi, I., Tan, H.L., Timms, E., Capparelli, C., Morony, S., Oliveira-dos-Santos, A.J., Van, G., Itie, A., et al. 1999. OPGL is a key regulator of osteo-clastogenesis, lymphocyte development and lymph-node organogenesis. *Nature* **397:** 315–323.

Kotake, S., Udagawa, N., Takahashi, N., Matsuzaki, K., Itoh, K., Ishiyama, S., Saito, S., Inoue, K., Kamatani, N., Gillespie, M.T., et al. 1999. IL-17 in synovial fluids from patients with rheumatoid arthritis is a potent stimulator of osteoclastogenesis. *J. Clin. Invest.* **109:** 1345–1352.

Lacey, D.L., Timms, E., Tan, H.L., Kelley, M.J., Dunstan, C.R., Burgess, T., Elliott, R.,

Colombero, A., Elliott, G., Scully, S., et al. 1998. Osteoprotegerin ligand is a cytokine that regulates osteoclast differentiation and activation. *Cell* **93:** 165–176.

Lagasse, E. and Weissman, I.L. 1997. Enforced expression of Bcl-2 in monocytes rescues macrophages and partially reverses osteopetrosis in *op/op* mice. *Cell* **89:** 1021–1031.

Lam, J., Takeshita, S., Barker, J.E., Kanagawa, O., Ross, F.P., and Teitelbaum, S.L. 2000. TNF-α induces osteoclastogenesis by direct stimulation of macrophages exposed to permissive levels of RANK ligand. *J. Clin. Invest.* **106:** 1481–1488.

Li, P., Schwarz, E.M., O'Keefe, R.J., Ma, L., Boyce, B.F., and Xing, L. 2004. RANK signaling is not required for TNFα-mediated increase in CD11^{hi} osteoclast precursors but is essential for mature osteoclast formation in TNFα-mediated inflammatory arthritis. *J. Bone Miner. Res.* **19:** 207–213.

Lomaga, M.A., Yeh, W.C., Sarosi, I., Duncan, G.S., Furlonger, C., Ho, A., Morony, S., Capparelli, C., Van, G., Kaufman, S., et al. 1999. TRAF6 deficiency results in osteopetrosis and defective interleukin-1, CD40, and LPS signaling. *Genes Dev.* **13:** 1015–1024.

Loser, K., Mehling, A., Loeser, S., Apelt, J., Kuhn, A., Grabbe, S., Schwarz, T., Penninger, J.M., and Beissert, S. 2006. Epidermal RANKL controls regulatory T-cell numbers via activation of dendritic cells. *Nat. Med.* **12:** 1372–1379.

McGill, G.G., Horstmann, M., Widlund, H.R., Du, J., Motyckova, G., Nishimura, E.K., Lin, Y.L., Ramaswamy, S., Avery, W., Ding, H.F., et al. 2002. Bcl2 regulation by the melanocyte master regulator Mitf modulates lineage survival and melanoma cell viability. *Cell* **109:** 707–718.

Mocsai, A., Humphrey, M.B., Van Ziffle, J.A., Hu, Y., Burghardt, A., Spusta, S.C., Majumdar, S., Lanier, L.L., Lowell, C.A., and Nakamura, M. C. 2004. The immunomodulatory adapter proteins DAP12 and Fc receptor γ-chain (FcRγ) regulate development of functional osteoclasts through the Syk tyrosine kinase. *Proc. Natl. Acad. Sci.* **101:** 6158–6163.

Naito, A., Azuma, S., Tanaka, S., Miyazaki, T., Takaki, S., Takatsu, K., Nakao, K., Nakamura, K., Katsuki, M., Yamamoto, T., et al. 1999. Severe osteopetrosis, defective interleukin-1 signalling and lymph node organogenesis in TRAF6-deficient mice. *Genes Cells* **4:** 353–362.

Paloneva, J., Kestilä, M., Wu, J., Salminen, A., Böhling, T., Ruotsalainen, V., Hakola, P., Bakker, A.B., Phillips, J.H., Pekkarinen, P., et al. 2000. Loss-of-function mutations in TYROBP (DAP12) result in a presenile dementia with bone cysts. *Nat. Genet.* **25:** 357–361.

Paloneva, J., Manninen, T., Christman, G., Hovanes, K., Mandelin, J., Adolfsson, R., Bianchin, M., Bird, T., Miranda, R., Salmaggi, A., et al. 2002. Mutations in two genes encoding different subunits of a receptor signaling complex result in an identical disease phenotype. *Am. J. Hum. Genet.* **71:** 656–662.

Ross, F.P. and Teitelbaum, S.L. 2005. α$_v$β$_3$ and macrophage colony-stimulating factor: Partners in osteoclast biology. *Immunol. Rev.* **208:** 88–105.

Sato, K., Suematsu, A., Nakashima, T., Takemoto-Kimura, S., Aoki, K., Morishita, Y., Asahara, H., Ohya, K., Yamaguchi, A., Takai, T., et al. 2006a. Regulation of osteoclast differentiation and function by the CaMK-CREB pathway. *Nat. Med.* **12:** 1410–1416.

Sato, K., Suematsu, A., Okamoto, K., Yamaguchi, A., Morishita, Y., Kadono, Y., Tanaka, S., Kodama, T., Akira, S., Iwakura, Y., et al. 2006b. Th17 functions as an osteoclastogenic helper T cell subset that links T cell activation and bone destruction. *J. Exp. Med.* **203:** 2673–2682.

Shinohara, M., Koga, T., Okamoto, K., Sakaguchi, S., Arai, K., Yasuda, H., Takai, T.,

Kodama, T., Morio, T., Geha, R.S., et al. 2008. Tyrosine kinases Btk and Tec regulate osteoclast differentiation by linking RANK and ITAM signals. *Cell* **132:** 794–806.

Simonet, W.S., Lacey, D.L., Dunstan, C.R., Kelley, M., Chang, M.S., Luthy, R., Nguyen, H.Q., Wooden, S., Bennett, L., Boone, T., et al. 1997. Osteoprotegerin: A novel secreted protein involved in the regulation of bone density. *Cell* **89:** 309–319.

Sobacchi, C., Frattini, A., Guerrini, M.M., Abinun, M., Pangrazio, A., Susani, L., Bredius, R., Mancini, G., Cant, A., Bishop, N., et al. 2007. Osteoclast-poor human osteopetrosis due to mutations in the gene encoding RANKL. *Nat. Genet.* **39:** 960–962.

Suda, T., Takahashi, N., Udagawa, N., Jimi, E., Gillespie, M.T., and Martin, T.J. 1999. Modulation of osteoclast differentiation and function by the new members of the tumor necrosis factor receptor and ligand families. *Endocr. Rev.* **20:** 345–357.

Takayanagi, H. 2005. Mechanistic insight into osteoclast differentiation in osteoimmunology. *J. Mol. Med.* **83:** 170–179.

Takayanagi, H. 2007. Osteoimmunology: Shared mechanisms and crosstalk between the immune and bone systems. *Nat. Rev. Immunol.* **7:** 292–304.

Takayanagi, H., Ogasawara, K., Hida, S., Chiba, T., Murata, S., Sato, K., Akinori, T., Yokochi, T., Oda, H., Tanaka, K., et al. 2000. T cell-mediated regulation of osteoclastogenesis by signalling cross-talk between RANKL and IFN-γ. *Nature* **408:** 600–605.

Takayanagi, H., Kim, S., Koga, T., Nishina, H., Isshiki, M., Yoshida, H., Saiura, A., Isobe, M., Yokochi, T., Inoue, J., et al. 2002a. Induction and activation of the transcription factor NFATc1 (NFAT2) integrate RANKL signaling for terminal differentiation of osteoclasts. *Dev. Cell* **3:** 889–901.

Takayanagi, H., Kim, S., Matsuo, K., Suzuki, H., Suzuki, T., Sato, K., Yokochi, T., Oda, H., Nakamura, K., Ida, N., et al. 2002b. RANKL maintains bone homeostasis through c-Fos-dependent induction of *interferon-β*. *Nature* **416:** 744–749.

Takeshita, S., Namba, N., Zhao, J.J., Jiang, Y., Genant, H.K., Silva, M.J., Brodt, M.D., Helgason, C.D., Kalesnikoff, J., Rauh, M.J., et al. 2002. SHIP-deficient mice are severely osteoporotic due to increased numbers of hyper-resorptive osteoclasts. *Nat. Med.* **8:** 943–949.

Theill, L.E., Boyle, W.J., and Penninger, J.M. 2002. RANK-L and RANK: T cells, bone loss, and mammalian evolution. *Annu. Rev. Immunol.* **20:** 795–823.

Tondravi, M.M., McKercher, S.R., Anderson, K., Erdmann, J.M., Quiroz, M., Maki, R., and Teitelbaum, S.L. 1997. Osteopetrosis in mice lacking haematopoietic transcription factor PU.1. *Nature* **386:** 81–84.

Tsuda, E., Goto, M., Mochizuki, S., Yano, K., Kobayashi, F., Morinaga, T., and Higashio, K. 1997. Isolation of a novel cytokine from human fibroblasts that specifically inhibits osteoclastogenesis. *Biochem. Biophys. Res. Commun.* **234:** 137–142.

Umeda, S., Beamer, W.G., Takagi, K., Naito, M., Hayashi, S., Yonemitsu, H., Yi, T., and Shultz, L.D. 1999. Deficiency of SHP-1 protein-tyrosine phosphatase activity results in heightened osteoclast function and decreased bone density. *Am. J. Pathol.* **155:** 223–233.

Wada, T., Nakashima, T., Oliveira-dos-Santos, A.J., Gasser, J., Hara, H., Schett, G., and Penninger, J.M. 2005. The molecular scaffold Gab2 is a crucial component of RANK signaling and osteoclastogenesis. *Nat. Med.* **11:** 394–399.

Wagner, E.F. and Eferl, R. 2005. Fos/AP-1 proteins in bone and the immune system. *Immunol. Rev.* **208:** 126–140.

Whyte, M.P. and Mumm, S. 2004. Heritable disorders of the RANKL/OPG/RANK signaling pathway. *J. Musculoskelet. Neuronal Interact.* **4:** 254–267.

Wong, B.R., Rho, J., Arron, J., Robinson, E., Orlinick, J., Chao, M., Kalachikov, S., Cayani, E., Bartlett III, F.S., Frankel, W.N., et al. 1997. TRANCE is a novel ligand of the tumor necrosis factor receptor family that activates c-Jun N-terminal kinase in T cells. *J. Biol. Chem.* **272:** 25190–25194.

Wong, B.R., Josien, R., Lee, S.Y., Vologodskaia, M., Steinman, R.M., and Choi, Y. 1998. The TRAF family of signal transducers mediates NF-κB activation by the TRANCE receptor. *J. Biol. Chem.* **273:** 28355–28359.

Yamashita, T., Yao, Z., Li, F., Zhang, Q., Badell, I.R., Schwarz, E.M., Takeshita, S., Wagner, E.F., Noda, M., Matsuo, K., et al. 2007. NF-κB p50 and p52 regulate receptor activator of NF-κB ligand (RANKL) and tumor necrosis factor-induced osteoclast precursor differentiation by activating c-Fos and NFATc1. *J. Biol. Chem.* **282:** 18245–18253.

Yang, S. and Li, Y.P. 2007. RGS10-null mutation impairs osteoclast differentiation resulting from the loss of $[Ca^{2+}]i$ oscillation regulation. *Genes Dev.* **21:** 1803–1816.

Yasuda, H., Shima, N., Nakagawa, N., Yamaguchi, K., Kinosaki, M., Mochizuki, S., Tomoyasu, A., Yano, K., Goto, M., Murakami, A., et al. 1998. Osteoclast differentiation factor is a ligand for osteoprotegerin/osteoclastogenesis-inhibitory factor and is identical to TRANCE/RANKL. *Proc. Natl. Acad. Sci.* **95:** 3597–3602.

10

Bone Mineralization

Lynda F. Bonewald, Sarah L. Dallas, and Jeff P. Gorski

Bone Biology Research Program
Department of Oral Biology
University of Missouri at Kansas City School of Dentistry
Kansas City, Missouri 64108

THE MECHANISMS BY WHICH MINERALIZED TISSUES such as bone acquire and regulate their mineral component are complex. The process of mineralization can be divided into the establishment of a primed, mineralizable matrix in which de novo mineral nucleation can occur, followed by the growth, expansion, and maturation of crystals. Perturbations in any of these events can result in bone disease and fragility. Elucidating the cellular and molecular mechanisms has been difficult because biomineralization is a combination of a physico-chemical process and a biological one. For example, phosphate participates directly in the formation of the hydroxyapatite crystals but recent studies have shown that phosphate can also directly regulate gene expression. This chapter summarizes current opinion within the field on key issues related to mineralization, such as whether mineralization is an active (cell-mediated) or a passive (physico-chemical) process and the role of cell-derived organelles/vesicles in mineralization. The molecular mediators and regulators of mineralization are reviewed and the question of whether there are unique mechanisms of mineralization in different types of bone tissue is addressed. Even though bone mineral density is currently the standard for predicting bone fragility, it is important to not only understand the inorganic component of bone, but also the organic component, as individuals with similar bone densities can have different susceptibility to fracture. These less clear properties of the skeleton that contribute to bone strength remain the focus of much investigation.

The Skeletal System ©2009 Cold Spring Harbor Laboratory Press 978-087969-825-6

THE CELLS RESPONSIBLE FOR BONE FORMATION
AND MINERALIZATION

Cells in the osteoblast/osteocyte lineage are responsible for bone formation and mineralization. The osteoblast is derived from a precursor stem
cell of mesenchymal origin. During its differentiation, the osteoblast goes
through three major phenotypically identifiable stages, including proliferation, matrix production, and maturation. It is during the maturation
phase that mineralization occurs. Cells in each of these phases have been
described as early preosteoblasts, proliferating osteoblasts, mature
osteoblasts, preosteocytes (or osteoid osteocytes), and finally mature
osteocytes within the mineralized matrix. Not every matrix-producing
osteoblast will differentiate ultimately into an embedded osteocyte. Some
may adopt a flattened morphology and remain on the bone surface as a
quiescent-lining cell. A proportion of osteoblasts also undergo pro-
grammed cell death or apoptosis (for review, see Bonewald 2008).

Osteoblasts, osteoid-osteocytes, and osteocytes each play unique roles
in the initiation and regulation of mineralization of bone. Osteoblasts on
the bone surface are responsible for production of the nonmineralized
matrix known as osteoid. These cells possess extensive Golgi and produce
large amounts of collagen either apically, in the case of lamellar bone, or
in three dimensions as in woven bone (see below). Osteoid-osteocytes
appear to play an important role in the initiation and control of matrix
calcification (Barragan-Adjemian et al. 2006). These cells actively secrete
matrix and are also responsible for calcifying this matrix. Like osteoblasts
in lamellar bone, their activity is polarized towards the mineralization front
to which their initial cellular processes or dendrites are oriented. Exciting
recent discoveries have shown that osteocytes play a major role in regulating phosphate homoeostasis, which further underscores their importance
in the regulation of mineralization (for review, see Bonewald 2008).

TWO DISTINCT FORMS OF ADULT BONE:
WOVEN AND LAMELLAR

New bone formation in the embryo occurs by two distinct condensation
processes: intramembranous to form the calvaria and jaws, and endochondral, which forms the long bones. With intramembranous bone formation, woven bone is rapidly formed without the preexistence of
cartilage, whereas endochondral bone uses a previously formed vascularized cartilage template (Olsen 2006). In the adult skeleton, there are two
distinct types of bone formation that may have common but also unique
molecular and cellular control mechanisms. Woven or primary bone can

be viewed as "de novo" bone and is defined as bone that forms where no bone existed previously (Fig. 1). In contrast, lamellar (or mature bone) can be viewed as "replacement bone" (Fig. 2). It is formed when previously formed bone (either woven or lamellar bone) is resorbed by osteoclasts and then replaced by new lamellar bone. The terms "modeling" and "remodeling" have also been applied to the formation of woven bone and lamellar bone, respectively.

Woven bone is generally formed in situations where a rapid deposition of bone is required, not only during embryonic bone formation, but also during fracture healing and adaptive bone gain following mechanical loading (generally greater than 3000 microstrain). It is perhaps not surprising, therefore, that the rate of deposition of woven bone has been estimated to be six times faster than that for lamellar bone (Kimmel and Jee 1980). The key difference in these two types of bone is their organizational structure. Because woven bone is formed rapidly and in a less polarized manner compared to lamellar bone, it has a much more random orientation of collagen fibers and is less organized (Fig. 1).

Lamellar bone derives its name from the appearance of its highly organized and layered appearance (Fig. 2). The structural hallmark of pri-

Woven Bone Formation

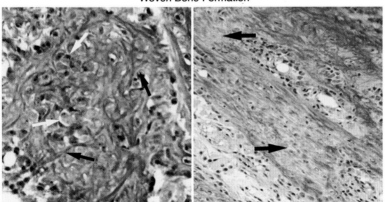

Figure 1. An example of two stages of woven bone produced during fracture healing of canine tibia. The higher power, undecalcified image on the *left* shows the formation of fibrous condensed mesenchymal matrix (black arrows) around the osteoblastic cells (white arrows) at 7 days postfracture. Note the lack of directionality. The image on the *right* depicts a strand of condensed matrix and osteoblastic cells at 14 days postfracture undergoing mineralization (arrows demark green-stained mineral deposits). Both sections were stained with Goldner trichrome stain.

Lamellar Bone Formation

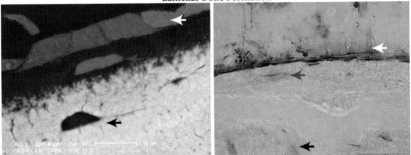

Figure 2. An example of lamellar bone formation. The image on the *left* is a scanning electron micrograph (SEM) of murine cortical bone, while the image on the *right* is immunohistochemical staining for alkaline phosphatase (brown), a marker for osteoblasts, counterstained with methyl green. The white arrows denote the osteoblasts on the bone surface, the red arrows denote newly embedding osteoid-osteocytes, and the black arrow indicates a fully embedded mature osteocyte. Note the mineralization front that is clearly visible in the SEM, showing a gradient in sizes of calcospherulites (white) that appear to fuse around the osteocyte.

mate, but not rodent lamellar bone, is the osteon, comprised of a series of concentric lamellar rings surrounding a central blood vessel. This organizational structure results from the action of sheets of osteoblastic cells that deposit bone matrix in a highly polarized manner from their basal surfaces (Fig. 2). The orientation of the collagen fibrils is highly ordered and consequently, lamellar bone has high birefringent properties compared to woven bone. This high degree of organization results in bone that is mechanically much stronger than woven bone.

It is clear that there is much commonality in the molecular pathways that control mineralization in woven and lamellar bone. Nevertheless, a growing literature suggests that there also may be unique differences in the cellular and molecular mechanisms regulating mineralization in these two types of bone (see below). From a biological point of view, this is not unexpected, given the profound differences in the rate of formation and the structural organization of the two types of bone.

PROPOSED MECHANISMS OF MINERAL NUCLEATION IN WOVEN AND LAMELLAR BONE

The biochemical mechanisms responsible for biomineralization of woven as compared to lamellar bone, endochondral compared to intramembranous bone, and embryonic compared to adult bone formation are predicted to share but also to have distinct molecular pathways and underlying mechanisms (Gorski 1998). In both woven and lamellar bone, initial hydroxyapatite crystal nucleation is mediated by organic molecules. There are two essential events in mineralization: The first is the nucleation of mineral to produce crystals approximately 100 nm in length (Simmons et al. 1991) and the second is mineral crystal propagation, also known as the crystal growth phase (for review, see Boskey 1998). These two stages occur in all mineralized tissues, but may have different mechanisms, regulatory proteins, and substrates in different tissues (Fig. 3).

Several cell-derived structures and/or organelles have been proposed to be responsible for the initiation and control of the mineral nucleation event. These include matrix vesicles (Bonucci 1967; Anderson 1969), biomineralization foci (BMF), and crystal ghosts (Bonucci 2002) or calcospherulites (Aaron et al. 1999), in addition to collagen fibrils (Glimcher 1989). The relative contribution of each of these to the initial biomineralization event remains unclear and it seems likely that their relative contribution to biomineralization may vary in different tissues or developmental stages. In the following sections, each of these proposed mechanisms are briefly reviewed.

Matrix Vesicles

These small 50–300-nm vesicles play a crucial role in the mineralization of epiphyseal cartilage, as described over 40 years ago by Anderson (1969) and Bonucci (1967). Their role in mineralization of bone has yet to be definitively established. Extracellular matrix vesicles are phospholipid membrane-delimited spheres that form by budding from the cell-surface membrane. The composition of matrix vesicles is clearly different from the plasma membrane as alkaline phosphatase and other factors are highly elevated in this organelle. These vesicles are initially devoid of mineral and are thought to be deposited or "left behind" in the newly forming matrix to act as nucleators of mineralization in a front distant from the chondrocyte or osteoblast (Anderson 1983). These spherical membrane vesicles appear to contain factors that are responsible for initiating subsequent apatite crystal formation. They contain nucleoside triphosphate pyrophosphohydrolase activities that are thought to dynamically balance the level of pyrophosphate inhibitor. When crystals form, they are randomly deposited in radial-shaped spherical clusters that then appear to rupture the vesicle membrane, forming calcospherites that associate with collagen. After the initial seeding of mineral crystals on collagen, it is thought that crystal growth and expansion can occur by other physico-chemical mechanisms. Whereas matrix vesicles can be

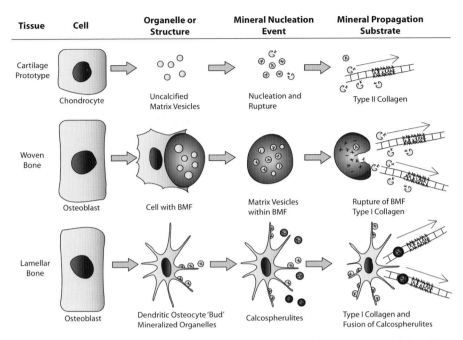

Figure 3. Nucleation and propagation of mineral in cartilage, woven, and lamellar bone, and variations on a theme. It is well accepted that chondrocytes generate uncalcified matrix vesicles that serve as sites of mineral nucleation in the matrix of cartilage. Upon nucleation of calcium and phosphate, the crystal growth ruptures the matrix vesicle, allowing the mineral to propagate along Type II collagen fibers. It is proposed that woven bone produces large structures called bone mineralization foci (BMF) that serve to sequester promoters (and not inhibitors) of mineralization within its domain. Matrix vesicles within the BMF serve to initiate nucleation, but in this tissue both the vesicles and the BMFs disintegrate to allow mineral propagation along Type I collagen. The late osteoblast-osteoid osteocyte most likely regulates the mineralization process in a tightly controlled manner. Osteoid osteocytes bud vesicular structures that calcify while associated with the cell membrane. These then dissociate from the cell membrane and become calcospherulites that associate with Type I collagen, grow in size, and fuse to surround and enclose the mature osteocyte.

readily isolated from cartilage using a protease digestion protocol, their function in bone is not as well defined and is based on morphological similarities with structures in bone sections (Boskey 1996, 1998).

Crystal Ghosts

Crystal ghosts were described by Bonucci as organic structures masked by mineral that only become recognizable after decalcification (Bonucci

2002). Calcified, spherically shaped structures with a lower size limit of 300–600 nm in diameter, termed calcospherulites or crystal ghost aggregates, have also been observed in bone (Boyde and Sela 1978) and in primary osteoblastic cultures (Nanci et al. 1996). These structures have also been called "calcifying nodules" and range in size from 1–300 nm and up to 3 μm in diameter in bone in remodeling Haversian canals (Lowe et al. 1983). They have also been described as "calcified microspheres" in bovine bone (Aaron et al. 1999). How crystal ghosts are formed is still the subject of investigation, although a shared presence of bone sialoprotein (BSP) within BMF (below) suggests a temporal and physical relationship between these structures.

Calcospherulites

Embedded osteoblasts, or osteocytes, most likely control the mineralization process in bone. Vesicular structures budding from the plasma membrane of both the cell body and dendrites have been described in a late osteoblast/early osteocyte-like cell line (Barragan-Adjemian et al. 2006). These structures appear to mineralize before leaving the cell. It was therefore proposed that in lamellar bone, it is the osteoid-osteocyte that regulates the mineralization process as the cell and its numerous dendritic processes become encased in mineral. The cell "buds" or "blebs" are small vesicular structures of 20–50 nm that nucleate crystals before being released from the cell membrane (see Fig. 3). These may form the basis for calcospherulites, which are calcified structures that have a collapsed or encased membrane vesicle in their center. Similar structures have also been described in cultures of primary chondrocytes (Wu et al. 1995). Consistent with these studies, it has recently been proposed that intracellular vesicles can initiate crystal formation before leaving the cell via an exocytic process (Rohde and Mayer 2007). These investigators used bone marrow stromal cells and proposed that mineral nucleation can begin intracellularly under certain circumstances.

BMF

Large structures, approximately the size of a cell, called "biomineralization foci," or BMF, were first described in three osteoblast cell lines (Midura et al. 2004). Analogous structures were then shown to occur in developing periosteal bone and in new bone formed following marrow ablation (Gorski et al. 2004). These structures may therefore play a specific role in initiating mineralization in primary bone formation, also known as intramembranous osteogenesis. These three-dimensional extracellular structures are 20–35 microns in diameter in young periosteum and in the marrow ablation model. In several different osteoblast culture models, these structures occur in two size ranges of 15–20 and 150–250 microns. Proteomic analyses on isolated BMF suggest that they may sequester acidic phosphoproteins within a protected environment for the purpose of mineral nucleation in association with matrix vesicles also located within these structures (see Fig. 3). The structure of BMF appears to include an organized assembly of thin fibrils and vesicles that include biomarkers such as bone acidic glycoprotein-75 (BAG-75) and BSP, which have been associated with mineralization. BMF calcify in vitro in osteoblastic cells upon addition of a source of phosphate. Collagen fibrils do not appear to be present within these BMF. Once nucleation occurs, the BMF structure appears to rupture and mineral propagation along collagen fibrils can then occur. Calcein-labeled calcospherulites resembling BMF have been isolated from young growing periosteum and are able to seed the calcification of collagen fibrils in vitro within collagen gels (Midura et al. 2007). Others (Sela et al. 1992) have described a continuum of mineralization foci within newly forming bone induced by marrow ablation, ranging from matrix vesicles to calcospherites.

Collagen Fibrils

Collagen fibrils may not only serve as a substrate for crystal propagation, but may also act as a nucleator. The factors that initiate nucleation of apatite on the collagen fibril are still unclear. Glimcher (1989) proposed that the initial mineral nucleation is mediated by phosphoproteins associated with the "hole region" of type I collagen fibrils. While this model remains under investigation, conclusive evidence for the colocalization of phosphoproteins to the hole regions of type I collagen fibrils is yet to be found. Collagen mineralization can be related to matrix vesicle deposition or can be independent (Arsenault and Kohler 1994). It has been proposed that matrix vesicles are responsible for providing the bond for the individual nucleation events and alignment of crystals onto the collagen (Landis 1999). Lipids associated with both collagen and matrix vesicles may provide the link between vesicle and collagen-mediated mineralization (Boskey 1996). In this way, the data suggest that collagen fibrils facilitate the propagation of newly formed mineral crystals nucleated in association with organelles/structures as previously described.

Questions remain about the relative contributions of each of the mechanisms to mineralization in different tissue types, and whether they function together or in combination. Another important question to be resolved is whether different mechanisms operate in conditions of normal versus pathological mineralization. Overall, there is a substantial body of evidence to suggest that initial nucleation of mineral crystals is likely an active, cell-mediated process.

MINERAL CRYSTAL PROPAGATION

After the initial nucleation of mineral crystals, the second phase of mineral crystal propagation can occur (see Fig. 3). Heteropolymers of collagens type I and XI for bone, heteropolymers of collagen type II and type XI for cartilage, and amelogenin for tooth enamel are the accepted templates or substrates for mineral propagation. It is thought that granular mineral is deposited in bone in a highly organized manner in the "hole" and overlap zones of collagen type I fibrils. Fibrillar mineral crystals, generally longer than the dimensions of the hole zone, are located mainly in the interfibrillary space (Bonucci 2007). Propagation of newly formed crystals seems to occur primarily through duplication of crystals rather than an increase in individual crystal size. In cortical bone, the newly formed crystals grow in length along their crystallographic c-axis, parallel to the collagen fibril axes (Glimcher 1989; Landis 1999). However, filamentous crystals are more prevalent in woven versus lamellar (cortical) bone, where the packing of collagen fibrils is looser and the width of interfibrillary spaces larger. In woven bone, the orientation of most crystals does not show a periodic banding as seen in lamellar bone (Bonucci 2007).

MOLECULAR MEDIATORS OF BIOMINERALIZATION

A diverse group of molecular mediators are responsible for and/or regulate biomineralization. These range from the mineral ions themselves, enzymes and proteins that regulate their homeostasis and/or local concentrations, protein and nonprotein inhibitors and nucleators of mineralization, extracellular matrix proteins that form a scaffold for mineral deposition, and enzymes that play a critical role in this process. In the following section, the major molecular mediators are briefly reviewed.

Calcium, Phosphate, and Pyrophosphate Ions

Calcium and phosphate ions are the main constituents of hydroxyapatite crystals, so it is not surprising that their levels in the circulation play a permissive but critical role in bone mineralization. Defective regulation of calcium and phosphate result in diseases of osteomalacia (i.e., unmineralized bone) (Harmey et al. 2004). The major tissues involved in regulating blood calcium and phosphate are the thyroid and parathyroid glands, bone, intestine, and kidney. The thyroid and parathyroid glands sense blood-ionized calcium and secrete calcitonin or parathyroid hormone (PTH) to maintain ion levels. The intestinal epithelial cell layer mediates calcium absorption from the gut. In addition, in response to a low blood calcium, osteoclasts resorb bone to release stored calcium and phosphorus into the circulation. Finally, the kidney generally recovers 98% of filtered calcium ions from the urine and returns it to the blood. Pyrophosphate, an inhibitor of mineral nucleation, is present in serum at low picomolar levels (Lust and Seegmiller 1976). The balance between local levels of phosphate and pyrophosphate may be key to the control of mineralization.

Several enzymes and transporters directly regulate the local concentrations of free phosphate and pyrophosphate within bone. These include nucleotide pyrophosphatase/phosphodiesterase (PC-1), tissue nonspecific alkaline phosphatase (TNAP), and phosphate transporters. A component of matrix vesicle and plasma cell membranes, PC-1 is an enzyme that cleaves nucleotides to generate pyrophosphate, thereby having a local anti-mineralization effect. This is believed to be balanced by the pro-mineralization function of TNAP, which cleaves pyrophosphate into inorganic phosphate. Consistent with these proposed functions, mutations in PC-1 lead to extraskeletal calcification, while mutations in TNAP lead to hypophosphatasia. In support of the concept that the balanced activities of TNAP and PC-1 are key determinants of mineralization, a double knockout of TNAP and PC-1 largely rescues the defective mineralization phenotypes of either of these single knockouts (Harmey et al. 2004).

The in vivo and in vitro actions of sodium-phosphate cotransporter inhibitors, such as foscarnet, demonstrate their importance for osteoblast-mediated mineralization. Osteoblasts express several different type II and type III sodium phosphate cotransporters (Lundquist et al. 2007). These likely play a role in controlling the phosphate concentrations locally, in matrix vesicles, or in other nucleating subcellular structures to elevate them to levels permissive for mineralization. Type II transporter Npt2 is also responsible for resorbing the bulk of filtered phosphate by kidney tubules. As a result, null mice display hypophosphatemia and defective mineralization (Beck et al. 1998).

Serum-ionized calcium levels are maintained within 10% of the normal level, primarily through the actions of PTH and activated vitamin D_3 (1,25-$(OH)_2$ vitamin D_3). Calcium is sensed by the Calcium Sensing

Receptor CasR of the parathyroid gland, which, along with serum phosphate level, regulates release of PTH. PTH stimulates release of calcium (and phosphorus) from bone and reabsorption of calcium by the kidney, and blocks reabsorption of phosphorus. $1,25$-$(OH)_2$ vitamin D_3, induced by PTH, increases intestinal absorption of calcium (and phosphorus) and bone resorption. Interestingly, both the CasR and Vitamin D receptor (VDR) knockout mice show phenotypes of osteomalacia with a hypomineralized skeleton, underscoring their importance in mineralization (Christakos and Holick 2006).

The Role of DMP-1, Phex, MEPE, and FGF23 in Regulation of Mineralization

Based on the fact that their inactivation causes either hyper- or hypophosphatemia, this diverse group of proteins are believed to comprise a pathway responsible for regulating serum phosphate concentration. Phosphate-regulating neutral endopeptidase on the X chromosome (PHEX) is a metalloendoprotease originally described on the plasma membrane of osteoblasts and osteocytes (Ruchon et al. 2000). Loss-of-function mutations in the corresponding gene, *Pex*, result in X-linked hypophosphatemic rickets (HYP Consortium 1995). Mutation of the *Pex* gene is responsible for the phenotype of the "hyp" mouse. The precise function and/or physiological substrate(s) of Phex are unclear but it plays an essential role in phosphate homeostasis and bone mineralization. As this molecule is highly expressed in osteocytes, Nijweide and coworkers (Nijweide and Mulder 1986) were the first to propose that the osteocyte network may function as a "gland" that regulates phosphate metabolism and bone mineralization through expression of Phex (for review, see Bonewald 2007).

MEPE/OF45 is also highly expressed in osteocytes as compared to osteoblasts. Deletion of this gene in mice results in increased bone formation and bone mass and less trabecular bone loss with age (Gowen et al. 2003). The MEPE protein is highly phosphorylated in a carboxy-terminal region called the ASARM peptide, which binds to the active site of Phex after its release and acts as a potent inhibitor of mineralization. Cathepsins can cleave MEPE, releasing the ASARM peptide in vitro (Rowe et al. 2004). High ASARM peptide generation by osteocytes is also detected in the X-linked rickets mouse model, the "hyp" mouse.

Dentin matrix protein-1 (DMP-1), an acidic phosphorylated extracellular matrix protein originally identified from a rat incisor cDNA library, has highest expression in odontocytes in teeth and osteocytes in bone, with much lower expression in osteoblasts and hypertrophic chondrocytes (for review, see Bonewald 2007). Autosomal Recessive Hypophosphatemic Rickets in humans is due to mutations in *Dmp1*. Both this human condition and mice lacking DMP1 have highly elevated circulating levels of fibroblast growth factor 23 (FGF23), which appears to derive primarily from osteocytes (Feng et al. 2006). Bone appears to be the source of FGF23, which targets the kidney to regulate phosphate reabsorption. The hypophosphatemic rickets and elevated FGF23 levels that occur in *Dmp1* null mice resemble the "hyp" mouse, which has a mutation in *Pex*. In both these animal models, the skeletal abnormalities can be largely but not completely rescued by providing the mice with a high phosphate diet, suggesting that Dmp1 and Phex may function within a common pathway that is essential for phosphate metabolism. All three genes, *Dmp1*, *Pex*, and *FGF23* are highly expressed in the osteocytes, suggesting that these cells may play a central role in phosphate metabolism and that bone may function as an endocrine organ to control phosphate metabolism (Liu et al. 2006).

The Role of BSP, BAG-75, and MGP in the Regulation of Mineralization

BSP has been shown to colocalize with calcospherulites in osteoid and in osteoblastic cultures. Based on the presence of polyacidic domains and a capacity to nucleate hydroxyapatite in vitro, BSP has been proposed to act as a nucleator in bone and cartilage. Interestingly, although BSP is associated with the first needle-like crystals produced in developing bone or culture models, it has not been reported to be a component of matrix vesicles. In this way, BSP appears to define a biochemically distinct population of particles (e.g., crystal ghost aggregates), which are larger than matrix vesicles and which can also become mineralized (for review, see Ganss et al. 1999).

BAG-75 is expressed at sites of mineral nucleation in osteoblastic cultures and primary bone cells during development, fracture healing, and bone modeling (Gorski 1998). BAG-75 has been shown to demark the sites of mineral deposition in vivo and in vitro, termed BMF, prior to the appearance of crystals (Gorski et al. 2004; Midura et al. 2004). BAG-75 self-associates into long, thin microfibrils that may serve to recruit other BMF components such as BSP through protein–protein interactions, as well as to attract calcium ions through its many phosphate groups and acidic side chains. Recently, it has been shown that cleaved fragments of BSP and of BAG-75 are enriched within BMF in an osteoblastic culture model (Huffman et al. 2007). The serine protease inhibitor, AEBSF, prevented this cleavage and also blocked mineral nucleation, suggesting that mineralization within BMF may be associated with proteolytic fragmentation of BAG-75 and BSP.

Matrix-Gla-protein (MGP) was so named because it contains the vitamin K dependent γ-carboxylated amino acid, Gla. MGP is found in both soft and hard tissues, but is more abundant in cartilage compared to bone. MGP was found to be essential for the inhibition of mineralization of soft tissue (Luo et al. 1997). Mice lacking MGP perinatally died due to massive calcification of cartilage and blood vessels.

A number of other noncollagenous extracellular matrix molecules play a role in regulation of mineralization, which are too numerous to review here. These include the proteins osteonectin, osteopontin, and fibronectin; the protoglycans aggrecan, versican, fibromodulin, biglycan, and decorin; and matrix metalloproteinases (for review, see Zhu et al. 2008). Targeted gene deletion has been used to determine the function of these ECM components. Because most of the resultant null strains have not displayed severe skeletal abnormalities, some functions may be redundant. Table 1 gives a brief list of additional components of the extracellular matrix that appear necessary for normal bone mineralization.

MINERALIZATION: AN ACTIVE OR A PASSIVE PROCESS?

A key question that is still debated within the field is whether mineralization is an active (cell-mediated) or passive (physico-chemical) event. There are a number of important rate-limiting factors in vivo that control the ability of tissues to mineralize and may provide a partial explanation for the spatial restriction of mineralization to skeletal tissues. The major chemical constituents of hydroxyapatite crystals are calcium and inorganic phosphate (Pi) ions, and current data suggest that the serum level of phosphate, as well as of calcium, directly influence bone mineralization (Christakos and Holick 2006; Feng et al. 2006). There are also several groups of physiological inhibitors of calcification, including small molecules and macromolecules. Perhaps the most significant small inhibitor appears to be inorganic pyrophosphate (PPi), which is found in all tissues. PPi inhibits mineralization by preventing incorporation of Pi into the nascent hydroxyapatite crystals. The local balance of available Pi and its inhibitor PPi appear to be key determinants as to whether mineralization will occur (Harmey et al. 2004). This balance is determined by the relative expression and activities of PC-1 and tissue nonspecific alkaline phosphatase, which converts PPi into Pi, thereby locally coun-

Table 1. Enzymes, proteins, and proteoglycans that contribute to bone mineralization

Marker	Expression in Skeletal Tissues	Function in Mineralization
Bone specific Alkaline Phosphatase, TNAP	Early to late osteoblast, early osteocyte	Generates inorganic phosphate from phosphate esters
nucleotide pyrophosphatase/ phosphodiesterase (PC-1)		Degrades pyrophosphate, an inhibitor of mineralization
Matrix-Gla-protein (MGP)	Expressed in all tissues	Inhibitor of mineralization
α_2-HS or fetuin A	Produced in liver, circulates in blood	Binds HA and inhibits growth
PHEX	Early and late osteocytes	Metalloproteinase essential for normal biomineralization and phosphate metabolism
DMP1	Early and mature osteocytes	Phosphate metabolism and mineralization
OF45/MEPE	Late osteoblast through osteocytes	Inhibitor of bone formation/ regulator of phosphate metabolism
FGF23	Early and mature osteocytes	Regulated by PHEX and DMP1, targets the kidney and induces hypophosphatemia
BSP	Osteoblasts	Nucleator of mineralization
BAG-75	Osteoblasts (primary bone)	Biomarker for biomineralization foci, calcium binding
Osteopontin, OPN	Many tissues	Inhibitor of mineralization
Decorin		Organizer of matrix
Biglycan		
Fibronectin		Scaffold or matrix template
Aggrecan		Inhibitor of crystal formation
Osteocalcin	Late osteoblasts, osteocytes	May regulate glucose metabolism, not biomineralization

teracting the inhibitory effect of its precursor. Macromolecular inhibitors of mineralization include MGP, osteopontin, and fetuin (Zhu et al. 2008). Finally, another major limiting factor for mineralization is the deposition and presence of an organized extracellular matrix that is appropriately "primed" for the nucleation of mineral.

Schinke and Karsenty (Schinke et al. 1999) proposed that calcification reactions are passive processes that are regulated locally. Based on their observations of a series of transgenic mouse models exhibiting abnormal mineralization, Murshed and coworkers (Murshed et al. 2005) proposed that calcification will occur in any tissue if the following conditions are met: (1) there is an adequate source of calcium and phosphate, (2) there is an available extracellular matrix of fibrillar collagen, and (3) there is an enzyme such as alkaline phosphatase that is able to remove the inhibitory PPi from the local microenvironment. These investigators provided elegant experimental evidence for this using a mouse model in which alkaline phosphatase was ectopically expressed in the skin, resulting in calcification of collagen fibrils in the skin.

The simplicity of this passive model is attractive, and it may provide a realistic molecular explanation for bone mineralization if this is defined merely as the deposition of hydroxyapatite crystals on collagen fibrils. However, such a model does not adequately explain the finely tuned processes of mineralization that form a highly organized tissue containing interconnected osteocyte networks and a hierarchical tissue structure and microstructure capable of remodeling itself in response to hormonal and mechanical stimuli. Although nucleation and propagation of mineral crystals could occur via a passive, physico-chemical process, active control of mineralization by the cell could occur at many levels. Specifically, regulation of expression of inhibitors and promoters of mineralization, regulation of the availability of calcium and phosphate ions in the local milieu, regulation of extracellular matrix assembly and modification, and/or regulation of proteolytic events that modify the matrix represent some potential active regulatory mechanisms. Cellular regulation may occur through cellular organelles, as described previously. Genetic inactivation of over thirty different genes leads to abnormal bone mineralization, suggesting a critical role of cells in this process. However, the resultant bone phenotypes are not interpretable in terms of a single mechanism involving phosphate, pyrophosphate, alkaline phosphatase, and collagen.

A view that is receiving increasing acceptance and support in the literature is that there may be both active and passive components to the mineralization process. It seems likely that the initial nucleation of mineral deposition is an active cell-mediated process. A second phase of mineral crystal propagation or duplication likely represents a passive, physico-chemical process that can occur once the initial mineral crystals have been seeded. Clearly, future research is needed to verify this and to further refine models of the molecular and cellular control of biomineralization.

SUMMARY AND PERSPECTIVE

In summary, bone mineralization is a complex, multi-step extracellular process that depends upon (1) the serum concentrations of ionic calcium and phosphate, (2) systemic regulatory factors such as PTH and FGF23, (3) osteoblast/osteocyte cells to regulate mineralization in a sequential, timed sequence of events through cellular organelles such as matrix vesicles, calcospherulties, and BMFs, and (4) the macromolecules of proteins, enzymes, and proteoglycans that function as scaffold, nucleators or inhibitors of nucleation, or as organizers of the ECM. The composition of bone matrix and the morphological form of mineral nucleation varies depending upon the type of bone (primary or lamellar). While these latter distinctions have received considerable emphasis in the past, we believe they do not imply fundamental differences, but rather mechanistic adaptations to a common pathway.

REFERENCES

Aaron, J.E., Oliver, B., Clarke, N., and Carter, D.H. 1999. Calcified microspheres as biological entities and their isolation from bone. *Histochem. J.* **31:** 455–470.

Anderson, H.C. 1969. Vesicles associated with calcification in the matrix of epiphyseal cartilage. *J. Cell Biol.* **41:** 59–72.

Anderson, H.C. 1983. Calcific diseases. A concept. *Arch. Pathol. Lab. Med.* **107:** 341–348.

Arsenault, A.L. and Kohler, D.M. 1994. Image analysis of the extracellular matrix. *Microsc. Res. Tech.* **28:** 409–421.

Barragan-Adjemian, C., Nicolella, D., Dusevich, V., Dallas, M.R., Eick, J.D., and Bonewald, L.F. 2006. Mechanism by which MLO-A5 late osteoblasts/early osteocytes mineralize in culture: Similarities with mineralization of lamellar bone. *Calcif. Tissue Int.* **79:** 340–353.

Beck, L., Karaplis, A.C., Amizuka, N., Hewson, A.S., Ozawa, H., and Tenenhouse, H.S. 1998. Targeted inactivation of Npt2 in mice leads to severe renal phosphate wasting, hypercalciuria, and skeletal abnormalities. *Proc. Natl. Acad. Sci.* **95:** 5372–5377.

Bonewald, L.F. 2007. Osteocytes as dynamic, multifunctional cells. *Ann. N.Y. Acad. Sci.* **1116:** 281–290.

Bonewald, L.F. 2008. *The osteocyte.* In *Osteoporosis,* 3rd ed. (ed. R. Marcus et al.), pp. 169–190. Elsevier, Amsterdam.

Bonucci, E. 1967. Fine structure of early cartilage calcification. *J. Ultrastruct. Res.* **20:** 33–50.

Bonucci, E. 2002. Crystal ghosts and biological mineralization: Fancy spectres in an old castle, or neglected structures worthy of belief? *J. Bone Miner. Metab.* **20:** 249–265.

Bonucci, E. 2007. *Biological calcification: Normal and pathological processes in the early stages,* pp. 511–512. Springer-Verlag, New York.

Boskey, A. 1996. Matrix proteins and mineralization: An overview. *Connect. Tissue Res.* **35:** 357–363.

Boskey, A. 1998. Biomineralization: Conflicts, challenges, and opportunities. *J. Cell. Biochem. Suppl.* **30–31:** 83–91.

Boyde, A. and Sela, J. 1978. Scanning electron microscope study of separated calcospherites from the matrices of different mineralizing systems. *Calcif. Tissue Res.* **26:** 47–49.

Christakos, S. and Holick, M.F., Eds. 2006. *Mineral homeostasis.* American Society for Bone and Mineral Research, Washington, D.C.

Feng, J.Q., Ward, L.M., Liu, S., Lu, Y., Xie, Y., Yuan, B., Yu, X., Rauch, F., Davis, S.I., Zhang, S., et al. 2006. Loss of DMP1 causes rickets and osteomalacia and identifies a role for osteocytes in mineral metabolism. *Nat. Genet.* **38:** 1310–1315.

Ganss, B., Kim, R.H., and Sodek, J. 1999. Bone sialoprotein. *Crit. Rev. Oral Biol. Med.* **10:** 79–98.

Glimcher, M.J. 1989. Mechanism of calcification: Role of collagen fibrils and collagen-phosphoprotein complexes in vitro and in vivo. *Anat. Rec.* **224:** 139–153.

Gorski, J.P. 1998. Is all bone the same? Distinctive distributions and properties of non-collagenous matrix proteins in lamellar vs. woven bone imply the existence of different underlying osteogenic mechanisms. *Crit. Rev. Oral Biol. Med.* **9:** 201–223.

Gorski, J.P., Wang, A., Lovitch, D., Law, D., Powell, K., and Midura, R.J. 2004. Extracellular bone acidic glycoprotein-75 defines condensed mesenchyme regions to be mineralized and localizes with bone sialoprotein during intramembranous bone formation. *J. Biol. Chem.* **279:** 25455–25463.

Gowen, L.C., Petersen, D.N., Mansolf, A.L., Qi, H., Stock, J.L., Tkalcevic, G.T., Simmons, H.A., Crawford, D.T., Chidsey-Frink, K.L., Ke, H.Z., et al. 2003. Targeted disruption of the osteoblast/osteocyte factor 45 gene (OF45) results in increased bone formation and bone mass. *J. Biol. Chem.* **278:** 1998–2007.

Harmey, D., Hessle, L., Narisawa, S., Johnson, K.A., Terkeltaub, R., and Millan, J.L. 2004. Concerted regulation of inorganic pyrophosphate and osteopontin by akp2, enpp1, and ank: An integrated model of the pathogenesis of mineralization disorders. *Am. J. Pathol.* **164:** 1199–1209.

Huffman, N.T., Keightley, J.A., Chaoying, C., Midura, R.J., Lovitch, D., Veno, P.A., Dallas, S.L., and Gorski, J.P. 2007. Association of specific proteolytic processing of bone sialoprotein and bone acidic glycoprotein-75 with mineralization within biomineralization foci. *J. Biol. Chem.* **282:** 26002–26013.

HYP Consortium. 1995. A gene (PEX) with homologies to endopeptidases is mutated in patients with X-linked hypophosphatemic rickets. The HYP Consortium. *Nat. Genet.* **11:** 130–136.

Kimmel, D.B. and Jee, W.S. 1980. A quantitative histologic analysis of the growing long bone metaphysis. *Calcif. Tissue Int.* **32:** 113–122.

Landis, W. 1999. An overview of vertebrate mineralization with emphasis on collagen-mineral interaction. *Gravit. Space Biol. Bull.* **12:** 15–26.

Lee, N.K., Sowa, H., Hinoi, E., Ferron, M., Ahn, J.D., Confavreux, C., Dacquin, R., Mee, P.J., McKee, M.D., Jung, D.Y., et al. 2007. Endocrine regulation of energy metabolism by the skeleton. *Cell* **130:** 456–469.

Liu, S., Zhou, J., Tang, W., Jiang, X., Rowe, D.W., and Quarles, L.D. 2006. Pathogenic role of Fgf23 in Hyp mice. *Am. J. Physiol. Endocrinol. Metab.* **291:** E38–E49.

Lowe, J., Bab, I., Stein, H., and Sela, J. 1983. Primary calcification in remodeling haversian systems following tibial fracture in rats. *Clin. Orthop. Relat. Res.* **176:** 291–297.

Luo, G., Ducy, P., McKee, M.D., Pinero, G.J., Loyer, E., Behringer, R.R., and Karsenty, G. 1997. Spontaneous calcification of arteries and cartilage in mice lacking matrix GLA protein. *Nature* **386:** 78–81.

Lundquist, P., Murer, H., and Biber, J. 2007. Type II Na^+-P_i cotransporters in osteoblast mineral formation: Regulation by inorganic phosphate. *Cell. Physiol. Biochem.* **19:** 43–56.

Lust, G. and Seegmiller, J.E. 1976. A rapid, enzymatic assay for measurement of inorganic pyrophosphate in biological samples. *Clin. Chim. Acta* **66:** 241–249.

Midura, R.J., Vasanji, A., Su, X., Wang, A., Midura, S.B., and Gorski, J.P. 2007. Calcospherulites isolated from the mineralization front of bone induce the mineralization of type I collagen. *Bone* **41:** 1005–1016.

Midura, R.J., Wang, A., Lovitch, D., Law, D., Powell, K., and Gorski, J.P. 2004. Bone acidic glycoprotein-75 delineates the extracellular sites of future bone sialoprotein accumulation and apatite nucleation in osteoblastic cultures. *J. Biol. Chem.* **279:** 25464–25473.

Murshed, M., Harmey, D., Millan, J.L., McKee, M.D., and Karsenty, G. 2005. Unique coexpression in osteoblasts of broadly expressed genes accounts for the spatial restriction of ECM mineralization to bone. *Genes Dev.* **19:** 1093–1104.

Nanci, A., Zalzal, S., Gotoh, Y., and McKee, M.D. 1996. Ultrastructural characterization and immunolocalization of osteopontin in rat calvarial osteoblast primary cultures. *Microsc. Res. Tech.* **33:** 214–231.

Nijweide, P.J. and Mulder, R.J. 1986. Identification of osteocytes and osteoblast-like cell cultures using a monoclonal antibody specifically directed against oocytes. *Histochemistry* **84:** 342–347.

Olsen, B.R. 2006. Bone embryology. In *Primer on the metabolic bone diseases and disorders of mineral metabolism*, 6th ed. (ed. M.J. Favus), pp. 1–6. Lippincott Williams and Wilkins, Philadelphia.

Rohde, M. and Mayer, H. 2007. Exocytotic process as a novel model for mineralization by osteoblasts in vitro and in vivo determined by electron microscopic analysis. *Calcif. Tissue Int.* **80:** 323–336.

Rowe, P.S., Kumagai, Y., Gutierrez, G., Garrett, I.R., Blacher, R., Rosen, D., Cundy, J., Navvab, S., Chen, D., Drezner, M.K., et al. 2004. MEPE has the properties of an osteoblastic phosphatonin and minhibin. *Bone* **34:** 303–319.

Ruchon, A.F., Tenenhouse, H.S., Marcinkiewicz, M., Siegfried, G., Aubin, J.E., DesGroseillers, L., Crine, P., and Boileau, G. 2000. Developmental expression and tissue distribution of Phex protein: Effect of the Hyp mutation and relationship to bone markers. *J. Bone Miner. Res.* **15:** 1440–1450.

Schinke, T., McKee, M.D., and Karsenty, G. 1999. Extracellular matrix calcification: Where is the action? *Nat. Genet.* **21:** 150–151.

Sela, J., Schwartz, Z., Swain, L., and Boyan, B.D. 1992. Role of matrix vesicles in calcification. In *Calcification in biological systems* (ed. E. Bonucci), pp. 73–105. CRC, Boca Raton, Florida.

Simmons, Jr., E.D., Pritzker, K.P., and Grynpas, M.D. 1991. Age-related changes in the human femoral cortex. *J. Orthop. Res.* **9:** 155–167.

Wu, L.N., Ishikawa, Y., Sauer, G.R., Genge, B.R., Mwale, F., Mishima, H., and Wuthier, R.E. 1995. Morphological and biochemical characterization of mineralizing primary cultures of avian growth plate chondrocytes: Evidence for cellular processing of Ca2+ and Pi prior to matrix mineralization. *J. Cell. Biochem.* **57:** 218–237.

Zhu, W., Robey, P.G., and Boskey, A.L. 2008. The regulatory role of matrix proteins in mineralization of bone. In *Osteoporosis,* 3rd ed. (ed. R. Marcus et al.), pp. 191–240. Elsevier, Amsterdam.

11

Bone Remodeling: Cellular and Molecular Events

T. John Martin and Natalie A. Sims
St Vincent's Institute of Medical Research and
University of Melbourne Department of Medicine
Fitzroy, 3065, Victoria, Australia

BONE REMODELING REFERS TO THE RENEWAL process whereby small packets of old trabecular and cortical bone, dispersed throughout the skeleton and separated from others geographically as well as chronologically, are replaced by new bone throughout adult life. A major feature of bone remodeling is that it does not occur uniformly throughout the skeleton, but takes place asynchronously in focal or discrete sites known as basic multicellular units (BMUs) of bone turnover (Frost 1964; Parfitt 1996). The BMU describes the cells within a packet of bone that is resorbed and then fully rebuilt. The resorption activity in a BMU in human bone takes approximately 3 weeks and the formation response takes 3–4 months. The process is such that remodeling replaces about 5%–10% of the skeleton each year, with the entire adult human skeleton replaced in 10 years. Understanding the tightly-controlled processes of bone resorption and formation that take place in individual BMUs throughout the skeleton requires appreciation of the many pathways that control cells of the osteoblast and osteoclast lineage and how they communicate among themselves.

SEQUENCE OF CELLULAR EVENTS IN BONE REMODELING: OSTEOCYTE INVOLVEMENT

Cancellous bone remodeling starts on the bone surface, initiated by any of several possible stimuli. Among these are pressure changes sensed by osteocytes, resulting in signals delivered to surface cells, and damage in the form of microcracks in bone that lead to osteocyte stimulation, or even apoptosis, and the release from other nearby bone cells of signals, some of which are likely to be known cytokines and prostanoids.

One of the major recent advances in bone biology is the realization of the importance of the osteocyte. Osteocytes make up over 90% of all bone cells, are viable for decades in bone, and have long dendritic processes that resemble those of neurons, equipping them ideally to communicate with cells on the surface. Most relevant to the initiation of remodeling, viable osteocytes were found to prevent osteoclast activation (Tomkinson et al. 1998) and evidence was also obtained that apoptotic osteocytes contribute to the recruitment of osteoclasts (Verborgt et al. 2000, 2002). Osteocytes sense when skeletal unloading occurs (e.g., immobilization) and respond with signals to surface cells that recruit osteoclasts for resorption, or sense skeletal loading or microcrack formation and provide signals favoring bone formation. There is increasing evidence that osteocyte-derived signals have profound effects on both bone formation (see below) and resorption. A significant advance in appreciation of the osteocyte's role in skeletal homeostasis came with work in which the dentin matrix protein-1 (DMP-1) promoter was used to prepare mice expressing diphtheria toxin (DT) receptor specifically in osteocytes, and DT was used to kill the osteocytes in the mice (Tatsumi et al. 2007). Shortly after osteocyte killing, increased expression of mRNA for RANKL was detected in bones, leading to increased osteoclast formation and bone resorption, supporting the idea that viable osteocytes prevent osteoclast recruitment and activation, whereas the reverse accompanies osteocyte death.

THE RESORPTION PHASE: OSTEOCLAST FORMATION IN THE BMU

The first essential step in the remodeling cycle is the generation of active osteoclasts from hemopoietic precursors. Regardless of the source of the initiation signal, osteoclasts are likely derived from early and late precursors available in marrow adjacent to activation sites, or could be recruited from blood available at the bone interface through a sinus structure of bone remodeling compartments (BRCs) (Hauge et al. 2001). This concept has been extended to provide a mechanism by which both osteoblast and osteoclast precursors circulate and so may arrive at the BRC via the circulation and via capillaries penetrating the canopy that overlies the BRC (Fujikawa et al. 1996; Eghbali-Fatourechi et al. 2007; Eriksen et al. 2007). Osteoblast progenitors are associated with vascular structures in the marrow, and several studies suggest there may also be common progenitors giving rise to cells forming the blood vessel and pluripotent perivascular cells (Doherty et al. 1998; Howson et al. 2005; Matsumoto et al. 2006; Modder and Khosla 2008; Otsuru et al. 2008). Osteoclast formation can take place rapidly in vivo, and it is appealing to consider that this might be due to possible niches of partially differentiated cells available in the BRC (Yamamoto et al. 2006).

Osteoclasts are formed by the attraction of haematopoietic myelomonocytic precursors to the resorption site, followed by their fusion, and attachment of the subsequent multinucleated cell to the bone surface (Fig. 1). Chemoattractants for osteoclast precursors may include factors from the bone matrix itself (Malone et al. 1982), factors produced by cells of the osteoblast lineage (Yu et al. 2003), or signals from apoptotic osteocytes (Noble et al. 2003). The precise regulation of the osteoclast fusion process is not well understood, although it is clear that the initial stimulus depends on the presence of receptor activator of NF-κB ligand (RANKL), a factor produced by the osteoblast. RANKL stimulates the mononuclear osteoclast precursors to express a number of fusion proteins, including Atp6v0d2 (Lee et al. 2006), DAP12, FcRγ, and DC-STAMP (Kukita et al. 2004; Mocsai et al. 2004). Although some aspects of osteoclast differentiation do not appear to be completely dependent on fusion, the resorption capability of mononuclear osteoclast-like cells is vastly impaired compared with multinucleated osteoclasts (Yagi et al. 2005).

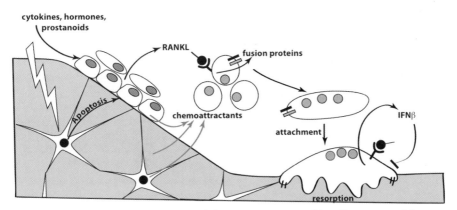

Figure 1. Initial events in bone remodeling. Hormones and locally generated factors, as well as osteocyte apoptosis resulting from microfractures, influence cells of the osteoblast lineage to produce RANKL and promote osteoclast differentiation and fusion. The osteoclasts attach to bone and resorb. Osteoclasts can limit their own formation by, for example, generation of interferon β.

When the remodeling cycle is initiated, say by mechanical strain that would generate cytokines or prostanoids (Parfitt 1996), it has been proposed that the thin layer of nonmineralized matrix (osteoid), under osteoblast-lineage lining cells, is initially digested by collagenase derived from lining cells to expose the mineralized matrix that osteoclasts can resorb (Chambers et al. 1985; Chambers and Fuller 1985). To initiate resorption, the multinucleated osteoclast attaches to the bone matrix via the interaction of integrins with arginine-glycine-aspartic acid (RGD) sequences in matrix proteins, including osteopontin and bone sialoprotein (Ross and Teitelbaum 2005), which were laid down by osteoblasts during the previous cycle of bone formation. The sites at which attachment occur, known as podosomes, are rapidly assembled and disassembled, allowing osteoclast movement and resorption across the bone surface. One of the critical signaling pathways through which this occurs is via the interaction of Src, Cbl, and Pyk2 tyrosine kinases (Horne et al. 2005; Gil-Henn et al. 2007).

The formation of osteoclasts from hematopoietic precursors in the monocyte-macrophage lineage needs to be regulated through direct interaction with cells of the osteoblast lineage, stimulated by hormones, cytokines, and prostanoids. Osteoblast lineage cells in bone, together with immune cells in the marrow microenvironment, play an essential role as a source of cytokines, including tumor necrosis factor-α (TNF-α), IL-1, and IL-6. Since the receptors for these and many other bone-resorbing paracrine factors are expressed in cells of the osteoblast lineage, they must first act on the osteoblast to modify osteoclast formation (Suda et al. 1999). Despite the several different signaling mechanisms that these agents use upon their target cells, they converge to a common mechanism in promoting osteoclast formation, which requires RANKL as the essential mediator of osteoclast formation in response to all known stimuli (Dougall et al. 1999). The osteoblast lineage also produces the decoy receptor, osteoprotegerin (OPG), a soluble member of the TNF receptor superfamily, which is a very effective inhibitor of osteoclast formation that acts locally as a "brake" on osteoclast formation (Bucay et al. 1998). Osteoblasts/stromal cells are also the source of M-CSF, which plays a crucial role in osteoclast formation by promoting the proliferation of precursors.

THE RESORPTION PHASE: ITS CESSATION IN THE BMU

Resorption in the BMU takes place in a sealed-off subosteoclastic resorbing compartment that is acidified through the active transport of protons driven by a V-type H+ATPase. At the same time, passive transport of chloride through chloride channel ClC-7 preserves electroneutrality. This results in dissolution of bone mineral, thereby exposing the organic matrix to proteolytic enzymes, particularly cathepsin K. These enzymes are responsible for the degradation of organic matrix. Inactivation of any of these (V-type H+ATPase, ClC-7 channel, or cathepsin K) by genetic or pharmacologic means results in failure of osteoclasts to resorb bone (Bruzzaniti and Baron 2006).

Once a BMU site is activated, resorption within that site must be limited. If it were to continue unchecked, resorption would be excessive. A novel aspect of signaling in osteoclast development that could be relevant to the control of osteoclasts in the BMU is the ability of RANKL to limit its own osteoclastogenic effect by promoting interferon-β (IFN-β) production by monocytic osteoclast precursors (Takayanagi et al. 2002). This inhibits osteoclast formation by preventing RANKL-induced expression of c-fos. The latter was known as an essential transcription factor in osteoclast differentiation, since c-fos null mice were osteopetrotic because of failed osteoclast formation (Grigoriadis et al. 1994). RANKL-induced IFN-β provides an appealing mechanism, contained within the osteoclast itself, and contributing to regulated osteoclast formation within the BMU. A further possible mechanism of osteoclast restraint within the BMU might come from reverse signaling through osteoblast-derived EphB4, acting on ephrinB2 in the osteoclast, to restrict osteoclast formation (Zhao et al. 2006).

An important unanswered question about osteoclast behavior and the control of resorption in the remodeling cycle is how each osteoclast knows when to stop resorbing. The process is likely to finish with osteoclast death, which has been studied in vitro to some extent, but its regulation in vivo remains obscure. Osteoclasts phagocytose osteocytes, which might provide a mechanism to remove the signal for resorption (Elmardi et al. 1990; Suzuki et al. 2000). TGF-β is available when it is resorbed from matrix and activated by acid pH, and can promote osteoclast apoptosis (Hughes and Boyce 1997). A direct effect of estrogen to enhance osteoclast apoptosis through the mediation of Fas ligand has been identified in mouse genetic experiments (Nakamura et al. 2007). Some insights into control of apoptosis arise from genetic and pharmacological studies, showing that inhibition of acidification of the resorption space by blockade of either ClC-7 or the V-type H+ATPase of the osteoclast results in prolonged osteoclast survival (Henriksen et al. 2004; Karsdal et al. 2005, 2007). This might suggest a role for acidification in determining osteoclast lifespan, perhaps even through TGF-β activation. In determining how resorption ends in the BMU, there is much to be learned of mechanisms regulating osteoclast apoptosis in that microenvironment.

THE REVERSAL PHASE

Toward the end of resorption, mononuclear cells are seen at the bottom of resorption pits, where they prepare the pits for the engagement of osteoblasts in bone formation (Villanueva et al. 1986). In this "reversal" phase of remodeling, macrophages had been long considered responsible for the postresorption digestion of collagen fragments in the BMU, but recent evidence suggests the involvement of osteoblasts and/or bone lining cells, with these cells identified cytologically at sites of resorption, both in calvaria and long bones, and shown engulfing the collagen fragments on the bone surface after the osteoclasts have resorbed (Everts et al. 2002; Perez-Amodio et al. 2004). This activity appears to be mediated by membrane matrix metalloproteinases, which carry out the task of cleaning out resorption pits after osteoclasts complete their task. They then go on to lay down a thin layer of collagen along the Howship's lacuna, closely associated with a cement line.

The reversal line (cement line) contains a large abundance of osteopontin (Chen et al. 1994), which is produced both by osteoclasts and osteoblasts. This is an RGD-containing extracellular matrix protein that interacts with integrin receptors $\alpha_v\beta_3$ in osteoclasts and primarily $\alpha_v\beta_5$ in osteoblasts. These integrin receptors were shown not only to mediate cell attachment of both cell types to the bone matrix, but also to act as signal transducing receptors (Hynes 1992). It is not yet fully established what influence osteopontin has on osteoclast or osteoblast activity, but it appears to be required for a normal response of bone to mechanical unloading or matrix mineralization (Ishijima et al. 2001). The presence of osteopontin on the reversal line raises the possibility that it may be one of the signals for cessation of osteoclast activity or initiation of osteoblastic bone formation, or possibly both.

COUPLING AND CELL COMMUNICATION IN REMODELING

The next major stage of the remodeling process is the recruitment and differentiation of mesenchymal precursors to osteoblasts, sufficient to synthesize the amount of bone lost in the resorption process at that site. Of the several potential sources of osteoblast precursors, one is that lining cells, the single layer of flattened cells that have ceased their bone-forming function, can revert to that activity. Other likely sources are adjacent marrow stromal cell precursors and blood-borne osteoblast precursors that could be presented to the BMU from capillaries within the sinus structure, provided by the BRC underneath its canopy (Hauge et al. 2001; Eghbali-Fatourechi et al. 2007).

Outstanding questions concerning bone remodeling relate to the mechanisms by which bone formation begins, and the amount of bone formed in a BMU is linked to the amount resorbed—the coupling mechanism (Fig. 2). Baylink and colleagues (Howard et al. 1981) suggested that "coupling" is due to bone-forming growth factors released from the bone matrix during bone resorption. Indeed, a large number of substances that are mitogenic to osteoblasts or stimulate bone formation in vivo could be extracted from bone matrix (Baylink et al. 1993). A new view supportive of the hypothesis of growth factor involvement comes from a study using inhibitors of cathepsin K that prevent degradation of bone matrix while allowing demineralization to occur. This blockade inhibits bone resorption without a coupled inhibition of formation, most likely due to adequate release of intact growth factors from the demineralized matrix (Fuller et al. 2008). Questions raised by this model of coupling, via growth-factor release from the matrix, relate to the time course and the distance between the resorption and formation processes, and whether activation can be controlled with sufficient precision in this way. The evidence of bone growth-factor involvement is increasing, and it is possible that they also play a role in modeling.

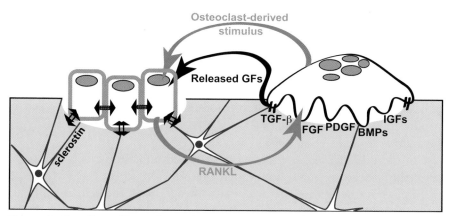

Figure 2. Coupling of bone formation to resorption in the BMU. Osteoblast differentiation and bone formation are promoted by growth factors from matrix and osteoclast-derived factors. Sclerostin production by the connecting osteocytes limits the expansion of osteoblasts, which is also regulated by intercellular communication.

Finally, observations made in genetically manipulated mice and human genetics suggest that the osteoclast itself could also be the source of an activity that contributes to the fine control of the coupling process in bone remodeling. In individuals with the osteopetrotic syndrome ADOII, due to inactivating mutations in the chloride-7 channel (ClC-7), bone resorption is deficient because of failure of the osteoclast acidification process. Bone formation in these patients is nevertheless normal, rather than diminished, as might be expected because of the greatly impaired resorption (Del Fattore et al. 2006). Furthermore, in mice deficient in either c-src (Soriano et al. 1991; Kornak et al. 2001) or tyrosine phosphatase epsilon (Chiusaroli et al. 2004), bone resorption is inhibited without inhibition of formation. In these three knockout mouse lines, osteoclast resorption is greatly reduced by the mutation, although osteoclast numbers are not reduced. Indeed, osteoclast numbers are actually increased because of reduced osteoclast apoptosis. A possibility is that these osteoclasts, although unable to resorb bone, are nevertheless capable of generating a factor (or factors) contributing to bone formation. On the other hand, mice lacking c-fos, which are unable to generate osteoclasts, have reduced bone formation and resorption (Grigoriadis et al. 1994). Studies with other mutant mice, with specific inactivation of each of the two alternative signaling pathways through gp130, led to the conclusion that resorption alone was insufficient to promote the coupled bone formation, but active osteoclasts are the likely source (Sims et al. 2004; Martin and Sims 2005).

In a search for osteoclast products that might contribute to the coupling process, evidence was obtained in mice, using genetic and pharmacological approaches, that osteoclast-derived ephrinB2 acts through a contact-dependent mechanism on EphB4, its receptor in the osteoblast, to promote osteoblast differentiation and bone formation (Zhao et al. 2006). At the time of writing, this ephrin-Eph communication process is the first specific candidate-coupling mechanism. There are likely to be several such factors, and another possibility is cardiotropin-1, which is produced by osteoclasts and promotes osteoblast differentiation and bone formation (Walker et al. 2008).

In remodeling, just as osteoclasts need to know how much to resorb, osteoblasts need to know how much bone to replace in the BMU. One possibility is that within the osteoblast lineage itself, the cells are able to sense spatial requirements for appropriate replacement in the BMU. There is evidence to support that possibility from the work of Boyde and colleagues (Gray et al. 1996), in which they showed in vitro that rat calvarial cells grown on bone slices with mechanically excavated crevices and grooves, made bone in those defects, filling them exactly to a flat surface, indicating that the participating cells themselves are able to sense spatial limits, and most likely do so by chemical communication that takes place between the developing osteoblasts. The communicating factors could be the growth factors that are made and stored in the matrix, as well as gap junction communication between the osteoblasts themselves (Stains and Civitelli 2005).

Secondly, among the newly recognized functions of the osteocyte, its role in limiting bone formation is having a major impact on thinking in bone biology. Mutations in the *Sost* gene were found responsible for the rare sclerosing bone dysplasias, sclerosteosis, and Van Buchem disease. Each is characterized by a greatly increased amount of bone. The *Sost* gene product, sclerostin, is produced exclusively in bone by osteocytes, and is a negative regulator of bone formation, inhibiting Wnt signaling through binding to the LRP5/6, thus preventing its participation in the receptor complex that activates the Wnt pathway and bone formation (van Bezooijen et al. 2004). Production of sclerostin by osteocytes is rapidly decreased by treatment with PTH or PTHrP (Bellido et al. 2005; Keller and Kneissel 2005), as well as by mechanical loading (Robling et al. 2006). For each of these stimuli, removal of sclerostin as a constitutive inhibitor of bone formation could at least partly explain the accompanying increased bone formation. Physiologically, rapid changes in osteocyte production of sclerostin could signal to surface cells to limit the filling of remodeling spaces by osteoblasts, in addition to keeping lining cells in a quiescent state on nonremodeling bone surfaces.

Finally, the first direct evidence that there could be controlling influences upon osteoblast differentiation from within the osteoblast lineage comes from a study that identified a novel mechanism in which mature osteoblasts directed the differentiation of early mesenchymal osteoblast precursors through activation of canonical Wnt signaling (Zhou et al. 2008). Such a mechanism has the added advantage that its regulation would be susceptible to inhibition by osteoblast-derived inhibitors of Wnt signaling, such as a dickkopf (Dkk) protein or secreted Frizzled-related protein (sFRP). Zhou et al. (2008) provided evidence for inhibition of the process by sFRP1.

CIRCULATING HORMONES AND CYTOKINES

The importance of estrogen as a hormone that inhibits bone resorption has long been recognized. Data from animal models and cell and tissue cultures suggest that estrogen deficiency increases and estrogen replacement decreases the activity of IL-1, TNF-α, and IL-6, as well as prostaglandins (Kawaguchi et al. 1995; Kimble et al. 1995; Manolagas and Jilka 1995), all of which promote osteoclast formation and bone resorption. In a study of women who were premenopausal, or postmenopausal with and without estrogen treatment (Eghbali-Fatourechi et al. 2003), serum levels of RANKL or OPG did not differ among the three groups, but analyses of bone marrow mononuclear cells showed that the production of RANKL per cell was increased up to threefold in untreated postmenopausal women, and correlated negatively with serum estradiol-17β-estradiol levels, and positively with bone resorption markers. The study draws attention to the close involvement of cells of the immune system in estrogen action on the skeleton, but without the increases in T- or B-cell numbers that occur in the mouse in estrogen deficiency. Although any or all of these mechanisms have been put forward to explain the increased numbers of osteoclasts and bone resorption that follows estrogen withdrawal, there still remains the possibility that estrogen acts directly on the osteoclast. Although estrogen can increase TGF-β, leading to osteoclast apoptosis (Kimble et al. 1995), evidence in support of a direct effect of estrogen-enhancing osteoclast apoptosis has been discovered through experiments in which the ERα was specifically depleted in osteoclasts (Nakamura et al. 2007). If this is confirmed in human osteoclasts, it might be an important contributor to the several mechanisms invoked in explaining estrogen action on bone.

Parathyroid hormone (PTH) has a central role in calcium homeostasis, and for that reason has been regarded as an important influence upon bone remodeling. Further support comes from the action of intermittent PTH injection as an anabolic for the skeleton (Neer et al. 2001), and the accompanying evidence for PTH treatment resulting in an increased number of active BMUs (Compston 2007). PTH-related protein (PTHrP) resembles PTH in its amino-terminal sequence (Suva et al. 1987), and the two have very similar structural requirements for activity on their common receptor, PTHR1 (Kemp et al. 1987). Further, a contribution to the anabolic effect of PTH and PTHrP could come from their ability to rapidly reduce the production of sclerostin as a constitutive inhibitor of bone formation (Bellido et al. 2005; Keller and Kneissel 2005). Rapidly induced reduction in osteocyte production of sclerostin could signal to surface cells to enhance the filling of remodeling spaces by osteoblasts, in addition to releasing lining cells from a quiescent state in nonremodeling bone surfaces.

Of great physiological interest is the recent evidence obtained through mouse genetics that osteoblast-derived PTHrP has a crucial role in bone remodeling (Miao et al. 2005). Although PTHrP circulates as a hormone that causes hypercalcemia in patients with certain cancers, it has no normal endocrine role except in the fetus, where it promotes the transfer of calcium across the placenta from mother to fetus, and during late pregnancy and lactation, where its function is still not fully defined (Philbrick et al. 1996; Martin et al. 1997). The physiological role of PTHrP in postnatal mammals appears to relate to its function as a paracrine effector in several tissues. The discovery of PTHrP production in bone (Moseley et al. 1991; Suda et al. 1996) pointed to possible local functions of this peptide, in which case it should be regarded as a cytokine rather than as a hormone. PTHrP production by cells of mesenchymal lineage that differentiate into osteoblasts was shown to occur very early during differentiation and to decrease with further cell maturation. The receptor, PTHR1, which is shared with PTH, is expressed at a later stage in differentiation, by committed preosteoblasts (Allan et al. 2003).

Ablation of *Pthrp* in mice revealed a role for PTHrP in endochondral bone formation (Amizuka et al. 1994, 1996). Although *Pthrp*$^{-/-}$ mice died at the time of birth because of a cartilage defect, *Pthrp*$^{+/-}$ mice survived, and haploinsufficient mice were markedly osteoporotic by 3 months of age, suggesting that PTHrP might be a local factor of some importance in maintaining bone. Importantly, osteoblast-specific ablation of *Pthrp* in mice mimicked this phenotype, with osteoporosis and impaired bone formation, resulting from an impaired capacity to generate osteoblasts from bone marrow precursors and increased osteoblast apoptosis—even in the presence of normal levels of circulating PTH. Furthermore, in both *Pthrp* haploinsufficient mice and osteoblast-specific *Pthrp* knockout mice, the number of osteoclasts was decreased (Miao et al. 2005), consistent with a requirement for PTHrP to enhance production of RANKL by PTHR1-positive preosteoblasts. Thus, PTHrP plays a critical physiological role in promoting recruitment and survival of osteoblasts, and in enhancing osteoclast formation.

In determining the relative contributions to bone physiology of the endocrine factor PTH and the paracrine factor PTHrP, Miao et al. (2004, 2005) concluded that postnatally, PTH is secreted as a hormone in response to a hypocalcemic signal to regulate calcium homeostasis by promoting bone resorption, whereas PTHrP functions as a bone cytokine to control bone mass. Thus, PTHrP may be the crucial ligand in bone for the PTHR1, with actions that reproduce the known effects of PTH: stimulation of bone formation by promoting the differentiation of committed osteoblast precursors (Dobnig and Turner 1995) and by inhibiting apoptosis of mature osteoblasts and osteocytes (Jilka et al. 1999; Jilka 2007). On the other hand, sustained elevation of PTH levels favor osteoclast formation through the generation of RANKL in target cells (Ma et al. 2001), which promotes osteoclast production from hemopoietic precursors. Thus, the local release of PTHrP from osteoblastic cells must be tightly regulated in time and space so that it is presented only briefly to target cells. For PTHrP to contribute to remodeling, control mechanisms must exist to ensure that only short-lived, high levels of PTHrP are available to local targets in order to favor bone formation, as persistently increased local PTHrP levels would favor increased osteoclast formation, through stimulation of RANKL production. We can only speculate about mechanisms that might achieve such fine control. They could involve cytokine-mediated and/or neural control of PTHrP gene transcription, mRNA stability, or proteolytic processing of the protein.

From an evolutionary point of view, PTHrP deficiency could contribute to the age-related decrease in bone formation that takes place at the BMU level. The maintenance of skeletal integrity is an essential survival function for mammals. Participation in this process by PTHrP might be sufficient to explain the remarkable conservation of PTHrP amino-acid sequence among mammalian species.

CENTRAL AND SYMPATHETIC NERVOUS SYSTEM REGULATION

In the last few years, there has been accumulating evidence that the nervous system plays a crucial role in bone remodeling and therefore in the maintenance of bone mass. Although this was surprising at first, such a relationship has been long suspected because of the loss of bone occurring in subjects with spinal injury, which accompanies sensory or sympathetic denervation. Among neuroendocrine controls that have been identified so far, leptin, an adipocyte gene product, was shown to inhibit bone formation (Ducy et al. 2000) through hypothalamic receptors, involving an action of CART (Cocaine and Amphetamine-Regulated Transcript) (Elefteriou et al. 2005) and peripherally through the sympathetic nervous system, with β-adrenergic agonists decreasing, and antagonists increasing bone mass in mice (Takeda et al. 2002). Another hypothalamic connection comes from experiments using hypothalamic deletion of the Y2 receptor, showing that neuropeptide Y or a related peptide can regulate bone formation, but independent of sympathetic control (Baldock et al. 2002). Furthermore, mice deficient in neuromedin U (NMU) were obese independent of leptin, but had increased bone mass similar to that in leptin-deficient mice. This work led to the conclusion that NMU is a key central mediator in the regulation of bone mass by the sympathetic nervous system (Sato et al. 2007).

The concept of an adipocyte gene product regulating bone is indeed an intriguing one, with further interest provided by evidence that bone-derived osteocalcin acts as a hormone that increases insulin-sensitive glucose utilization and pancreatic β-cell mass, and reduces visceral fat (Lee et al. 2007). The translation of these exciting findings to human skeletal physiology and the pathogenesis of bone loss, as well as their use in approaches to treatment, are awaited with great interest. There is evidence from an epidemiological and case-control study (Reid et al. 2005) for protection of the skeleton against bone loss through the use of β-adrenergic antagonists. Resolution of these issues will require carefully planned studies.

CONCLUSION

The strict coupling of bone resorption and formation that is essential for the maintenance of skeletal integrity takes place in BMUs distributed throughout the skeleton. Cellular communication processes determine the initiation of resorption in the BMU, with osteoclasts generated and activated, and their numbers and life limited by signals generated from the osteoblast lineage, the matrix, and from within the osteoclast lineage. When osteoblast precursors are recruited, they differentiate to a mature form that equips them to replace the required amount of bone in the BMU. Their differentiation is controlled by matrix- and osteoclast-derived signals, and likely also from regulatory mechanisms within the osteoblast lineage itself. Control of the process is achieved predominantly through actions of cytokines and contact-dependent signals, influenced in turn by endocrine and neural regulation.

REFERENCES

Allan, E.H., Ho, P.W., Umezawa, A., Hata, J., Makishima, F., Gillespie, M.T., and Martin, T.J. 2003. Differentiation potential of a mouse bone marrow stromal cell line. *J. Cell. Biochem.* **90:** 158–169.

Amizuka, N., Warshawsky, H., Henderson, J.E., Goltzman, D., and Karaplis, A.C. 1994. Parathyroid hormone-related peptide-depleted mice show abnormal epiphyseal cartilage development and altered endochondral bone formation. *J. Cell Biol.* **126:** 1611–1623.

Amizuka, N., Karaplis, A.C., Henderson, J.E., Warshawsky, H., Lipman, M.L., Matsuki, Y., Ejiri, S., Tanaka, M., Izumi, N., Ozawa, H., et al. 1996. Haploinsufficiency of parathyroid hormone-related peptide (PTHrP) results in abnormal postnatal bone development. *Dev. Biol.* **175:** 166–176.

Baldock, P.A., Sainsbury, A., Couzens, M., Enriquez, R.F., Thomas, G.P., Gardiner, E.M., and Herzog, H. 2002. Hypothalamic Y2 receptors regulate bone formation. *J. Clin. Invest.* **109:** 915–921.

Baylink, D.J., Finkelman, R.D., and Mohan, S. 1993. Growth factors to stimulate bone formation. *J. Bone Miner. Res.* (suppl. 2) **8:** S565–S572.

Bellido, T., Ali, A.A., Gubrij, I., Plotkin, L.I., Fu, Q., O'Brien, C.A., Manolagas, S.C., and Jilka, R.L. 2005. Chronic elevation of parathyroid hormone in mice reduces expression of sclerostin by osteocytes: A novel mechanism for hormonal control of osteoblastogenesis. *Endocrinology* **146:** 4577–4583.

Bruzzaniti, A. and Baron, R. 2006. Molecular regulation of osteoclast activity. *Rev. Endocr. Metab. Disord.* **7:** 123–139.

Bucay, N., Sarosi, I., Dunstan, C.R., Morony, S., Tarpley, J., Capparelli, C., Scully, S., Tan, H.L., Xu, W., Lacey, D.L., et al. 1998. *osteoprotegerin*-deficient mice develop early onset osteoporosis and arterial calcification. *Genes Dev.* **12:** 1260–1268.

Chambers, T.J. and Fuller, K. 1985. Bone cells predispose bone surfaces to resorption by exposure of mineral to osteoclastic contact. *J. Cell Sci.* **76:** 155–165.

Chambers, T.J., Darby, J.A., and Fuller, K. 1985. Mammalian collagenase predisposes bone surfaces to osteoclastic resorption. *Cell Tissue Res.* **241:** 671–675.

Chen, J., McKee, M.D., Nanci, A., and Sodek, J. 1994. Bone sialoprotein mRNA expression and ultrastructural localization in fetal porcine calvarial bone: Comparisons with osteopontin. *Histochem. J.* **26:** 67–78.

Chiusaroli, R., Knobler, H., Luxenburg, C., Sanjay, A., Granot-Attas, S., Tiran, Z., Miyazaki, T., Harmelin, A., Baron, R., and Elson, A. 2004. Tyrosine phosphatase epsilon is a positive regulator of osteoclast function in vitro and in vivo. *Mol. Biol. Cell* **15:** 234–244.

Compston, J.E. 2007. Skeletal actions of intermittent parathyroid hormone: Effects on bone remodelling and structure. *Bone* **40:** 1447–1452.

Del Fattore, A., Peruzzi, B., Rucci N., Recchia, I., Cappariello, A., Longo, M., Fortunati, D., Ballanti, P., Iacobini, M., Luciani M., et al. 2006. Clinical, genetic, and cellular analysis of 49 osteopetrotic patients: Implications for diagnosis and treatment. *J. Med. Genet.* **43:** 315–325.

Dobnig, H. and Turner, R.T. 1995. Evidence that intermittent treatment with parathyroid hormone increases bone formation in adult rats by activation of bone lining cells. *Endocrinology* **136:** 3632–3638.

Doherty, M.J., Ashton, B.A., Walsh, S., Beresford, J.N., Grant, M.E., and Canfield, A.E. 1998. Vascular pericytes express osteogenic potential in vitro and in vivo. *J. Bone Miner. Res.* **13:** 828–838.

Dougall, W.C., Glaccum, M., Charrier, K., Rohrbach, K., Brasel, K., De Smedt, T., Daro, E., Smith, J., Tometsko, M.E., Maliszewski, C.R., et al. 1999. RANK is essential for osteoclast and lymph node development. *Genes Dev.* **13:** 2412–2424.

Ducy, P., Amling, M., Takeda, S., Priemel, M., Schilling, A.F., Beil, F.T., Shen, J., Vinson, C., Rueger, J.M., and Karsenty, G. 2000. Leptin inhibits bone formation through a hypothalamic relay: A central control of bone mass. *Cell* **100:** 197–207.

Eghbali-Fatourechi, G., Khosla, S., Sanyal, A., Boyle, W.J., Lacey, D.L., and Riggs, B.L. 2003. Role of RANK ligand in mediating increased bone resorption in early postmenopausal women. *J. Clin. Invest.* **111:** 1221–1230.

Eghbali-Fatourechi, G.Z., Modder, U.I., Charatcharoenwitthaya, N., Sanyal, A., Undale, A.H., Clowes, J.A., Tarara, J.E., and Khosla, S. 2007. Characterization of circulating osteoblast lineage cells in humans. *Bone* **40:** 1370–1377.

Elefteriou, F., Ahn, J.D., Takeda, S., Starbuck, M., Yang, X., Liu, X., Kondo, H., Richards,

W.G., Bannon, T.W., Noda, M., et al. 2005. Leptin regulation of bone resorption by the sympathetic nervous system and CART. *Nature* **434:** 514–520.

Elmardi, A.S., Katchburian, M.V., and Katchburian, E. 1990. Electron microscopy of developing calvaria reveals images that suggest that osteoclasts engulf and destroy osteocytes during bone resorption. *Calcif. Tissue Int.* **46:** 239–245.

Eriksen, E.F., Eghbali-Fatourechi, G.Z., and Khosla, S. 2007. Remodeling and vascular spaces in bone. *J. Bone Miner. Res.* **22:** 1–6.

Everts, V., Delaisse, J.M., Korper, W., Jansen, D.C., Tigchelaar-Gutter, W., Saftig, P., and Beertsen, W. 2002. The bone lining cell: Its role in cleaning Howship's lacunae and initiating bone formation. *J. Bone Miner. Res.* **17:** 77–90.

Frost H.M. 1964. Dynamics of bone remodeling. In *Bone biodynamics*, pp. 315–333. Little, Brown, Boston.

Fujikawa, Y., Quinn, J.M., Sabokbar, A., McGee, J.O., and Athanasou, N.A. 1996. The human osteoclast precursor circulates in the monocyte fraction. *Endocrinology* **137:** 4058–4060.

Fuller, K., Lawrence, K.M., Ross, J.L., Grabowska, U.B., Shiroo, M., Samuelsson, B., and Chambers, T.J. 2008. Cathepsin K inhibitors prevent matrix-derived growth factor degradation by human osteoclasts. *Bone* **42:** 200–211.

Gil-Henn, H., Destaing, O., Sims, N.A., Aoki, K., Alles, N., Neff, L., Sanjay, A., Bruzzaniti, A., De Camilli, P., Baron, R., et al. 2007. Defective microtubule-dependent podosome organization in osteoclasts leads to increased bone density in Pyk2$^{-/-}$ mice. *J. Cell Biol.* **178:** 1053–1064.

Gray, C., Boyde, A., and Jones, S.J. 1996. Topographically induced bone formation in vitro: Implications for bone implants and bone grafts. *Bone* **18:** 115–123.

Grigoriadis, A.E., Wang, Z.Q., Cecchini, M.G., Hofstetter, W., Felix, R., Fleisch, H.A., and Wagner, E.F. 1994. c-Fos: A key regulator of osteoclast-macrophage lineage determination and bone remodeling. *Science* **266:** 443–448.

Hauge, E.M., Qvesel, D., Eriksen, E.F., Mosekilde, L., and Melsen, F. 2001. Cancellous bone remodeling occurs in specialized compartments lined by cells expressing osteoblastic markers. *J. Bone Miner. Res.* **16:** 1575–1582.

Henriksen, K., Gram, J., Schaller, S., Dahl, B.H., Dziegiel, M.H., Bollerslev, J., and Karsdal, M.A. 2004. Characterization of osteoclasts from patients harboring a G215R mutation in ClC-7 causing autosomal dominant osteopetrosis type II. *Am. J. Pathol.* **164:** 1537–1545.

Horne, W.C., Sanjay, A., Bruzzaniti, A., and Baron, R. 2005. The role(s) of Src kinase and Cbl proteins in the regulation of osteoclast differentiation and function. *Immunol. Rev.* **208:** 106–125.

Howard, G.A., Bottemiller, B.L., Turner, R.T., Rader, J.I., and Baylink, D.J. 1981. Parathyroid hormone stimulates bone formation and resorption in organ culture: Evidence for a coupling mechanism. *Proc. Natl. Acad. Sci.* **78:** 3204–3208.

Howson, K.M., Aplin, A.C., Gelati, M., Alessandri, G., Parati, E.A., and Nicosia, R.F. 2005. The postnatal rat aorta contains pericyte progenitor cells that form spheroidal colonies in suspension culture. *Am. J. Physiol. Cell Physiol.* **289:** C1396–C1407.

Hughes, D.E. and Boyce, B.F. 1997. Apoptosis in bone physiology and disease. *Mol. Pathol.* **50:** 132–137.

Hynes, R.O. 1992. Integrins: Versatility, modulation, and signaling in cell adhesion. *Cell* **69:** 11–25.

Ishijima, M., Rittling, S.R., Yamashita, T., Tsuji, K., Kurosawa, H., Nifuji, A., Denhardt, D.T., and Noda, M. 2001. Enhancement of osteoclastic bone resorption and suppression of osteoblastic bone formation in response to reduced mechanical stress do not occur in the absence of osteopontin. *J. Exp. Med.* **193:** 399–404.

Jilka, R.L. 2007. Molecular and cellular mechanisms of the anabolic effect of intermittent PTH. *Bone* **40:** 1434–1446.

Jilka, R.L., Weinstein, R.S., Bellido, T., Roberson, P., Parfitt, A.M., and Manolagas, S.C. 1999. Increased bone formation by prevention of osteoblast apoptosis with parathyroid hormone. *J. Clin. Invest.* **104:** 439–446.

Karsdal, M.A., Henriksen, K., Sorensen, M.G., Gram, J., Schaller, S., Dziegiel, M.H., Heegaard, A.M., Christophersen, P., Martin, T.J., Christiansen, C., et al. 2005. Acidification of the osteoclastic resorption compartment provides insight into the coupling of bone formation to bone resorption. *Am. J. Pathol.* **166:** 467–476.

Karsdal, M.A., Martin, T.J., Bollerslev, J., Christiansen, C., and Henriksen, K. 2007. Are non-resorbing osteoclasts sources of bone anabolic activity? *J. Bone Miner. Res.* **22:** 487–494.

Kawaguchi, H., Pilbeam, C.C., Harrison, J.R., and Raisz, L.G. 1995. The role of prostaglandins in the regulation of bone metabolism. *Clin. Orthop. Relat. Res.* **313:** 36–46.

Keller, H. and Kneissel, M. 2005. SOST is a target gene for PTH in bone. *Bone* **37:** 148–158.

Kemp, B.E., Moseley, J.M., Rodda, C.P., Ebeling, P.R., Wettenhall, R.E., Stapleton, D., Diefenbach-Jagger, H., Ure, F., Michelangeli, V.P., Simmons, H.A., et al. 1987. Parathyroid hormone-related protein of malignancy: Active synthetic fragments. *Science* **238:** 1568–1570.

Kimble, R.B., Matayoshi, A.B., Vannice, J.L., Kung, V.T., Williams, C., and Pacifici, R. 1995. Simultaneous block of interleukin-1 and tumor necrosis factor is required to completely prevent bone loss in the early postovariectomy period. *Endocrinology* **136:** 3054–3061.

Kornak, U., Kasper, D., Bosl, M.R., Kaiser, E., Schweizer, M., Schulz, A., Friedrich, W., Delling, G., and Jentsch, T.J. 2001. Loss of the ClC-7 chloride channel leads to osteopetrosis in mice and man. *Cell* **104:** 205–215.

Kukita, T., Wada, N., Kukita, A., Kakimoto, T., Sandra, F., Toh, K., Nagata, K., Iijima, T., Horiuchi, M., Matsusaki, H., et al. 2004. RANKL-induced DC-STAMP is essential for osteoclastogenesis. *J. Exp. Med.* **200:** 941–946.

Lee, S.H., Rho, J., Jeong, D., Sul, J.Y., Kim, T., Kim, N., Kang, J.S., Miyamoto, T., Suda, T., Lee, S.K., et al. 2006. v-ATPase V0 subunit d2-deficient mice exhibit impaired osteoclast fusion and increased bone formation. *Nat. Med.* **12:** 1403–1409.

Lee, N.K., Sowa, H., Hinoi, E., Ferron, M., Ahn, J.D., Confavreux, C., Dacquin, R., Mee, P.J., McKee, M.D., Jung, D.Y., et al. 2007. Endocrine regulation of energy metabolism by the skeleton. *Cell* **130:** 456–469.

Ma, Y.L., Cain, R.L., Halladay, D.L., Yang, X., Zeng, Q., Miles, R.R., Chandrasekhar, S., Martin, T.J., and Onyia, J.E. 2001. Catabolic effects of continuous human PTH (1–38) in vivo is associated with sustained stimulation of RANKL and inhibition of osteoprotegerin and gene-associated bone formation. *Endocrinology* **142:** 4047–4054.

Malone, J.D., Kahn, A.J., and Teitelbaum, S.L. 1982. Dissociation of organic acid secretion from macrophage mediated bone resorption. *Biochem. Biophys. Res. Commun.* **108:** 468–473.

Manolagas, S.C. and Jilka, R.L. 1995. Bone marrow, cytokines, and bone remodeling. Emerging insights into the pathophysiology of osteoporosis. *N. Engl. J. Med.* **332:** 305–311.

Martin, T.J. and Sims, N.A. 2005. Osteoclast-derived activity in the coupling of bone formation to resorption. *Trends Mol. Med.* **11:** 76–81.

Martin, T.J., Moseley, J.M., and Williams, E.D. 1997. Parathyroid hormone-related protein: Hormone and cytokine. *J. Endocrinol.* (suppl.) **154:** S23–S37.

Matsumoto, T., Kawamoto, A., Kuroda, R., Ishikawa, M., Mifune, Y., Iwasaki, H., Miwa, M., Horii, M., Hayashi, S., Oyamada, A., et al. 2006. Therapeutic potential of vasculogenesis and osteogenesis promoted by peripheral blood CD34-positive cells for functional bone healing. *Am. J. Pathol.* **169:** 1440–1457.

Miao, D., Li, J., Xue, Y., Su, H., Karaplis, A.C., and Goltzman, D. 2004. Parathyroid hormone-related peptide is required for increased trabecular bone volume in parathyroid hormone-null mice. *Endocrinology* **145:** 3554–3562.

Miao, D., He, B., Jiang, Y., Kobayashi, T., Soroceanu, M.A., Zhao, J., Su, H., Tong, X., Amizuka, N., Gupta, A., et al. 2005. Osteoblast-derived PTHrP is a potent endogenous bone anabolic agent that modifies the therapeutic efficacy of administered PTH 1-34. *J. Clin. Invest.* **115:** 2402–2411.

Mocsai, A., Humphrey, M.B., Van Ziffle, J.A., Hu, Y., Burghardt, A., Spusta, S.C., Majumdar, S., Lanier, L.L., Lowell, C.A., and Nakamura, M.C. 2004. The immunomodulatory adapter proteins DAP12 and Fc receptor γ-chain (FcRγ) regulate development of functional osteoclasts through the Syk tyrosine kinase. *Proc. Natl. Acad. Sci.* **101:** 6158–6163.

Modder, U.I. and Khosla, S. 2008. Skeletal stem/osteoprogenitor cells: Current concepts, alternate hypotheses, and relationship to the bone remodeling compartment. *J. Cell. Biochem.* **103:** 393–400.

Moseley, J.M., Hayman, J.A., Danks, J.A., Alcorn, D., Grill, V., Southby, J., and Horton, M.A. 1991. Immunohistochemical detection of parathyroid hormone-related protein in human fetal epithelia. *J. Clin. Endocrinol. Metab.* **73:** 478–484.

Nakamura, T., Imai, Y., Matsumoto, T., Sato, S., Takeuchi, K., Igarashi, K., Harada, Y., Azuma, Y., Krust, A., Yamamoto, Y., et al. 2007. Estrogen prevents bone loss via estrogen receptor α and induction of Fas ligand in osteoclasts. *Cell* **130:** 811–823.

Neer, R.M., Arnaud, C.D., Zanchetta, J.R., Prince, R., Gaich, G.A., Reginster, J.Y., Hodsman, A.B., Eriksen, E.F., Ish-Shalom, S., Genant, H.K., et al. 2001. Effect of parathyroid hormone (1-34) on fractures and bone mineral density in postmenopausal women with osteoporosis. *N. Engl. J. Med.* **344:** 1434–1441.

Noble, B.S., Peet, N., Stevens, H.Y., Brabbs, A., Mosley, J.R., Reilly, G.C., Reeve, J., Skerry, T.M., and Lanyon, L.E. 2003. Mechanical loading: Biphasic osteocyte survival and targeting of osteoclasts for bone destruction in rat cortical bone. *Am. J. Physiol. Cell Physiol.* **284:** C934–C943.

Otsuru, S., Tamai, K., Yamazaki, T., Yoshikawa, H., and Kaneda, Y. 2008. Circulating bone marrow-derived osteoblast progenitor cells are recruited to the bone-forming site by the CXCR4/stromal cell-derived factor-1 pathway. *Stem Cells* **26:** 223–234.

Parfitt, A.M. 1996. Skeletal heterogeneity and the purposes of bone remodeling: Implications for the understanding of osteoporosis. In *Osteoporosis* (ed. R. Marcus et al.), pp. 315–339. Academic, San Diego.

Perez-Amodio, S., Beertsen, W., and Everts, V. 2004. (Pre-)osteoclasts induce retraction of osteoblasts before their fusion to osteoclasts. *J. Bone Miner. Res.* **19:** 1722–1731.

Philbrick, W.M., Wysolmerski, J.J., Galbraith, S., Holt, E., Orloff, J.J., Yang, K.H., Vasavada, R.C., Weir, E.C., Broadus, A.E., and Stewart, A.F. 1996. Defining the roles of parathyroid hormone-related protein in normal physiology. *Physiol. Rev.* **76:** 127–173.

Reid, I.R., Lucas, J., Wattie, D., Horne, A., Bolland, M., Gamble, G.D., Davidson, J.S., and

Grey, A.B. 2005. Effects of a β-blocker on bone turnover in normal postmenopausal women: A randomized controlled trial. *J. Clin. Endocrinol. Metab.* **90:** 5212–5216.

Robling, A.G., Bellido, T., and Turner, C.H. 2006. Mechanical stimulation in vivo reduces osteocyte expression of sclerostin. *J. Musculoskelet. Neuronal Interact.* **6:** 354.

Ross, F.P. and Teitelbaum, S.L. 2005. $\alpha_v\beta_3$ and macrophage colony-stimulating factor: Partners in osteoclast biology. *Immunol. Rev.* **208:** 88–105.

Sato, S., Hanada, R., Kimura, A., Abe, T., Matsumoto, T., Iwasaki, M., Inose, H., Ida, T., Mieda, M., Takeuchi, Y., et al. 2007. Central control of bone remodeling by neuromedin U. *Nat. Med.* **13:** 1234–1240.

Sims, N.A., Jenkins, B.J., Quinn, J.M., Nakamura, A., Glatt, M., Gillespie, M.T., Ernst, M., and Martin, T.J. 2004. Glycoprotein 130 regulates bone turnover and bone size by distinct downstream signaling pathways. *J. Clin. Invest.* **113:** 379–389.

Soriano, P., Montgomery, C., Geske, R., and Bradley, A. 1991. Targeted disruption of the c-src proto-oncogene leads to osteopetrosis in mice. *Cell* **64:** 693–702.

Stains, J.P. and Civitelli, R. 2005. Gap junctions in skeletal development and function. *Biochim. Biophys. Acta* **1719:** 69–81.

Suda, N., Gillespie, M.T., Traianedes, K., Zhou, H., Ho, P.W., Hards, D.K., Allan, E.H., Martin, T.J., and Moseley, J.M. 1996. Expression of parathyroid hormone-related protein in cells of osteoblast lineage. *J. Cell. Physiol.* **166:** 94–104.

Suda, T., Takahashi, N., Udagawa, N., Jimi, E., Gillespie, M.T., and Martin, T.J. 1999. Modulation of osteoclast differentiation and function by the new members of the tumor necrosis factor receptor and ligand families. *Endocr. Rev.* **20:** 345–357.

Suva, L.J., Winslow, G.A., Wettenhall, R.E., Hammonds, R.G., Moseley, J.M., Diefenbach-Jagger, H., Rodda, C.P., Kemp, B.E., Rodriguez, H., Chen, E.Y., et al. 1987. A parathyroid hormone-related protein implicated in malignant hypercalcemia: Cloning and expression. *Science* **237:** 893–896.

Suzuki, R., Domon, T., and Wakita, M. 2000. Some osteocytes released from their lacunae are embedded again in the bone and not engulfed by osteoclasts during bone remodeling. *Anat. Embryol.* **202:** 119–128.

Takayanagi, H., Kim, S., Matsuo, K., Suzuki, H., Suzuki, T., Sato, K., Yokochi, T., Oda, H., Nakamura, K., Ida, N., et al. 2002. RANKL maintains bone homeostasis through c-Fos-dependent induction of interferon-β. *Nature* **416:** 744–749.

Takeda, S., Elefteriou, F., Levasseur, R., Liu, X., Zhao, L., Parker, K.L., Armstrong, D., Ducy, P., and Karsenty, G. 2002. Leptin regulates bone formation via the sympathetic nervous system. *Cell* **111:** 305–317.

Tatsumi, S., Ishii, K., Amizuka, N., Li, M., Kobayashi, T., Kohno, K., Ito, M., Takeshita, S., and Ikeda, K. 2007. Targeted ablation of osteocytes induces osteoporosis with defective mechanotransduction. *Cell Metab.* **5:** 464–475.

Tomkinson, A., Gevers, E.F., Wit, J.M., Reeve, J., and Noble, B.S. 1998. The role of estrogen in the control of rat osteocyte apoptosis. *J. Bone Miner. Res.* **13:** 1243–1250.

van Bezooijen, R.L., Roelen, B.A., Visser, A., van der Wee-Pals, L., de Wilt, E., Karperien, M., Hamersma, H., Papapoulos, S.E., ten Dijke, P., and Lowik, C.W. 2004. Sclerostin is an osteocyte-expressed negative regulator of bone formation, but not a classical BMP antagonist. *J. Exp. Med.* **199:** 805–814.

Verborgt, O., Gibson, G.J., and Schaffler, M.B. 2000. Loss of osteocyte integrity in association with microdamage and bone remodeling after fatigue in vivo. *J. Bone Miner. Res.* **15:** 60–67.

Verborgt, O., Tatton, N.A., Majeska, R.J., and Schaffler, M.B. 2002. Spatial distribution of

Bax and Bcl-2 in osteocytes after bone fatigue: Complementary roles in bone remodeling regulation? *J. Bone Miner. Res.* **17:** 907–914.

Villanueva, A.R., Sypitkowski, C., and Parfitt, A.M. 1986. A new method for identification of cement lines in undecalcified, plastic embedded sections of bone. *Stain Technol.* **61:** 83–88.

Walker, E.C., McGregor, N.E., Poulton, I.J., Pompolo, S., Allan, E.A., Quinn, J.M.W., Gillespie, M.T., Martin, T.J., and Sims, N.A. 2008. Cardiotrophin-1 is a locally-acting regulator of bone formation produced by osteoblasts. *J. Bone Miner. Res.* (in press).

Yagi, M., Miyamoto, T., Sawatani, Y., Iwamoto, K., Hosogane, N., Fujita, N., Morita, K., Ninomiya, K., Suzuki, T., Miyamoto, K., et al. 2005. DC-STAMP is essential for cell-cell fusion in osteoclasts and foreign body giant cells. *J. Exp. Med.* **202:** 345–351.

Yamamoto, Y., Udagawa, N., Matsuura, S., Nakamichi, Y., Horiuchi, H., Hosoya, A., Nakamura, M., Ozawa, H., Takaoka, K., Penninger, J.M., et al. 2006. Osteoblasts provide a suitable microenvironment for the action of receptor activator of nuclear factor-κB ligand. *Endocrinology* **147:** 3366–3374.

Yu, X., Huang, Y., Collin-Osdoby, P., and Osdoby, P. 2003. Stromal cell-derived factor-1 (SDF-1) recruits osteoclast precursors by inducing chemotaxis, matrix metalloproteinase-9 (MMP-9) activity, and collagen transmigration. *J. Bone Miner. Res.* **18:** 1404–1418.

Zhao, C., Irie, N., Takada, Y., Shimoda, K., Miyamoto, T., Nishiwaki, T., Suda, T., and Matsuo, K. 2006. Bidirectional ephrinB2-EphB4 signaling controls bone homeostasis. *Cell Metab.* **4:** 111–121.

Zhou, H., Mak, W., Zheng, Y., Dunstan, C.R., and Seibel, M.J. 2008. Osteoblasts directly control lineage commitment of mesenchymal progenitor cells through Wnt signaling. *J. Biol. Chem.* **283:** 1936–1945.

12

Genetics of Human Skeletal Disease

Bjorn R. Olsen
Department of Developmental Biology
Harvard School of Dental Medicine
Boston, Massachusetts 02115

DURING THE PAST 25 YEARS, the use of genetic approaches has contributed substantially to the understanding of skeletal development and growth. Identification of mutations responsible for a large number of human osteochondrodysplasias and dysostoses (Mundlos and Olsen 1997a,b) has provided insights into the roles not only of individual genes, but also of entire developmental pathways. The correlation of clinical phenotypes with molecular alterations has allowed analyses of structure-function relationships. Coupled with studies of the phenotypic consequences of gene mutations in inbred mouse strains, and more recently also in zebrafish, such analyses have resulted in deep insights into the genetic mechanisms that underlie skeletal assembly, growth, maintenance, and functions.

SKELETAL DEVELOPMENT AND GENETIC DISORDERS

The vertebrate skeleton is the product of mesenchymal cells (osteochondroprogenitors of cartilage-forming chondrocytes and bone-forming osteoblasts) derived from cranial neural crest, paraxial mesoderm, and lateral plate mesoderm (Olsen et al. 2000). Bone marrow-derived myeloid cells are the progenitors for bone- and cartilage-resorbing cells, called osteoclasts.

Neural crest cells give rise to the branchial arch derivatives of the craniofacial skeleton, paraxial mesoderm contributes to both the craniofacial and the axial skeleton, and the lateral plate mesoderm supplies progenitor cells for the limb skeleton (Fig. 1). Progenitor cells from these

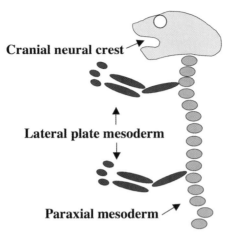

Figure 1. Diagram illustrating the three sources of osteochondroprogenitor cells for the craniofacial (light gray), axial (intermediate gray), and limb skeleton (dark gray).

sources migrate into the regions in which future bones are formed, condense into elements of high cellular density, and differentiate into either osteoblasts or chondrocytes. Osteoblastic differentiation, followed by synthesis of bone extracellular matrix, occurs in regions of membranous ossification, such as the calvarium of the skull, the maxilla, mandible, and a portion of the clavicle. Chondrocytic differentiation and formation of cartilage models (anlagen) of future bones occur in the remaining skeleton. These cartilage models are replaced by bone and bone marrow in a process called endochondral ossification.

Mutations in genes that control the migration, proliferation, and condensation of osteochondroprogenitor cells in the regions of future skeletal elements may result in abnormalities in specific skeletal elements, leaving the rest of the skeleton largely unaffected. Such inherited abnormalities have traditionally been described as dysostoses (Mundlos and Olsen 1997a). In contrast, mutations in genes that are primarily involved in osteoblastic or chondrocytic differentiation and bone and cartilage matrix production will affect the development and growth of skeletal elements more broadly. These disorders have been classified as osteochondrodysplasias. However, since many genes can be important for controlling both early and late stages of skeletal development, many inherited syndromes exhibit both defects in patterning of specific skeletal elements, as well as the differentiation and growth of cartilage and/or bone. Finally, many of the genes that are crucial for skeletal development also have essential roles in the development of nonskeletal organs. Therefore, skeletal genetic defects are frequently components of syndromes with significant nonskeletal phenotypic features (for classification of genetic skeletal disorders, see Rimoin et al. 2007; Superti-Furga and Unger 2007).

GENETIC DISORDERS OF BONE AND CARTILAGE-CELL DIFFERENTIATION

Two examples of inherited defects that include aspects of both dysostoses and osteochondrodysplasias are Cleidocranial dysplasia (CCD) and Campomelic dysplasia (CD). CCD is a dominantly inherited condition characterized by defects in both endochondral and membranous bone formation. Affected individuals have small or absent clavicles, cranial fontanelles and sutures that only slowly and incompletely are filled in with bone, short stature, and supernumerary teeth in the permanent dentition. Patients with CCD are heterozygous for loss-of-function mutations in the transcription factor *RUNX2* (Mundlos et al. 1997). RUNX2 is a member of a small family of Runx factors that form heterodimers with a common subunit, CBF-β (Ducy et al. 1997; Kundu et al. 2002). Both RUNX2 and CBF-β are crucial for osteoblastic differentiation. The complete loss of *Runx2* gene function in mice results in embryos with no bones and a skeleton consisting entirely of cartilage (Komori et al. 1997; Otto et al. 1997).

A critical mediator of *Runx2* action in osteoblastic cells is a zinc-finger transcription factor called Osterix (Osx) (Fig. 2). As in the case of *Runx2*, *Osx*-deficient mice are born with no differentiated osteoblasts and no bone tissue (Nakashima et al. 2002). *Runx2* and *Osx* are not necessary for the differentiation of chondrocytes and for formation of the endochondral cartilage skeleton. However, *Runx2* is up-regulated and required for normal differentiation of chondrocytes to hypertrophy, a process that is crucial for initiating endochondral ossification and allowing longitudinal bone growth (see next page).

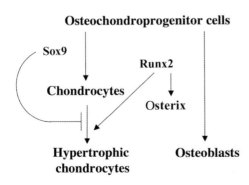

Figure 2. Diagram showing how chondrocytic differentiation is mediated by the transcription factor SOX9, while osteoblastic differentiation requires Runx2 and Osterix. Sox9 inhibits, while Runx2 induces, maturation of chondrocytes to hypertrophic chondrocytes.

Runx2 expression in hypertrophic chondrocytes is essential for the up-regulated expression of vascular endothelial growth factor (VEGF) in these cells (Zelzer et al. 2001, 2004). Expression of VEGF serves to stimulate sprouting angiogenesis and migration of osteoblast progenitors, hematopoietic progenitors, and osteoclasts into hypertrophic cartilage during endochondral ossification, and into the hypertrophic region of growth plates as long bones grow longitudinally. Therefore, mice that are deficient in *Runx2* function not only lack bone tissues, but they also have defects in chondrocyte hypertrophy and angiogenic invasion in the cartilage models of endochondral bones.

Individuals with CD have mutations that affect the expression or the coding sequence of the transcription factor *SOX9* (Foster et al. 1994; Wagner et al. 1994). SOX9 is essential for chondrocytic differentiation (Akiyama et al. 2002), and infants with heterozygous *SOX9* mutations die during the perinatal period or in early infancy with skeletal defects, defective airway cartilage, decreased number of ribs, and small rib cage and skull base. Similar defects are seen in patients with mutations in the cartilage-restricted collagen type II. This is not surprising since SOX9 directly regulates the *COL2A1* gene (Bell et al. 1997).

Heterozygous *Sox9* null mice display most of the abnormalities seen in CD patients, although some of the more severe defects in CD suggest that CD-associated mutations may not be simple loss-of-function mutations but instead have dominant-negative effects. Mutant mice die at birth with cleft palate, abnormal bending of limb bones, and alterations in growth plate cartilage. Expanded regions of hypertrophic chondrocytes in the mutant growth plates are consistent with the finding that SOX9 inhibits differentiation of chondrocytes to hypertrophy (see Fig. 2) (Bi et al. 2001; Akiyama et al. 2002).

GENETIC DISORDERS OF PATTERNING

Genetic skeletal patterning defects can be divided into three groups, depending on which of the three major sources of osteochondroprogenitor cells are primarily affected. Mutations in genes that control migration and differentiation of neural crest-derived cells will result in alterations that affect craniofacial skeletal development. Gene mutations that affect somite formation and differentiation will lead to abnormalities in the axial skeleton, and mutations in genes that are expressed primarily in lateral plate mesoderm-derived cells can cause limb skeletal deformities. However, many of the genes involved are required for the early stages of skeletal patterning in more than one of these "divisions," and their associated mutations can cause widespread abnormalities.

A good example is the Waardenburg syndrome with manifestations in both craniofacial and limb skeletal elements. In two of the clinical subtypes of this syndrome, missense mutations in the transcription factor *PAX3* cause abnormalities in craniofacial bone and soft tissues, cleft lip/palate and spina bifida, fusion of wrist bones, fusion of fingers (syndactyly), and small/missing phalangeal bones combined with deafness and defects in pigment cells (Tassabehji et al. 1992). During development, *PAX3* is expressed in the somites, in the dorsal region of the neural tube, and in neural crest cells. In developing limbs, *PAX3* regulates the migration of muscle progenitor cells (Bober et al. 1994; Mansouri et al. 2001). Most *PAX3* mutations are within one of the two DNA binding domains of the protein (paired and homeo domains). Interestingly, the same mutation can cause different abnormalities, even within the same family. One of the target genes for *PAX3* is the transcription factor *Microphthalmia* (*MITF*), and mutant PAX3 proteins have been shown to exhibit reduced activity as activators of the *MITF* promoter. Interestingly, mutations in *MITF* have been identified in patients with another clinical subtype of Waardenburg syndrome, causing various degrees of deafness and pigmentation abnormalities in combination with congenital megacolon (Tassabehji et al. 1994).

In mice, but apparently not in humans, strong dominant mutations in *Mitf* can cause defects in osteoclast formation and increased deposition of bone (osteopetrosis). The explanation for this comes from studies showing that Mitf and its homolog Tfe3, when activated by phosphorylation during osteoclast differentiation from myeloid cells, associate with the transcriptional coactivator p300/CBP to stimulate the formation of osteoclasts (Steingrimsson et al. 2002). In addition, MITF is a transcriptional activator of the gene for cathepsin K, an enzyme that is essential for the ability of osteoclasts to degrade the organic bone matrix (Motyckova et al. 2001). Mutations in *MITF* can therefore affect both osteoclast formation and the resorptive function of these cells.

Abnormalities in the segmentation of the vertebral column and formation of the sacral bone are caused by mutations in genes that regulate somite development. Among such genes are components of a Notch signaling pathway that controls the process of segmentation of axial mesoderm into somites (Conlon et al. 1995). Mutations in the Notch-ligand *Delta-like 3* (*Dll3*) cause disruption of this process both in humans and in mice (Kusumi et al. 1998; Bulman et al. 2000). In humans, loss-of-function mutations in *DLL3* cause spondylocostal dysostosis, characterized by short stature and vertebral and rib abnormalities. Similar defects are seen in mice with *Dll3* deficiency (Dunwoodie et al. 2002). Mutations in a gene encoding another Notch-ligand, *JAGGED 1* (*JAG 1*), result in a human syndrome (Alagille syndrome) with abnormally shaped vertebral bodies, combined with liver, heart, eye, and facial defects (Oda et al. 1997).

GENETIC DEFECTS OF LIMB DEVELOPMENT

In the skeleton of the distal limbs, patterning defects such as polydactylies (extra fingers), syndactylies (fusion of fingers), and growth and joint fusion defects such as brachydactylies (short fingers) and symphalangism (loss of finger joints), have been shown to be the result of mutations in components of several major signaling pathways. In fact, studies of these limb defects have provided significant insights into the developmental roles and physiological functions of the pathways involved (Table 1). Striking examples are the studies of conditions such as Greig cephalopolysyndactyly, Townes-Brocks syndrome, and polysyndactyly. In the first of these disorders, a combination of craniofacial anomalies, broad thumbs, polydactyly, and syndactyly is a consequence of deletions or truncation mutations in the gene for the transcriptional repressor *GLI3* (Vortkamp et al. 1991). In the presence of hedgehog signaling (see next page), proteolytic processing of the full-length GLI3 protein to the repressor form is inhibited while the expression of the transcription factors *GLI1* and *GLI2* is stimulated. Thus, mutations in *GLI3* in Greig

Table 1. Select list of genetic disorders of limb development

Disorders	Gene	MIM*
Greig cephalopolysyndactyly	GLI3	175700
Townes-Brocks syndrome	SALL1	107480
Preaxial polydactyly	SHH	174400
Acheiropodia	SHH	200500
Synpolydactyly type 1	HOXD13	186000
Brachydactyly A1	IHH	112500
Brachydactyly B	ROR2	113000
Robinow syndrome	ROR2	268310
Brachydactyly C	NOGGIN, GDF5	113100
Hunter-Thompson acromesomelic dysplasia	GDF5	201250
Proximal symphalangism	NOGGIN	185800
Multiple dysostoses syndrome	NOGGIN	186500

*MIM numbers in this and subsequent tables refer to entries in the Online Mendelian Inheritance in Man® database.

cephalopolysyndactyly mimick the effects of enhanced hedgehog signaling (but without GLI1/GLI2-dependent effects). Hedgehog signaling also controls the expression of the zinc-finger transcription factor *SALL1*. In Townes-Brocks syndrome, mutations in *SALL1* result in craniofacial, hand, renal, and anal anomalies, all areas in which hedgehog signaling plays a role during development (Kohlhase et al. 1998).

These human diseases provide strong genetic evidence for the importance of hedgehog signaling in the development of the distal limb skeleton. The importance is further underscored by abnormalities caused by mutations that affect Sonic hedgehog (Shh) expression in the autopod (hand and foot) region of the developing limb (Hill et al. 2003). Several forms of preaxial polydactyly (extra digits on the anterior [thumb] side of hands and feet)-affected individuals are heterozygous for mutations within an enhancer region for *SHH* (Furniss et al. 2008; Sun et al. 2008). This enhancer is located within another gene, *LMBR1*, far away from the *SHH* proximal promoter, and is required for localized expression of SHH in the zone of polarizing activity (ZPA) at the posterior margin of the developing autopod. Gain-of-function mutations in the enhancer result in ectopic expression of *SHH* (Lettice et al. 2008). In contrast, loss-of-function mutation in the *SHH* control region is the basis for a rare recessive disorder called acheiropodia, characterized by absence of both hands and feet.

Both positive and negative interactions connect *SHH* and *GLI3* with genes in the *HOX* gene clusters. SHH signaling inhibits formation of the GLI3 repressor protein, and GLI3 in turn acts as a repressor of *HOX* gene expression. HOXD12 and HOXD13 proteins can bind to GLI3 and convert its function from a transcriptional repressor to an activator (Chen et al. 2004). Finally, the *HOX* genes are required for initiating Shh expression in the ZPA of the developing limb. These interactions provide a molecular framework for understanding digit anomalies caused by mutations in *HOXD13*. In synpolydactyly type 1 (SPD1), expansion of a polyalanine region in HOXD13 results in misfolding and cytoplasmic aggregation of the mutant protein (Muragaki et al. 1996; Albrecht et al. 2004). This results in reduced chondrocyte differentiation and proliferation in distal limb skeletal elements (Johnson et al. 1998; Albrecht et al. 2002). The pathogenetic mechanism explains the correlation between penetrance and severity of the phenotype and the length of the polyalanine expansions in individuals with SPD; the longer expansions cause a more severe phenotype. It also partly explains the partial phenotypic overlap between SPD and individuals who carry mutations resulting in deletions or reduced DNA-binding activity of the homeodomain of HOXD13 (Goodman et al. 1998). However, there are also significant differences among the phenotypes caused by different kinds of mutations in *HOXD13*. For example, missense mutations within the homeodomain that appear to increase the affinity for some target sequences and reduce it for others are associated with forms of brachydactylies (Caronia et al. 2003; Johnson et al. 2003). Other missense mutations in the same domain are associated with a combination of synpolydactyly and brachydactyly (Zhao et al. 2007). These differences are likely explained by differences in the stability of the mutant proteins and how they interact with other transcription factors and target promoter sequences.

Several clinical forms of brachydactylies are caused by mutations in components of hedgehog, Wnt, and BMP-related pathways. For example, loss-of-function mutations in Indian hedgehog (*IHH*) result in brachydactyly type A1, characterized by short middle fingers, stature, and thumbs (Gao et al. 2001; McCready et al. 2002). IHH is a major stimulator of chondrocyte proliferation in long bone growth plates (Long et al. 2001) and the phenotypic features of patients with *IHH* mutations can be traced back to reduced proliferation of chondrocytes during long bone growth. Patients with brachydactyly type B have dominant loss-of-function mutations in the Wnt-binding tyrosine kinase receptor ROR2 (Oldridge et al. 2000; Billiard et al. 2005; Mikels and Nusse 2006). Homozygosity for mutations in this receptor results in Robinow syndrome, characterized by short-limbed dwarfism and defects in ribs and spine, in addition to brachydactyly (Afzal et al. 2000; van Bokhoven et al. 2000). Brachydactyly type C is caused by heterozygous mutations in a member of the BMP/TGF-β family of cytokines, GDF5 (Polinkovsky et al. 1997). Homozygosity for loss-of-function mutations in GDF5 are associated with a condition of severe shortening of arms, legs, and fingers (toes), called Hunter-Thompson acromesomelic dysplasia (Thomas et al. 1996). Finally, mutations in NOGGIN, a secreted BMP antagonist, have been identified in patients with brachydactyly type C, proximal symphalangism, and multiple dysostoses syndrome (Gong et al. 1999).

OSTEOCHONDRODYSPLASIAS AFFECTING SKELETAL GROWTH

In addition to the conditions with short fingers discussed above, a number of osteochondrodysplasias in humans affect the structure and function of epiphyseal growth regions, and therefore result in short stature (Fig. 3). Genetic mapping and identification of mutations in the responsible genes have in many cases been crucial for elucidating the signaling pathways that control proliferation and differentiation of growth plate

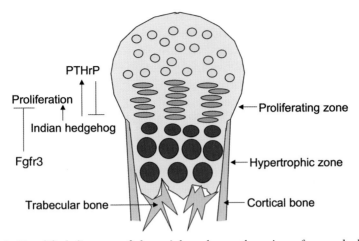

Figure 3. Simplified diagram of the epiphyseal growth region of an endochondral bone. When the secondary ossification center is established above the proliferating zone, the growth region becomes a growth plate. On the *left*, the negative and positive effects of Fgfr3 (expressed by chondrocytes in the hypertrophic zone) and Indian hedgehog (expressed in the boundary region between the proliferating and hypertrophic zones of chondrocytes) on chondrocyte proliferation are indicated. Also indicated is the feedback loop between Indian hedgehog and PTHrP. Indian hedgehog stimulates synthesis of PTHrP, which in turn (via the PTHrP receptor on chondrocytes) inhibits differentiation to hypertrophy and Indian hedgehog expression.

cartilage and formation of the associated trabecular bone. For example, the first realization that signaling through FGF receptors is essential in growth plates came from the discovery of dominant amino acid missense mutations in FGFR3 in achondroplasia, the most common form of dwarfism in humans (Shiang et al. 1994). Remarkably, most of the (sporadic) cases of the disorder are due to the same activating missense mutation in the transmembrane domain of the receptor. Fgfr3 is expressed in the zones of proliferating and hypertrophic chondrocytes in growth plates. Activation of the receptor by its ligand, Fgf18, serves to repress proliferation and regulate (via Ihh) the cellular transition to hypertrophy (Ohbayashi et al. 2002). Less common mutations in other parts of the FGFR3 receptor cause other anomalies, ranging from the very severe thanatophoric dysplasia to the least severe hypochondroplasia (Bellus et al. 1995; Tavormina et al. 1995).

While Fgfr3 signaling represents a negative regulator of growth plate chondrocyte proliferation, Ihh provides positive control (Lai and Mitchell 2005). Together with parathyroid hormone-related peptide, PTHrP, Ihh also forms part of a negative feedback loop that controls maturation of growth plate chondrocytes to hypertrophy (see Fig. 3) (Vortkamp et al. 1996). Mutations in the PTHrP receptor affect both chondrocyte prolif- eration and maturation. Activating mutations in the receptor are found in Jansen's metaphyseal chondrodysplasia (short limbs and increased bone turnover) (Schipani et al. 1996), and recessive loss-of-function mutations result in Blomstrand chondrodysplasia, characterized by accel- erated maturation of growth plate cartilage and advanced bone matura- tion (Jobert et al. 1998).

Chondrocyte proliferation and maturation to hypertrophy in limb growth plates is also regulated by the short stature homeobox gene (SHOX). SHOX and its homolog SHOX2 appear to be present only in vertebrates, and mutations in SHOX coding or regulatory gene sequences are associated with short stature in patients with Turner syndrome (loss of an X chromosome in females), patients with idiopathic short stature, and in patients with Lori-Weill dyschondrosteosis and Langer mesomelic dysplasia (Shears et al. 1998; Rao et al. 2001; Rappold et al. 2002). The SHOX gene is located at the tip of both X and Y chromosomes. Loss-of- function mutations in both alleles result in Langer mesomelic dysplasia, characterized by severe short stature and shortening of limbs, whereas loss-of-function mutations in one allele is the basis for a milder pheno- type with short stature and shortening of forearms and lower legs. The mechanism by which SHOX regulates growth plate function has not been fully characterized, but recent evidence has led to the identification of one target gene, the brain natriuretic peptide (BNP) gene (Marchini et al. 2007). Both SHOX and BNP are expressed in the late proliferative, pre- hypertrophic and hypertrophic zones of limb bone growth plates (Marchini et al. 2004). Since these are regions that also express the tran- scription factors Runx2 and Runx3, both required for chondrocyte hyper- trophy, it is possible that SHOX and BNP are part of a Runx2/ 3-dependent mechanism that controls chondrocyte hypertrophy (Yu et al. 2007). Consistent with such a possibility is the observation that mice overexpressing Bnp exhibit a phenotype of skeletal overgrowth with enlarged hypertrophic chondrocytes (Suda et al. 1998).

DISORDERS AFFECTING SKULL GROWTH

Several genetic disorders affect the functions of cranial sutures, the growth regions between the bones in the skull that allow the braincase to expand as the brain grows during development and postnatal growth (Rice 2005). Such disorders include CCD (see p. 319), characterized by delayed and decreased suture ossification resulting from heterozygous loss-of-function mutations in *RUNX2*, and several types of craniosynostoses, caused by mutations in the receptor tyrosine kinases FGFR1, -2, and -3, resulting in premature ossification of cranial sutures (Wilkie 1997; Rice 2005). The FGFR-associated conditions include disorders that have craniosynostosis as the primary characteristic feature, as well as syndromes with widespread anomalies and premature closure of cranial sutures as one component. Identification of the responsible FGFR mutations has provided important insights into processes that regulate the function of cranial sutures, and established a human genetic context for interpreting the results of further experimental studies in mice. However, the lack of a tight relationship between FGFR genotypes and phenotypes has in most cases made it very difficult to use the data to define pathogenetic mechanisms. For example, mutations in FGFR2 can result in phenotypes that are sufficiently different clinically to be known under different names, such as Jackson-Weiss (Jabs et al. 1994), Crouzon (Reardon et al. 1994), Pfeiffer (Rutland et al. 1995), Apert (Wilkie et al. 1995), and Beare-Stevenson syndrome (Przylepa et al. 1996). It is possible that this phenotypic variability is (at least in part) due to the different effects the different mutations have on interactions with other proteins. A good example suggesting that this is indeed the case is the nonsyndromic craniosynostosis of the coronal suture, seen in patients with Muenke syndrome (Muenke et al. 1997). This is caused by a proline-to-arginine substitution (at position 250) in the extracellular domain of FGFR3. In contrast, other mutations in the same receptor cause hypochondroplasia, achondroplasia, thanatophoric dysplasia, or Crouzon syndrome with acanthosis nigricans (Shiang et al. 1994; Bellus et al. 1995; Meyers et al. 1995; Tavormina et al. 1995). However, even in the case of the pro250, Arg substitution can result in a range of clinical presentations (Reardon et al. 1997).

The mechanisms by which Fgfr mutations cause premature closure of cranial sutures are currently being worked out by analyses of genetically modified mice and experimental manipulation of cell and organ cultures in vitro (Rice 2005). Many Fgf ligands and splice variants of Fgfr1, -2, and -3 are expressed in developing sutures. During suture fusion, the levels of Fgf2 increase, and adding Fgf2 or Fgf4 to calvarial organ cultures accelerates suture ossification. A stimulatory effect of Fgf signaling on suture ossification is consistent with the finding that human craniosynostoses are gain-of-function mutations. Expression of Fgfr2 in proliferating osteoprogenitor cells, and Fgfr1 in differentiated osteoblasts suggests that Fgfr2 signaling primarily controls proliferation, while Fgfr1 regulates osteoblastic differentiation.

Signaling through BMP receptors is also important for regulation of suture growth and ossification. The effects of BMP2 and BMP4 on suture cells are in part mediated by the transcription factor *MSX2*. A gain-of-function mutation in *MSX2* causes Boston type craniosynostosis, resulting from increased proliferation of osteoprogenitor cells (Jabs et al. 1993). Loss-of-function mutations in the transcription factor are associated with parietal bone defects (parietal foramina) (Wilkie et al. 2000). *MSX2* cooperates with the trascription factor *TWIST-1* in controlling suture-cell proliferation and differentiation. Heterozygous loss-of-function mutations in *TWIST-1* result in Saethre-Chotzen syndrome, characterized by craniosynostosis, facial anomalies, and abnormalities of the hands and feet (Howard et al. 1997). Twist-1 up-regulates transcription of *Fgfr2* and *Runx2* but inhibits Runx2 activity by binding to Runx2 (Yousfi et al. 2002; Bialek et al. 2004; Guenou et al. 2005). In addition, it appears that Twist-1 represses PI3K/Akt-dependent osteoblastic proliferation (Guenou et al. 2006). Thus, haploinsuffiency for *TWIST-1* in patients with Saethre-Chotzen syndrome may have complex effects on suture cells, with the net result being increased osteoblastic proliferation and differentiation.

Differential signaling through TGF-β isoforms represents another crucial mechanism for regulating suture growth. TGF-β3 is important for maintaining suture patency, while TGF-β1 has the opposite effect and is increased in ossifying sutures. Mutations in TGF-βR1 and -R2 cause Loeys-Dietz and Shprintzen-Goldberg syndromes (Dietz et al. 1995; Loeys et al. 2005). These are disorders in which craniosynostosis is a component of complex Marfan syndrome-like phenotypes. Many of the characteristics of Marfan syndrome, caused by mutations in the gene *FNB1*, which encodes the extracellular matrix molecule fibrillin-1 (Dietz et al. 1991; Loeys et al. 2004), are now recognized as being consequences of increased TGF-β signaling (Neptune et al. 2003; Habashi et al. 2006). This suggests that the mutations in TGF-βR1 and -R2 increase TGF-β signaling in cranial sutures. In further support of this is the finding that missense mutations in *FBN1* have been identified in some patients with Shprintzen-Goldberg syndrome (Dietz et al. 1995).

Finally, mutations in the gene *EFNB1*, which codes for the receptor tyrosine kinase ephrin B1, cause the craniofrontonasal syndrome (Wallis et al. 2008). This is an X-linked disorder with clinical manifestations of coronal synostosis and various other skeletal anomalies. The pathogenetic mechanisms are not entirely clear, but ephrin B1 is important for specifying the position of the future coronal suture during development.

GENETIC DISORDERS OF ALTERED REGULATION OF BONE MASS

Genetic disorders that affect accrual or maintenance of bone mass during development, postnatal growth, and in adulthood, can be divided into two groups, depending on whether they affect osteoblasts or osteoclasts. In the first group are disorders such as osteoporosis pseudoglioma and high bone mass, caused by mutations in the Wnt-binding receptor LRP5 (Gong et al. 2001; Boyden et al. 2002; Little et al. 2002). In osteoporosis pseudoglioma, affected individuals are homozygous for loss-of-function mutations in *LRP5*. This results in a phenotype of severe osteoporosis due to decreased bone formation and blindness, resulting from abnormal vascular processes in the vitreous and retina. Individuals with high bone mass carry dominant gain-of-function missense mutations in *LRP5*. Osteoblast-dependent high bone mass is also seen in syndromes (sclerosteosis and van Buchem disease) caused by loss-of-function mutations in the gene *SOST*, which encodes the protein sclerostin or deletion within its transcriptional control region (Balemans et al. 2001, 2002; Brunkow et al. 2001). This protein, primarily produced by osteoblasts in developing bone and osteocytes in mature bone, belongs to a class of extracellular BMP antagonists that includes Noggin and Chordin. While sclerostin can bind to BMP with low affinity, its negative effect on bone formation appears to be primarily through blocking Wnt signaling by binding to LRP5 (Semenov et al. 2005).

Numerous genetic disorders caused by mutations that affect osteoclast function result in increased bone density (Fig. 4). They can be divided into four groups, comprising disorders with (1) increased bone density without changes in bone shape, (2) increased bone density with diaphyseal involvement, (3) increased bone density with metaphyseal involvement, and (4) neonatal osteosclerotic dysplasias (Cundy et al. 2002). Table 2 summarizes these conditions and the genes involved. Mutations that result in increased osteoclast formation and therefore

Table 2. Select list of genetic disorders of abnormal bone density caused by abnormal osteoclastic function

Disorders	Genes	MIM
Infantile and delayed forms of osteopetrosis	*TCIRG1, CLCL7*	259700, 166600
Juvenile Paget's disease	*TNFRSF11B*	239000
Pyknodysostosis	*Cathepsin K*	265800
Cherubism	*SH3BP2*	118400

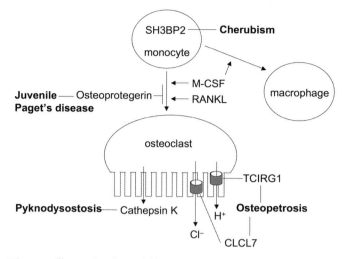

Figure 4. Diagram illustrating how differentiation of monocytes to macrophages and osteoclasts requires M-CSF and RANKL. Osteoprotegerin is a soluble decoy receptor for RANKL that prevents RANKL binding to the receptor RANK on monocytes/macrophages and inhibits osteoclast formation. Gain-of-function mutations in SH3BP2 in cherubism increases the response of monocytes to M-CSF and RANKL. Loss of osteoprotegerin, and therefore increased osteoclastogenesis, is seen in juvenile Paget's disease. Defects in osteoclast function are seen in pyknodysostosis (loss of the protease cathepsin K) and in forms of osteopetrosis (homozygous or heterozygous loss of the chloride channel protein, CLCL7, or an osteoclast-specific subunit, TCIRG1, of the vacuolar proton pump).

bone loss are found in cherubism and juvenile Paget's disease. Cherubism is a dominant disorder of progressive postnatal bone loss and inflammation in the jaws. The causative mutations are gain-of-function missense mutations within a small region of the signaling molecule SH3BP2 (Ueki et al. 2001). The mutations make bone marrow-derived monocytes, the precursors of macrophages and osteoclasts, highly sensitive to cytokines (M-CSF and RANKL) that induce their differentiation (Ueki et al. 2007). As a result, tissues of individuals with cherubism (particularly the jaws for reasons that are unclear) are invaded by large numbers of bone-destroying macrophages and osteoclasts. In patients with juvenile Paget's disease, homozygous loss of the gene *TNFRSF11B*, which encodes osteoprotegerin, has been identified (Cundy et al. 2002). Osteoprotegerin functions as a decoy receptor for RANKL, the cytokine that is crucial for osteoclast formation, and loss of the protein therefore leads to increased osteoclastogenesis (Simonet et al. 1997).

GENETIC DISORDERS CAUSED BY MUTATIONS THAT AFFECT THE EXTRACELLULAR MATRIX IN CARTILAGE AND BONE

Synthesis and assembly of extracellular matrices by chondrocytes and osteoblasts are crucial aspects of skeletal development. These matrices provide the skeleton not only with the mechanical properties that allow it to function as a rigid support system, but they also constitute an environment for transmission of mechanical and chemical signals, molecular structures for storage of growth factors and ions, and macromolecular scaffolds for attachment and migration of skeletal cells. Mutations in genes that affect the structure and function of skeletal extracellular matrix molecules can therefore result in a broad range of clinical disorders. However, since many of the major building blocks of extracellular matrices are expressed throughout the vertebrate skeleton, mutations in the corresponding genes usually result in osteochondrodysplasias, in which skeletal tissues are affected in a generalized fashion. The classical examples of such disorders are osteogenesis imperfecta and what has been called "type II collagenopathies" (Mundlos and Olsen 1997b). Also, since extracellular matrix molecules generally are components of multimolecular complexes, mutations in genes encoding different components will often result in similar phenotypes. Finally, in cases where the molecules are composed of multiple domains with different functional properties, different mutations in the same gene can result in different clinical manifestations. Osteochondrodysplasias caused by mutations in extracellular matrix components are therefore similar to disorders caused by mutations in FGF receptors in that mutations in one gene can be associated with several distinct clinical disorders, and a distinct clinical disorder can be associated with mutations in several different genes. These features are reflected in the classification of genetic skeletal disorders released by the International Skeletal Dysplasia Society (Superti-Furga and Unger 2007). A select subset of groups of these disorders involving proteins in the extracellular matrix and their associated genes are listed in Table 3.

Osteogenesis imperfecta (brittle bone disease) represents a group of disorders with increased fragility of bone and hearing loss combined in some forms with blue sclerae and dentinogenesis imperfecta associated with reduced levels of collagen I in bone. In the majority of cases, the disease is caused by mutations in COL1A1 or COL1A2, the genes that encode the two distinct polypeptide chains of collagen I molecules. Except for very rare cases of mutations in COL1A2 that result in clinical

Table 3. Groups of skeletal dysplasias caused by mutations in extracellular matrix genes

Groups of disorders	Genes	Proteins
Type II collagen disorders	COL2A1	collagen II
Type II XI collagen disorders	COL11A1, COL11A2	collagen XI
Perlecan disorders	PLC (HSPG2)	perlecan
Multiple epiphyseal dysplasia and pseudoachondroplasia	COMP, COL9A1 COL9A2, COL9A3 MATN3	cartilage oligomeric matrix protein, collagen IX, matrilin3
Metaphyseal dysplasia	COL10A1 MMP13	collagen X, matrix metallo- proteinase 13
Spondylo-epi-metaphyseal dysplasia	AGC1 MATN3 MMP13	aggrecan, matrilin 3, matrix metalloproteinase 13
Acromelic dysplasia	IHH GDF5 ADAMTS10 FBN1	Indian hedgehog, growth and differentiation factor 5, metalloproteinase with thrombospondin-like repeats 10, fibrillin 1
Acromesomelic dysplasia	NPR2 GDF5 BMPR1B	natriuretic peptide receptor 2, growth and differen- tiation factor 5, bone morphogenetic protein receptor 1B
Decreased bone density disorders	COL1A1, COL1A2 CRTAP P3H1/LEPRE1 LRP5	collagen I, cartilage associ- ated protein, prolyl 3- hydroxylase1/leprecan, LDL-receptor related protein 5

disease at homozygosity, these mutations are dominant, causing disease in heterozygous individuals. The most common mutations are missense mutations replacing a glycine residue with another amino acid within one of the Gly-Xxx-Yyy triplet sequences that are critical for the stability of the triple-helical structure of collagen molecules (Marini et al. 2007). Since the folding of collagen I molecules are initiated by trimerization of the large carboxyl propeptide regions of the type I procollagen chains (the biosynthetic precursor form of the collagen) and the triple-helical folding process proceeds in the amino-terminal direction, such missense mutations in the triple-helical region have dominant-negative effects on assembly of collagen I molecules.

For this reason, triple-helical glycine substitutions in *COL1A1* or *COL1A2* have more severe effects on the level of normal collagen I molecules in bone than a large gene deletion or mutations that compromise the trimerization property of the carboxyl propeptides. However, the phenotypic consequences of collagen I mutations in osteogenesis imperfecta are unlikely to be fully explained by their effects on collagen I levels in bone. Collagen I molecules contain multiple sites for interactions with multiple other extracellular structural and signaling components, and clustering of mutations when mapped onto collagen fibrils suggest that the mutations may disrupt interactions with other matrix molecules (Marini et al. 2007; Sweeney et al. 2008). Consistent with this conclusion are the findings that autosomal recessive forms of osteogenesis imperfecta are caused by mutations in proteins (prolyl 3-hydroxylase and cartilage associated protein, CRTAP) that are essential for generation of a small number of 3-hydroxyprolyl residues in collagen I (Baldridge et al. 2008). The clinical manifestations resulting from the loss of these residues are most likely the consequences of effects on processes other than triple-helical molecular folding.

ACKNOWLEDGMENTS

The original research that was carried out in the laboratory of Dr. B.R. Olsen and reviewed in this chapter was supported by National Institutes of Health grants AR36819 and AR36820 (to B.R.O.)

REFERENCES

Afzal, A.R., Rajab, A., Fenske, C.D., Oldridge, M., Elanko, N., Ternes-Pereira, E., Tuysuz, B., Murday, V.A., Patton, M.A., Wilkie, A.O., et al. 2000. Recessive Robinow syndrome, allelic to dominant brachydactyly type B, is caused by mutation of ROR2. *Nat. Genet.* **25:** 419–422.

Akiyama, H., Chaboissier, M.C., Martin, J.F., Schedl, A., and de Crombrugghe, B. 2002. The transcription factor Sox9 has essential roles in successive steps of the chondrocyte differentiation pathway and is required for expression of Sox5 and Sox6. *Genes Dev.* **16:** 2813–2828.

Albrecht, A.N., Schwabe, G.C., Stricker, S., Böddrich, A., Wanker, E.E., and Mundlos, S. 2002. The synpolydactyly homolog (spdh) mutation in the mouse—A defect in patterning and growth of limb cartilage elements. *Mech. Dev.* **112:** 53–67.

Albrecht, A.N., Kornak, U., Böddrich, A., Suring, K., Robinson, P.N., Stiege, A.C., Lurz, R., Stricker, S., Wanker, E.E., and Mundlos, S. 2004. A molecular pathogenesis for transcription factor associated poly-alanine tract expansions. *Hum. Mol. Genet.* **13:** 2351–2359.

Baldridge, D., Schwarze, U., Morello, R., Lennington, J., Bertin, T.K., Pace, J.M., Pepin,

M.G., Weis, M., Eyre, D.R., Walsh, J., et al. 2008. CRTAP and LEPRE1 mutations in recessive osteogenesis imperfecta. *Hum. Mutat.* **29:** 1435–1442.

Balemans, W., Ebeling, M., Patel, N., Van Hul, E., Olson, P., Dioszegi, M., Lacza, C., Wuyts, W., Van Den Ende, J., Willems, P., et al. 2001. Increased bone density in sclerosteosis is due to the deficiency of a novel secreted protein (SOST). *Hum. Mol. Genet.* **10:** 537–543.

Balemans, W., Patel, N., Ebeling, M., Van Hul, E., Wuyts, W., Lacza, C., Dioszegi, M., Dikkers, F.G., Hildering, P., Willems, P.J., et al. 2002. Identification of a 52 kb deletion downstream of the *SOST* gene in patients with van Buchem disease. *J. Med. Genet.* **39:** 91–97.

Bell, D.M., Leung, K.K., Wheatley, S.C., Ng, L.J., Zhou, S., Ling, K.W., Sham, M.H., Koopman, P., Tam, P.P., and Cheah, K.S. 1997. Sox9 directly regulates the type-II collagen gene. *Nat. Genet.* **16:** 174–178.

Bellus, G.A., McIntosh, I., Smith, E.A., Aylsworth, A.S., Kaitila, I., Horton, W.A., Greenhaw, G.A., Hecht, J.T., and Francomano, C.A. 1995. A recurrent mutation in the tyrosine kinase domain of fibroblast growth factor receptor 3 causes hypochondroplasia. *Nat. Genet.* **10:** 357–359.

Bi, W., Huang, W., Whitworth, D.J., Deng, J.M., Zhang, Z., Behringer, R.R., and de Crombrugghe, B. 2001. Haploinsufficiency of *Sox9* results in defective cartilage primordia and premature skeletal mineralization. *Proc. Natl. Acad. Sci.* **98:** 6698–6703.

Bialek, P., Kern, B., Yang, X., Schrock, M., Sosic, D., Hong, N., Wu, H., Yu, K., Ornitz, D.M., Olson, E.N., et al. 2004. A twist code determines the onset of osteoblast differentiation. *Dev. Cell* **6:** 423–435.

Billiard, J., Way, D.S., Seestaller-Wehr, L.M., Moran, R.A., Mangine, A., and Bodine, P.V. 2005. The orphan receptor tyrosine kinase Ror2 modulates canonical Wnt signaling in osteoblastic cells. *Mol. Endocrinol.* **19:** 90–101.

Bober, E., Franz, T., Arnold, H.H., Gruss, P., and Tremblay, P. 1994. *Pax-3* is required for the development of limb muscles: A possible role for the migration of dermomyotomal muscle progenitor cells. *Development* **120:** 603–612.

Boyden, L.M., Mao, J., Belsky, J., Mitzner, L., Farhi, A., Mitnick, M.A., Wu, D., Insogna, K., and Lifton, R.P. 2002. High bone density due to a mutation in LDL-receptor-related protein 5. *N. Engl. J. Med.* **346:** 1513–1521.

Brunkow, M.E., Gardner, J.C., Van Ness, J., Paeper, B.W., Kovacevich, B.R., Proll, S., Skonier, J.E., Zhao, L., Sabo, P.J., Fu, Y., et al. 2001. Bone dysplasia sclerosteosis results from loss of the *SOST* gene product, a novel cystine knot-containing protein. *Am. J. Hum. Genet.* **68:** 577–589.

Bulman, M.P., Kusumi, K., Frayling, T.M., McKeown, C., Garrett, C., Lander, E.S., Krumlauf, R., Hattersley, A.T., Ellard, S., and Turnpenny, P.D. 2000. Mutations in the human *Delta* homologue, *DLL3*, cause axial skeletal defects in spondylocostal dysostosis. *Nat. Genet.* **24:** 438–441.

Caronia, G., Goodman, F.R., McKeown, C.M., Scambler, P.J., and Zappavigna, V. 2003. An I47L substitution in the HOXD13 homeodomain causes a novel human limb malformation by producing a selective loss of function. *Development* **130:** 1701–1712.

Chen Y., Knezevic V., Ervin V., Hutson R., Ward Y., and Mackem S. 2004. Direct interaction with Hoxd proteins reverses Gli3-repressor function to promote digit formation downstream of Shh. *Development* **131:** 2339–2347.

Conlon, R.A., Reaume, A.G., and Rossant, J. 1995. *Notch1* is required for the coordinate segmentation of somites. *Development* **121:** 1533–1545.

Cundy, T., Hegde, M., Naot, D., Chong, B., King, A., Wallace, R., Mulley, J., Love, D.R.,

Seidel, J., Fawkner, M., et al. 2002. A mutation in the gene *TNFRSF11B* encoding osteo-protegerin causes an idiopathic hyperphosphatasia phenotype. *Hum. Mol. Genet.* **11:** 2119–2127.

Dietz, H.C., Cutting, G.R., Pyeritz, R.E., Maslen, C.L., Sakai, L.Y., Corson, G.M., Puffenberger, E.G., Hamosh, A., Nanthakumar, E.J., Curristin, S.M., et al. 1991. Marfan syndrome caused by a recurrent *de novo* missense mutation in the fibrillin gene (see comments). *Nature* **352:** 337–339.

Dietz, H.C., Sood, I., and McIntosh, I. 1995. The phenotypic continuum associated with *FBN1* mutations includes the Shprintzen-Goldberg syndrome. *Am. J. Hum. Genet.* **57:** A211. (Abstr.)

Ducy, P., Zhang, R., Geoffroy, V., Ridall, A.L., and Karsenty, G. 1997. Osf2/Cbfa1: A transcriptional activator of osteoblast differentiation. *Cell* **89:** 747–754.

Dunwoodie, S.L., Clements, M., Sparrow, D.B., Sa, X., Conlon, R.A., and Beddington, R.S. 2002. Axial skeletal defects caused by mutation in the spondylocostal dysplasia/pudgy gene *Dll3* are associated with disruption of the segmentation clock within the presomitic mesoderm. *Development* **129:** 1795–1806.

Foster, J.W., Dominguez-Steglich, M.A., Guioli, S., Kwok, C., Weller, P.A., Stevanović, M., Weissenbach, J., Mansour, S., Young, I.D., Goodfellow, P.N., et al. 1994. Campomelic dysplasia and autosomal sex reversal caused by mutations in an *SRY*-related gene. *Nature* **372:** 525–530.

Furniss, D., Lettice, L.A., Taylor, I.B., Critchley, P.S., Giele, H., Hill, R.E., and Wilkie, A.O. 2008. A variant in the sonic hedgehog regulatory sequence (ZRS) is associated with triphalangeal thumb and deregulates expression in the developing limb. *Hum. Mol. Genet.* **17:** 2417–2423.

Gao, B., Guo, J., She, C., Shu, A., Yang, M., Tan, Z., Yang, X., Guo, S., Feng, G., and He, L. 2001. Mutations in *IHH*, encoding Indian hedgehog, cause brachydactyly type A-1. *Nat. Genet.* **28:** 386–388.

Gong, Y., Krakow, D., Marcelino, J., Wilkin, D., Chitayat, D., Babul-Hirji, R., Hudgins, L., Cremers, C.W., Cremers, F.P., Brunner, H.G., et al. 1999. Heterozygous mutations in the gene encoding noggin affect human joint morphogenesis. *Nat. Genet.* **21:** 302–304.

Gong, Y., Slee, R.B., Fukai, N., Rawadi, G., Roman-Roman, S., Reginato, A.M., Wang, H., Cundy, T., Glorieux, F.H., Lev, D., et al. 2001. LDL receptor-related protein 5 (LRP5) affects bone accrual and eye development. *Cell* **107:** 513–523.

Goodman, F., Giovannucci-Uzielli, M.-L., Hall, C., Reardon, W., Winter R., and Scambler, P. 1998. Deletions in *HOXD13* segregate with an identical, novel foot malformation in two unrelated families. *Am. J. Hum. Genet.* **63:** 992–1000.

Guenou H., Kaabeche K., Mee S.L., and Marie P.J. 2005. A role for fibroblast growth factor receptor-2 in the altered osteoblast phenotype induced by Twist haploinsufficiency in the Saethre-Chotzen syndrome. *Hum. Mol. Genet.* **14:** 1429–1439.

Guenou, H., Kaabeche, K., Dufour, C., Miraoui, H., and Marie, P.J. 2006. Down-regulation of ubiquitin ligase Cbl induced by twist haploinsufficiency in Saethre-Chotzen syndrome results in increased PI3K/Akt signaling and osteoblast proliferation. *Am. J. Pathol.* **169:** 1303–1311.

Habashi, J.P., Judge, D.P., Holm, T.M., Cohn, R.D., Loeys, B.L., Cooper, T.K., Myers, L., Klein, E.C., Liu, G., Calvi, C., et al. 2006. Losartan, an AT1 antagonist, prevents aortic aneurysm in a mouse model of Marfan syndrome. *Science* **312:** 117–121.

Hill, R.E., Heaney, S.J., and Lettice, L.A. 2003. Sonic hedgehog: Restricted expression and limb dysmorphologies. *J. Anat.* **202:** 13–20.

Howard, T.D., Paznekas, W.A., Green, E.D., Chiang, L.C., Ma, N., Ortiz de Luna, R.I., Garcia Delgado, C., Gonzalez-Ramos, M., Kline, A.D., and Jabs, E.W. 1997. Mutations in *TWIST*, a basic helix-loop-helix transcription factor, in Saethre-Chotzen syndrome (see comments). *Nat. Genet.* **15**: 36–41.

Jabs, E.W., Muller, U., Li, X., Ma, L., Luo, W., Haworth, I.S., Klisak, I., Sparkes, R., Warman, M.L., Mulliken, J.B., et al. 1993. A mutation in the homeodomain of the human *MSX2* gene in a family affected with autosomal dominant craniosynostosis. *Cell* **75**: 443–450.

Jabs, E.W., Li, X., Scott, A.F., Meyers, G., Chen, W., Eccles, M., Mao, J.I., Charnas, L.R., Jackson, C.E., and Jaye, M. 1994. Jackson-Weiss and Crouzon syndromes are allelic with mutations in fibroblast growth factor receptor 2. *Nat. Genet.* **8**: 275–279.

Jobert, A.S., Zhang, P., Couvineau, A., Bonaventure, J., Roume, J., Le Merrer, M., and Silve, C. 1998. Absence of functional receptors for parathyroid hormone and parathyroid hormone-related peptide in Blomstrand chondrodysplasia. *J. Clin. Invest.* **102**: 34–40.

Johnson, K.R., Sweet, H.O., Donahue, L.R., Ward-Bailey, P., Bronson, R.T., and Davisson, M.T. 1998. A new spontaneous mouse mutation of *Hoxd13* with a polyalanine expansion and phenotype similar to human synpolydactyly. *Hum. Mol. Genet.* **7**: 1033–1038.

Johnson, D., Kan, S.H., Oldridge, M., Trembath, R.C., Roche, P., Esnouf, R.M., Giele, H., and Wilkie, A.O. 2003. Missense mutations in the homeodomain of HOXD13 are associated with brachydactyly types D and E. *Am. J. Hum. Genet.* **72**: 984–997.

Kohlhase, J., Wischermann, A., Reichenbach, H., Froster, U., and Engel, W. 1998. Mutations in the SALL1 putative transcription factor gene cause Townes-Brocks syndrome. *Nat. Genet.* **18**: 81–83.

Komori, T., Yagi, H., Nomura, S., Yamaguchi, A., Sasaki, K., Deguchi, K., Shimizu, Y., Bronson, R.T., Gao, Y.H., Inada, M., et al. 1997. Targeted disruption of *Cbfa1* results in a complete lack of bone formation owing to maturational arrest of osteoblasts. *Cell* **89**: 755–764.

Kundu, M., Javed, A., Jeon, J.P., Horner, A., Shum, L., Eckhaus, M., Muenke, M., Lian, J.B., Yang, Y., Nuckolls, G.H., et al. 2002. Cbfβ interacts with Runx2 and has a critical role in bone development. *Nat. Genet.* **32**: 639–644.

Kusumi, K., Sun, E.S., Kerrebrock, A.W., Bronson, R.T., Chi, D.C., Bulotsky, M.S., Spencer, J.B., Birren, B.W., Frankel, W.N., and Lander, E.S. 1998. The mouse pudgy mutation disrupts *Delta* homologue *Dll3* and initiation of early somite boundaries. *Nat. Genet.* **19**: 274–278.

Lai, L.P. and Mitchell, J. 2005. Indian hedgehog: Its roles and regulation in endochondral bone development. *J. Cell. Biochem.* **96**: 1163–1173.

Lettice, L.A., Hill, A.E., Devenney, P.S., and Hill, R.E. 2008. Point mutations in a distant sonic hedgehog *cis*-regulator generate a variable regulatory output responsible for preaxial polydactyly. *Hum. Mol. Genet.* **17**: 978–985.

Little, R.D., Carulli, J.P., Del Mastro, R.G., Dupuis, J., Osborne, M., Folz, C., Manning, S.P., Swain, P.M., Zhao, S.C., Eustace, B., et al. 2002. A mutation in the LDL receptor-related protein 5 gene results in the autosomal dominant high-bone-mass trait. *Am. J. Hum. Genet.* **70**: 11–19.

Loeys, B., De Backer, J., Van Acker, P., Wettinck, K., Pals, G., Nuytinck, L., Coucke, P., and De Paepe, A. 2004. Comprehensive molecular screening of the *FBN1* gene favors locus homogeneity of classical Marfan syndrome. *Hum. Mutat.* **24**: 140–146.

Loeys, B.L., Chen, J., Neptune, E.R., Judge, D.P., Podowski, M., Holm, T., Meyers, J., Leitch, C.C., Katsanis, N., Sharifi, N., et al. 2005. A syndrome of altered cardiovascular, craniofacial, neurocognitive and skeletal development caused by mutations in *TGFBR1* or

TGFBR2. *Nat. Genet.* **37**: 275–281.

Long, F., Zhang, X.M., Karp, S., Yang, Y., and McMahon, A.P. 2001. Genetic manipulation of hedgehog signaling in the endochondral skeleton reveals a direct role in the regulation of chondrocyte proliferation. *Development* **128**: 5099–5108.

Mansouri, A., Pla, P., Larue, L., and Gruss, P. 2001. *Pax3* acts cell autonomously in the neural tube and somites by controlling cell surface properties. *Development* **128**: 1995–2005.

Marchini, A., Marttila, T., Winter, A., Caldeira, S., Malanchi, I., Blaschke, R.J., Hacker, B., Rao, E., Karperien, M., Wit, J.M., et al. 2004. The short stature homeodomain protein SHOX induces cellular growth arrest and apoptosis and is expressed in human growth plate chondrocytes. *J. Biol. Chem.* **279**: 37103–37114.

Marchini, A., Hacker, B., Marttila, T., Hesse, V., Emons, J., Weiss, B., Karperien, M., and Rappold, G. 2007. BNP is a transcriptional target of the short stature homeobox gene *SHOX*. *Hum. Mol. Genet.* **16**: 3081–3087.

Marini, J.C., Forlino, A., Cabral, W.A., Barnes, A.M., San Antonio, J.D., Milgrom, S., Hyland, J.C., Korkko, J., Prockop, D.J., De Paepe, A., et al. 2007. Consortium for osteogenesis imperfecta mutations in the helical domain of type I collagen: Regions rich in lethal mutations align with collagen binding sites for integrins and proteoglycans. *Hum. Mutat.* **28**: 209–221.

McCready, M.E., Sweeney, E., Fryer, A.E., Donnai, D., Baig, A., Racacho, L., Warman, M.L., Hunter, A.G., and Bulman, D.E. 2002. A novel mutation in the *IHH* gene causes brachydactyly type A1: A 95- year-old mystery resolved. *Hum Genet* **111**: 368–375.

Meyers, G.A., Orlow, S.J., Munro, I.R., Przylepa, K.A., and Jabs, E.W. 1995. Fibroblast growth factor receptor 3 (FGFR3) transmembrane mutation in Crouzon syndrome with acanthosis nigricans. *Nat. Genet.* **11**: 462–464.

Mikels, A.J. and Nusse, R. 2006. Purified Wnt5a protein activates or inhibits β-catenin-TCF signaling depending on receptor context. *PLoS Biol.* **4**: e115.

Motyckova, G., Weilbaecher, K.N., Horstmann, M., Rieman, D.J., Fisher, D.Z., and Fisher, D.E. 2001. Linking osteopetrosis and pycnodysostosis: Regulation of cathepsin K expression by the microphthalmia transcription factor family. *Proc. Natl. Acad. Sci.* **98**: 5798–5803.

Muenke, M., Gripp, K.W., McDonald-McGinn, D.M., Gaudenz, K., Whitaker, L.A., Bartlett, S.P., Markowitz, R.I., Robin, N.H., Nwokoro N., Mulvihill J.J., et al. 1997. A unique point mutation in the fibroblast growth factor receptor 3 gene (*FGFR3*) defines a new craniosynostosis syndrome. *Am. J. Hum. Genet.* **60**: 555–564.

Mundlos, S. and Olsen, B.R. 1997a. Heritable diseases of the skeleton. Part I. Molecular insights into skeletal development-transcription factors and signaling pathways. *FASEB J.* **11**: 125–132.

Mundlos, S. and Olsen, B.R. 1997b. Heritable diseases of the skeleton. Part II. Molecular insights into skeletal development-matrix components and their homeostasis. *FASEB J.* **11**: 227–233.

Mundlos, S., Otto, F., Mundlos, C., Mulliken, J.B., Aylsworth, A.S., Albright, S., Lindhout, D., Cole, W.G., Henn, W., Knoll, J.H., et al. 1997. Mutations involving the transcription factor CBFA1 cause cleidocranial dysplasia. *Cell* **89**: 773–779.

Muragaki, Y., Mundlos, S., Upton, J., and Olsen, B.R. 1996. Altered growth and branching patterns in synpolydactyly caused by mutations in *HOXD13*. *Science* **272**: 548–551.

Nakashima, K., Zhou, X., Kunkel, G., Zhang, Z., Deng, J.M., Behringer, R.R., and de Crombrugghe, B. 2002. The novel zinc finger-containing transcription factor osterix is required for osteoblast differentiation and bone formation. *Cell* **108:** 17–29.

Neptune, E.R., Frischmeyer, P.A., Arking, D.E., Myers, L., Bunton, T.E., Gayraud, B., Ramirez, F., Sakai, L.Y., and Dietz, H.C. 2003. Dysregulation of TGF-β activation contributes to pathogenesis in Marfan syndrome. *Nat. Genet.* **33:** 407–411.

Oda, T., Elkahloun, A.G., Pike, B.L., Okajima, K., Krantz, I.D., Genin, A., Piccoli, D.A., Meltzer, P.S., Spinner, N.B., Collins, F.S., et al. 1997. Mutations in the human *Jagged1* gene are responsible for Alagille syndrome. *Nat. Genet.* **16:** 235–242.

Ohbayashi, N., Shibayama, M., Kurotaki, Y., Imanishi, M., Fujimori, T., Itoh, N., and Takada, S. 2002. FGF18 is required for normal cell proliferation and differentiation during osteogenesis and chondrogenesis. *Genes Dev.* **16:** 870–879.

Oldridge, M., Fortuna, A.M., Maringa, M., Propping, P., Mansour, S., Pollitt, C., DeChiara, T.M., Kimble, R.B., Valenzuela, D.M., Yancopoulos, G.D., et al. 2000. Dominant mutations in *ROR2*, encoding an orphan receptor tyrosine kinase, cause brachydactyly type B. *Nat. Genet.* **24:** 275–278.

Olsen, B.R., Reginato, A.M., and Wang, W. 2000. Bone development. *Annu. Rev. Cell Dev. Biol.* **16:** 191–220.

Otto, F., Thornell, A.P., Crompton, T., Denzel, A., Gilmour, K.C., Rosewell, I.R., Stamp, G.W., Beddington, R.S., Mundlos, S., Olsen, B.R., et al. 1997. *Cbfa1*, a candidate gene for cleidocranial dysplasia syndrome, is essential for osteoblast differentiation and bone development. *Cell* **89:** 765–771.

Polinkovsky, A., Robin, N.H., Thomas, J.T., Irons, M., Lynn, A., Goodman, F.R., Reardon, W., Kant, S.G., Brunner, H.G., van der Burgt I., et al. 1997. Mutations in *CDMP1* cause autosomal dominant brachydactyly type C (letter). *Nat. Genet.* **17:** 18–19.

Przylepa, K.A., Paznekas, W., Zhang, M., Golabi, M., Bias, W., Bamshad, M.J., Carey, J.C., Hall, B.D., Stevenson, R., Orlow, S., et al. 1996. Fibroblast growth factor receptor 2 mutations in Beare-Stevenson cutis gyrata syndrome. *Nat. Genet.* **13:** 492–494.

Rao, E., Blaschke, R.J., Marchini, A., Niesler, B., Burnett, M., and Rappold, G.A. 2001. The Leri-Weill and Turner syndrome homeobox gene SHOX encodes a cell-type specific transcriptional activator. *Hum. Mol. Genet.* **10:** 3083–3091.

Rappold, G.A., Fukami, M., Niesler, B., Schiller, S., Zumkeller, W., Bettendorf, M., Heinrich, U., Vlachopapadoupoulou, E., Reinehr, T., Onigata, K., et al. 2002. Deletions of the homeobox gene *SHOX* (short stature homeobox) are an important cause of growth failure in children with short stature. *J. Clin. Endocrinol. Metab.* **87:** 1402–1406.

Reardon, W., Winter, R.M., Rutland, P., Pulleyn, L.J., Jones, B.M., and Malcolm, S. 1994. Mutations in the fibroblast growth factor receptor 2 gene cause Crouzon syndrome. *Nat. Genet.* **8:** 98–103.

Reardon, W., Wilkes, D., Rutland, P., Pulleyn, L.J., Malcolm, S., Dean, J.C., Evans, R.D., Jones, B.M., Hayward, R., Hall, C.M., et al. 1997. Craniosynostosis associated with *FGFR3* pro250arg mutation results in a range of clinical presentations including unisutural sporadic craniosynostosis. *J. Med. Genet.* **34:** 632–636.

Rice, D.P. 2005. Craniofacial anomalies: From development to molecular pathogenesis. *Curr. Mol. Med.* **5:** 699–722.

Rimoin, D.L., Cohn, D., Krakow, D., Wilcox, W., Lachman, R.S., and Alanay, Y. 2007. The skeletal dysplasias: Clinical-molecular correlations. *Ann. N.Y. Acad. Sci.* **1117:** 302–309.

Rutland, P., Pulleyn, L.J., Reardon, W., Baraitser, M., Hayward, R., Jones, B., Malcolm, S., Winter, R.M., Oldridge, M., Slaney, S.F., et al. 1995. Identical mutations in the *FGFR2* gene cause both Pfeiffer and Crouzon syndrome phenotypes. *Nat. Genet.* **9:** 173–176.

Schipani, E., Langman, C.B., Parfitt, A.M., Jensen, G.S., Kikuchi, S., Kooh, S.W., Cole, W.G., and Juppner, H. 1996. Constitutively activated receptors for parathyroid hormone and parathyroid hormone-related peptide in Jansen's metaphyseal chondrodysplasia. *N. Engl. J. Med.* **335:** 708–714.

Semenov, M., Tamai, K., and He, X. 2005. SOST is a ligand for LRP5/LRP6 and a Wnt signaling inhibitor. *J. Biol. Chem.* **280:** 26770–26775.

Shears, D.J., Vassal, H.J., Goodman, F.R., Palmer, R.W., Reardon, W., Superti-Furga, A., Scambler, P.J., and Winter, R.M. 1998. Mutation and deletion of the pseudoautosomal gene *SHOX* cause Leri-Weill dyschondrosteosis. *Nat. Genet.* **19:** 70–73.

Shiang, R., Thompson, L.M., Zhu, Y.Z., Church, D.M., Fielder, T.J., Bocian, M., Winokur, S.T., and Wasmuth, J.J. 1994. Mutations in the transmembrane domain of FGFR3 cause the most common genetic form of dwarfism, achondroplasia. *Cell* **78:** 335–342.

Simonet, W.S., Lacey, D.L., Dunstan, C.R., Kelley, M., Chang, M.S., Luthy, R., Nguyen, H.Q., Wooden, S., Bennett, L., Boone, T., et al. 1997. Osteoprotegerin: A novel secreted protein involved in the regulation of bone density. *Cell* **89:** 309–319.

Steingrimsson, E., Tessarollo, L., Pathak, B., Hou, L., Arnheiter, H., Copeland, N.G., and Jenkins, N.A. 2002. Mitf and Tfe3, two members of the Mitf-Tfe family of bHLH-Zip transcription factors, have important but functionally redundant roles in osteoclast development. *Proc. Natl. Acad. Sci.* **99:** 4477–4482.

Suda, M., Ogawa, Y., Tanaka, K., Tamura, N., Yasoda, A., Takigawa, T., Uehira, M., Nishimoto, H., Itoh, H., Saito, Y., et al. 1998. Skeletal overgrowth in transgenic mice that overexpress brain natriuretic peptide. *Proc. Natl. Acad. Sci.* **95:** 2337–2342.

Sun, M., Ma, F., Zeng, X., Liu, Q., Zhao, X.L., Wu, F.X., Wu, G.P., Zhang, Z.F., Gu, B., Zhao, Y.F., et al. 2008. Triphalangeal thumb-polysyndactyly syndrome and syndactyly type IV are caused by genomic duplications involving the long range, limb-specific SHH enhancer. *J. Med. Genet.* **45:** 589–595.

Superti-Furga, A. and Unger, S. 2007. Nosology and classification of genetic skeletal disorders: 2006 revision. *Am. J. Med. Genet. A* **143:** 1–18.

Sweeney, S.M., Orgel, J.P., Fertala, A., McAuliffe, J.D., Turner, K.R., Di Lullo, G.A., Chen, S., Antipova, O., Perumal, S., Ala-Kokko, L., et al. 2008. Candidate cell and matrix interaction domains on the collagen fibril, the predominant protein of vertebrates. *J. Biol. Chem.* **283:** 21187–21197.

Tassabehji, M., Read, A.P., Newton, V.E., Harris, R., Balling, R., Gruss, P., and Strachan, T. 1992. Waardenburg's syndrome patients have mutations in the human homologue of the *Pax-3* paired box gene. *Nature* **355:** 635–636.

Tassabehji, M., Newton, V.E., and Read, A.P. 1994. Waardenburg syndrome type 2 caused by mutations in the human microphthalmia (*MITF*) gene. *Nat. Genet.* **8:** 251–255.

Tavormina, P.L., Shiang, R., Thompson, L.M., Zhu, Y.Z., Wilkin, D.J., Lachman, R.S., Wilcox, W.R., Rimoin, D.L., Cohn, D.H., and Wasmuth, J.J. 1995. Thanatophoric dysplasia (types I and II) caused by distinct mutations in fibroblast growth factor receptor 3. *Nat. Genet.* **9:** 321–328.

Thomas, J.T., Lin, K., Nandedkar, M., Camargo, M., Cervenka, J., and Luyten, F.P. 1996. A human chondrodysplasia due to a mutation in a TGF-β superfamily member. *Nat. Genet.* **12:** 315–317.

Ueki, Y., Tiziani, V., Santanna, C., Fukai, N., Maulik, C., Garfinkle, J., Ninomiya, C., doAmaral, C., Peters, H., Habal, M., et al. 2001. Mutations in the gene encoding c-Abl-binding protein SH3BP2 cause cherubism. *Nat. Genet.* **28:** 125–126.

Ueki, Y., Lin, C.Y., Senoo, M., Ebihara, T., Agata, N., Onji, M., Saheki, Y., Kawai, T., Mukherjee, P.M., Reichenberger, E., et al. 2007. Increased myeloid cell responses to M-

CSF and RANKL cause bone loss and inflammation in SH3BP2 "cherubism" mice. *Cell* **128:** 71–83.

van Bokhoven, H., Celli, J., Kayserili, H., van Beusekom, E., Balci, S., Brussel, W., Skovby, F., Kerr, B., Percin, E.F., Akarsu, N., et al. 2000. Mutation of the gene encoding the ROR2 tyrosine kinase causes autosomal recessive Robinow syndrome. *Nat. Genet.* **25:** 423–426.

Vortkamp, A., Gessler, M., and Grzeschik, K.H. 1991. GLI3 zinc-finger gene interrupted by translocations in Greig syndrome families. *Nature* **352:** 539–540.

Vortkamp, A., Lee, K., Lanske, B., Segre, G.V., Kronenberg, H.M., and Tabin, C.J. 1996. Regulation of rate of cartilage differentiation by Indian hedgehog and PTH-related protein. *Science* **273:** 613–622.

Wagner, T., Wirth, J., Meyer, J., Zabel, B., Held, M., Zimmer, J., Pasantes, J., Bricarelli, F., Keutel, J., Hustert, E., et al. 1994. Autosomal sex reversal and campomelic dysplasia are caused by mutations in and around the *SRY*-related gene *SOX9*. *Cell* **79:** 1111–1120.

Wallis, D., Lacbawan, F., Jain, M., Der Kaloustian, V.M., Steiner, C.E., Moeschler, J.B., Losken, H.W., Kaitila, I.I., Cantrell, S., Proud, V.K., et al. 2008. Additional *EFNB1* mutations in craniofrontonasal syndrome. *Am. J. Med. Genet. A* **146A:** 2008–2012.

Wilkie, A.O. 1997. Craniosynostosis: Genes and mechanisms. *Hum. Mol. Genet.* **6:** 1647-1656.

Wilkie, A.O., Slaney, S.F., Oldridge, M., Poole, M.D., Ashworth, G.J., Hockley, A.D., Hayward, R.D., David, D.J., Pulleyn, L.J., Rutland, P., et al. 1995. Apert syndrome results from localized mutations of *FGFR2* and is allelic with Crouzon syndrome. *Nat. Genet.* **9:** 165–172.

Wilkie, A.O., Tang, Z., Elanko, N., Walsh, S., Twigg, S.R., Hurst, J.A., Wall, S.A., Chrzanowska, K.H., and Maxson Jr., R.E. 2000. Functional haploinsufficiency of the human homeobox gene *MSX2* causes defects in skull ossification. *Nat. Genet.* **24:** 387–390.

Yousfi, M., Lasmoles, F., and Marie, P.J. 2002. TWIST inactivation reduces CBFA1/RUNX2 expression and DNA binding to the osteocalcin promoter in osteoblasts. *Biochem. Biophys. Res. Commun.* **297:** 641–644.

Yu, L., Liu, H., Yan, M., Yang, J., Long, F., Muneoka, K., and Chen, Y. 2007. *Shox2* is required for chondrocyte proliferation and maturation in proximal limb skeleton. *Dev. Biol.* **306:** 549–559.

Zelzer, E., Glotzer, D.J., Hartmann, C., Thomas, D., Fukai, N., Shay, S., and Olsen, B.R. 2001. Tissue specific regulation of VEGF expression by Cbfa1/Runx2 during bone development. *Mech. Dev.* **106:** 97–106.

Zelzer, E., Mamluk, R., Ferrara, N., Johnson, R.S., Schipani, E., and Olsen, B.R. 2004. VEGFA is necessary for chondrocyte survival during bone development. *Development* **131:** 2161–2171.

Zhao, X., Sun, M., Zhao, J., Leyva, J.A., Zhu, H., Yang, W., Zeng, X., Ao, Y., Liu, Q., Liu, G., et al. 2007. Mutations in *HOXD13* underlie syndactyly type V and a novel brachydactyly-syndactyly syndrome. *Am. J. Hum. Genet.* **80:** 361–371.

13

Extracellular Matrix in the Skeleton

Francesco Ramirez

Department of Pharmacology and Systems Therapeutics
Mount Sinai School of Medicine,
New York, New York 10029

T HE EXTRACELLULAR MATRIX (ECM) is a highly heterogeneous amalgam of multidomain molecules that are intimately involved in the development, growth, function, and homeostasis of every organ system, including the skeleton. Similar to other connective tissues, bone and cartilage matrices consist of collagens, proteoglycans (PGs), and noncollagenous (NC) proteins, in addition to including enzymes involved in matrix assembly and degradation. That the vast majority of these molecules are also found in other tissues indicates that relative differences in ECM composition specify form and function at discrete anatomical locations of the developing and adult skeleton. This chapter provides an introduction to ECM composition and organization in the skeleton, and a brief review of the contribution of selected matrix molecules to bone formation and remodeling that is mostly based on genetic evidence from loss-of-function studies in mice. Similar topics are also covered in other chapters of this book, and a number of excellent reviews are available that describe various aspects of ECM biology in greater detail.

ECM COMPOSITION AND ORGANIZATION

Collagens

Collagens are the most abundant and diverse components of the connective tissue (Mecham 1998; Birk and Bruckner 2005). All collagens possess at least one triple helical (or collagenous [COL]) domain and NC domains of variable length and composition. Most collagens give rise to morphologically diverse suprastructures that are also referred to as molecular composites because they include additional collagens and NC proteins (Birk and Bruckner 2005). For example, tissue-specific organization of collagen I or II networks is largely regulated by copolymerization with collagens V or XI, direct binding to PGs, and indirect association with other ECM components via collagens attached to the fibril surface. Additionally, collagens can interact with cells through sequences recognized by different receptors, such as integrins and tyrosine kinase discoidin domain receptors (DDRs) (Heino 2007).

Fibril-forming collagens are the major architectural components of skeletal tissues (Morris et al. 2002; Schenk et al. 2002; von der Mark 2006). Their distinguishing feature is the presence of a long uninterrupted COL domain flanked at both ends by globular propeptides, which are removed by BMP1/TLD and ADAMTS endopeptidases prior to collagen self-assembly into cross-linked fibrils. Nearly 95% of the bone organic matrix consists of collagen I fibrils that afford multidirectional strength, in addition to serving as substrate for tissue mineralization (Schenk et al. 2002; von der Mark 2006). Collagen I fibrils of woven bone display a random organization, whereas those of lamellar bone are arranged in parallel and alternatively oriented layers. Similarly, collagen II fibrils are the architectural determinants of tensile strength in all cartilaginous tissues (Morris et al. 2002; von der Mark 2006). Elastic fibers and collagen I fibrils contribute to the distinct mechanical properties of elastic cartilage and fibrocartilage, respectively, whereas collagen X networks are uniquely found in the hypertrophic zone of growth plates. Collagen II fibrils are thin and oriented randomly in the resting zone, and increasingly larger and aligned longitudinally in the lower layers of growth plates. They are organized as a fine and tightly woven meshwork in the pericellular matrix of chondrocytes, which also includes collagen VI, hyaluronan (HA), PGs, and NC proteins.

Proteoglycans

PGs represent a special class of NC proteins containing one or more glycosaminoglycan (GAG) side chains, covalently attached to a core protein (Mecham 1998; Heinegard et al. 2006; Hardingham 2006). GAG groups include chondroitin sulfate (CS), dermatan sulfate (DS), heparan sulfate (HS), keratan sulfate (KS), and HA. Structural considerations segregate PGs into small leucine-rich PGs (SLRPs; e.g., decorin and biglycan) and modular PGs, which are in turn divided into molecules bound to HA or lectin (e.g., aggrecan) or other GAG chains (e.g., perlecan).

PG content varies greatly in bone and cartilage, reflecting distinct physiological and biomechanical demands (Hardingham 2006). A case in point is the integration within cartilage fibrils of the aggrecan, link protein, and HA complex to form a "fiber-reinforced composite matrix swollen with water that can withstand both tensional and compressive loads" (Hardingham 2006). Whereas the HS-PG perlecan is only found in the inter-territorial region of cartilage, SLRPs that bind collagen fibrils (decorin) or modulate TGF-β activity (biglycan) are differentially expressed in both bone and cartilage (Morris et al. 2002; Heinegard et al. 2006).

NC Proteins

This heterogeneous group of ECM constituents includes a wide variety of structurally and functionally diverse proteins (Mecham 1998; Gunderberg and Nishimoto 2006; Heinegard et al. 2006). Amongst them, elastic fibers and 10-nm microfibrils represent the NC scaffold of skeletal tissues (Mecham 1998; Ramirez et al. 2008). In contrast to the restricted distribution of elastic fibers in the perichondrium and elastic cartilage, 10-nm microfibrils are widely dispersed throughout the skeleton, either associated with elastin or as elastin-free assemblies. They are made of polymers of fibrillins 1 and 2, cysteine-rich glycoproteins that contain a single integrin-binding (Arg-Gly-Asp; RGD) sequence, and heparin-binding sites that interact with extracellular and cell-surface PGs. Fibrillins can associate with several other ECM molecules, including latent TGF-β binding proteins (LTBPs). LTBP binding to fibrillin-rich microfibrils promotes sequestration and storage of the large latent TGF-β complex (LLC) into the ECM, from which it is released and activated during tissue formation or remodeling (Rifkin 2005). A comparable function has been proposed for BMP interaction with fibrillin-rich microfibrils (Gregory et al. 2005).

Osteocalcin is a mineral-binding Gla (γ-carboxylated glutamic acid) protein exclusively found in hard tissues and serum that is often grouped together with matrix-Gla-protein (MGP), which is instead produced by both chondrocytes and vascular smooth muscle cells (Gunderberg and Nishimoto 2006). Osteopontin and osteonectin are calcium-binding, RGD-containing molecules that belong to the matricellular family of ECM proteins (Mecham 1998). Osteopontin is secreted by osteoblasts and osteoclasts, and in a variety of injured tissues or during cancer progression. Similarly, osteonectin is a prominent component of mineralized matrices that is also expressed in tissues undergoing morphogenesis or remodeling.

ECM Proteinases

Enzymes involved in ECM assembly and degradation include members of the MMP, ADAMTS, and BMP1/TLD families of metalloproteinases (Flannery 2006; Ge and Greenspan 2006; Ra and Parks 2007). MMPs collectively cleave molecules as diverse as structural and enzymatic components of the ECM, growth factors and cytokines, and cell receptors and adhesion molecules (Flannery 2006; Ra and Parks 2007). MMPs are secreted into the matrix or anchored to the cell surface as proenzymes (zymogens) that are activated by proteolytic and nonproteolytic mechanisms. In contrast to in vitro evidence associating tissue inhibitors of metalloproteinases (TIMPs) with MMP silencing, in vivo data have implicated TIMP2 in proMMP2 activation through the formation of a ternary complex in which interaction of TIMP2 with proMMP2 and cell surface-anchored MMP14 tethers the zymogen close to the endopeptidase (Wang et al. 2000). That MMP2 activity is unaffected in MMP14 null mice suggests involvement of other membrane-type MMPs and/or additional mechanisms of zymogen activation (Holmbeck et al. 1999). Whereas some of the ADAMTSs participate in aggrecan proteolysis and collagen maturation, BMP1/TLDs are involved in both processing structural and enzymatic ECM components and regulating the activity of TGF-β super-family members (Flannery 2006; Ge and Greenspan 2006).

ECM FUNCTION

Endochondral Bone Formation

Growth plate differentiation is accompanied by substantial changes in ECM morphology and composition (Morris et al. 2002). Increased matrix deposition and directional cell division lead to the emergence of longitudinal septa (inter-territorial matrix) between chondrocytic columns, and transverse septa (territorial matrix) between individual cells. As chondrocytes mature, matrix vesicles appear on the longitudinal septa and release mineral clusters that eventually coalesce into a calcified matrix. Additional changes include down-regulation of collagen II production, degradation of collagen II and aggrecan, deposition of collagen X, chondrocyte apoptosis, and vascular invasion of transverse septa. This last event introduces osteoblasts that deposit bone-specific products, as well as osteoclasts and chondroclasts that secrete MMPs and other matrix-degrading enzymes.

Collagen II deficient mice have demonstrated the absolute requirement of this matrix component for bone formation, in addition to yielding mechanistic insights into key interactions with other matrix molecules and differentiating chondrocytes (Li et al. 1995a; Aszodi et al. 1998). Ectopic expression of collagens I and III by collagen II null chondrocytes, which otherwise produce all other cartilage-specific products and undergo terminal differentiation, has been interpreted to suggest that collagen II deposition normally suppresses the default expression of non-cartilaginous collagens in these cells (Aszodi et al. 1998). Additional genetic evidence supports the notion that reciprocal interactions between chondrocytes and collagen II stimulate outside-in and inside-out signals that instruct cell behavior and organize matrix assembly. Cartilage-specific loss of β_1 integrin has been reported to perturb the ability of chondrocytes to arrange into columns, as a result of defective adhesion to collagen II, and to proliferate, due in part to elevated FGFR3 expression (Aszodi et al. 2003). Similarly, dwarfism of *Ddr2* null mice has implicated this collagen-binding receptor in promoting chondrocyte proliferation (Labrador et al. 2001). Moreover, the reduced density and highly irregular morphology of collagen II fibrils in β_1 or α_{10}-deficient growth plates has strongly suggested that integrins may directly participate in assembling the collagen network (Aszodi et al. 2003; Bengtsson et al. 2004). Collagen II null mice have also revealed that this molecule establishes a mechanically suitable environment for notochord removal during intervertebral disc formation (Aszodi et al. 1998). Finally, the severely compromised growth plate of collagen XI mutant (*cho/cho*) mice has further reiterated the importance of collagen II interacting molecules in fibrillogenesis (Li et al. 1995b).

Structural and instructive functions have been attributed to aggrecan as well, based on the abnormal growth plates and deregulated expression of cartilage genes in *cmd/cmd* mice (Watanabe et al. 1994; Wai et al. 1998). Similar findings from perlecan null mice have been related to potential roles of this HS-PG in inhibiting ECM degradation, modulating FGF signals, and/or regulating Hedgehog diffusion (Arikawa-Hirasawa et al. 1999; Costell et al. 1999; Olsen 1999). This last hypothesis is consistent with results from glycosyltransferase-deficient mice, indicating that HS is required to sequester Hedgehog into the cartilage matrix, in addition to regulating the range of Hedgehog signaling negatively and in a concentration-dependent manner (Koziel et al. 2004; Stickens et al. 2005). Along the same lines, it has been proposed that chondroitin-4-sulfotransferase may direct chondrogenesis in the growth plate by coordinating TGF-β and BMP signals (Kluppel et al. 2005).

Perturbed matrix organization and bone mineralization in mice lacking collagen X have established its function as a tissue-specific factor that compartmentalizes the distribution of PGs and matrix vesicles within the growth plate matrix (Kwan et al. 1997). Similarly, ectopic calcification of growth plate cartilage and arteries in *Mgp* null mice have demonstrated its role as a potent local inhibitor of ECM mineralization (Luo et al. 1997; Murshed et al. 2004). While MGP deposition into the territorial matrix participates in restricting cartilage calcification to the longitudinal septa, matrix proteolysis is required to eliminate the transverse septa around late hypertrophic chondrocytes, thus enabling vascular invasion and bone formation. Analyses of MMP-deficient mice have revealed their discrete contributions to this process. Selective loss of MMP13 production by chondrocytes has associated this enzyme with the removal of transverse septa and the invasion of the ossification front (Inada et al. 2004; Stickens et al. 2004). MMP9 secretion by osteo-chondroclasts, on the other hand, has been shown to coordinate cartilage resorption, chondrocyte apoptosis, vascular invasion, and osteoblast differentiation (Vu et al. 1998; Engsig et al. 2000). In vivo and in vitro data strongly suggest that MMP9 control of vascular invasion is mostly exerted through the negative regulation of VEGF bioavailability in hypertrophic cartilage (Engsig et al. 2000). Finally, the phenotype of MMP14-deficient mice has indicated a specialized function in collagenolysis during modeling of soft issues, including unmineralized epiphyseal cartilage (Holmbeck et al. 1999, 2003). That mice with MMP-resistant aggrecan or lacking various aggrecanases display unremarkable levels of the PG has led to questioning whether aggrecan removal is a more complex process than previously suspected or not as critical as collagen degradation for bone formation (Little et al. 2005a,b).

Bone Remodeling

Dynamic changes in ECM composition underpin the locally coupled process of bone remodeling by matrix-producing osteoblasts and matrix-degrading osteoclasts (Schenk et al. 2002). During the catabolic phase, activated bone lining cells degrade the unmineralized matrix (osteoid) and expose (and perhaps deposit too) cell-binding signals that promote osteoclast adhesion and polarization. In the anabolic phase, pre-osteoblasts adhere to the eroded bone surface (through endogenous and/or newly deposited cell-binding proteins) and lay down a mineralization-competent osteoid.

Mouse models of osteogenesis imperfecta have firmly established collagen I role in bone formation and integrity, in addition to implicating other molecules in this function (Aszodi et al. 2006). Examples include the demonstration that PCOLCE1 enhancement of BMP1/TDL processing of procollagen I contributes to bone mechanical properties, and that prolyl 3-hydroxylation of collagen I is required for bone mass maintenance (Morello et al. 2006; Steiglitz et al. 2008). Other mouse studies have demonstrated that physiological restriction of matrix mineralization to bone is predicated on the proper assembly of a collagen I network, the relative concentration of extracellular phosphate, and the absence of inhibitors otherwise present in soft tissues (Murshed et al. 2005).

Several NC proteins have been implicated in supporting bone mass maintenance through distinct mechanisms. Osteopenia in osteonectin-deficient mice has been accounted for by decreased bone formation and fewer osteoblasts, a finding the latter correlated with a reduced pool of precursor cells in the marrow stroma and a relatively greater number of adipocytes in osteoblast cultures (Delaney et al. 2000, 2003). These and additional data have been interpreted to suggest that osteonectin may contribute to the formation of a differentiation-competent matrix by favoring expression of pro-osteoblastic factors while inhibiting expression of proadipocytic factors (Delaney et al. 2003).

Reduced bone formation and fewer osteoblasts characterize age-dependent osteopenia of biglycan-deficient mice as well (Xu et al. 1998). This phenotype has been associated with increased apoptosis and decreased response to TGF-β-stimulation of bone marrow stromal cells, as well as with impaired BMP4-induced differentiation of preosteoblasts (Chen et al. 2002, 2004; Bi et al. 2005). It has been proposed from this and additional studies that competitive interactions of matrix-bound biglycan and decorin with TGF-β and BMPs may modulate the respective signals on osteoblast differentiation (Chen et al. 2004).

Our unpublished data indicate that fibrillin 1 and fibrillin 2 null mice are both osteopenic due to promiscuous activation of latent TGF-β (H. Nistala, unpubl.). In contrast to fibrillin 1-deficient osteoblasts, those from fibrillin 2-null mice deposit less matrix and fail to differentiate properly. On the other hand, both mutant osteoblasts display higher RANKL levels and stimulate osteoclastogenesis in coculture experiments. Hence, fibrillins appear to modulate local TGF-β bioavailability in forming bone differently, perhaps as a result of discrete contributions to matrix formation, LLC binding, and/or cell-matrix interactions. This conclusion is consistent with the previous finding that loss of fibrillin 2, but not of fibrillin 1, causes a bone-patterning defect even though both proteins are coexpressed in the forming skeleton (Arteaga-Solis et al. 2001; Carta et al. 2006). Furthermore, fibrillin's role in controlling local TGF-β bioavailability is also in line with the argument that osteopetrosis in LTBP-3-deficient mice is the result of decreased TGF-β levels (Dabovic et al. 2005).

In vivo and in vitro studies have indicated that osteopontin participates in bone resorption by promoting osteoclast motility and adhesion (Chellaiah et al. 2003; Franzén et al. 2008). Age-dependent increase of bone mass in mice lacking $\alpha_v\beta_3$ integrin, which is selectively expressed by differentiating osteoclasts upon RANKL induction, has further underscored the importance of cell-matrix interactions in bone remodeling (McHugh et al. 2000). Contrary to expectations, loss of MMP2 activity in mice has been related with bone loss and deficiencies in osteoblast proliferation, osteoblast-dependent osteoclastogenesis, and osteocyte processes (Inoue et al. 2006; Mosig et al. 2007). Relevant to the last finding, collagen I degradation and formation of osteocyte processes are impaired in mice lacking MMP14, which is part of the cell surface complex involved in MMP2 activation (Holmbeck et al. 2004). The importance of collagen I degradation in remodeling is further supported by the perturbed activity of osteoblasts and osteoclasts in mice producing MMP-resistant collagen I fibrils (Zhao et al. 2000; Chiusaroli et al. 2003).

Seminal work by the Karsenty's group has recently demonstrated that osteocalcin is a bone-specific hormone that regulates energy metabolism. These investigators have in fact shown that loss of osteocalcin leads to an age-dependent increase in bone mass associated with visceral fat hypoplasia, and that circulating γ-carboxylated osteocalcin controls adipogenesis by regulating pancreatic β-cell proliferation, insulin production, and energy expenditure (Ducy et al. 1996; Lee et al. 2007). Together with previous evidence of leptin involvement in the control of bone mass (Takeda et al. 2002), these findings have implicated the skeleton as an integral component of the endocrine system that orchestrates bone homeostasis, fat uptake, and energy metabolism.

CONCLUSIONS

Studies highlighted in this review demonstrate that the ECM is a key extrinsic determinant of skeletal function that confers mechanical and physiological properties, specifies tissue boundaries, instructs cell behavior, and modulates morphogenetic and homeostatic events. As the genetic evidence indicates, this multiplicity of functions is exerted through the combinatorial organization of supramolecular structures, reciprocal cell-matrix interactions, complex and often redundant enzymatic processes, and local control of signaling molecules. While a genetic catalog of ECM function is gradually emerging, its highly insoluble nature still represents a formidable obstacle to understand how the orderly process of ECM assembly translates into the formation of higher order skeletal structures and acquisition of organ function. There is, however, good reason to believe that mouse genetics combined with new proteomics and imaging modalities may eventually yield a comprehensive in vivo picture of the ECM dynamics that drive bone formation and homeostasis.

ACKNOWLEDGMENTS

Studies from the author's laboratory were supported by grants from the National Institutes of Health (AR-049698 and AR-42044). Ms. Karen Johnson provided invaluable assistance with organizing the manuscript.

REFERENCES

Arikawa-Hirasawa, E., Watanabe, H., Takami, H., Hassell, J.R., and Yamada, Y. 1999. Perlecan is essential for cartilage and cephalic development. *Nature* **23:** 354–358.

Arteaga-Solis, E., Gayraud, B., Lee, S.Y., Shum, L., Sakai, L.Y., and Ramirez, F. 2001. Regulation of limb patterning by extracellular microfibrils. *J. Cell Biol.* **154:** 275–281.

Aszodi, A., Chan, D., Hunziker, E., Bateman, J.F., and Fassler, R. 1998. Collagen II is essential for the removal of the notochord and the formation of intervertebral discs. *J. Cell Biol.* **143:** 1399–1412.

Aszodi, A., Hunziker, E.B., Brakebusch, C., and Fassler, R. 2003. β1 integrins regulate chondrocyte rotation, G1 progression, and cytokinesis. *Genes Dev.* **17:** 2465–2479.

Aszodi, A., Legate, K.R., Nakchband, I., and Fassler, R. 2006. What mouse mutants teach us about extracellular matrix function. *Annu. Rev. Cell Dev. Biol.* **22:** 591–621.

Bengtsson, T., Aszodi, A., Nicolae, C., Hunziker, E.B., Lundgren-Akerlund, E., and Fassler, R. 2004. Loss of α10β1 integrin expression leads to moderate dysfunction of growth plate chondrocytes. *J. Cell Sci.* **118:** 929–936.

Bi, Y., Stueltens, C.H., Kilts, T., Wadhwa, S., Iozzo, R.V., Robey, P.G., Chen, X.D., and Young, M.F. 2005. Extracellular matrix proteoglycans control the fate of bone marrow stromal cells. *J. Biol. Chem.* **280:** 30481–30489.

Birk, D.E. and Bruckner, P. 2005. Collagen suprastructures. *Top. Curr. Chem.* **247:** 185–205.

Carta, L., Pereira, L., Arteaga-Solis, E., Lee-Arteaga, S.Y., Lenart, B., Starcher, B., Merkel, C.A., Sukoyan, M., Kerkis, A., Hazeki, N., et al. 2006. Fibrillins 1 and 2 perform partially overlapping functions during aortic development. *J. Biol. Chem.* **281:** 8016–8023.

Chellaiah, M.A., Kizer, N., Biswas, R., Alvarez, U., Strauss-Schoenberger, J., Rifas, L., Rittling, S.R., Denhardt, D.T., and Hruska, K.A. 2003. Osteopontin deficiency produces osteoclast dysfunction due to reduced CD44 surface expression. *Mol. Biol. Cell* **14:** 173–189.

Chen, X.-D., Shi, S., Xu, T., Robey, P.G., and Young, M.F. 2002. Age-related osteoporosis in biglycan-deficient mice is related to defects in bone marrow stromal cells. *J. Bone Miner. Res.* **17:** 331–340.

Chen, X.-D., Fisher, L.W., Robey, P.G., and Young, M.F. 2004. The small leucine-rich proteoglycan biglycan modulates BMP-4-induced osteoblast differentiation. *FASEB J.* **18:** 948–958.

Chiusaroli, R., Maier, A., Knight, M.C., Byrne, M., Calvi, L.M., Baron, R., Krane, S.M., and Schipani, E. 2003. Collagenase cleavage of type I collagen is essential for parathyroid hormone (PTH)/PTH-related peptide receptor-induced osteoclast activation and has differential effects on discrete bone compartments. *Endocrinology* **144:** 4106–4116.

Costell, M., Gustafason, E., Aszodi, A., Morgelin, M., Bloch, W., Hunziker, E., Addicks, K., Timpl, R., and Fassler, R. 1999. Perlecan maintains the integrity of cartilage and some basement membranes. *J. Cell Biol.* **147:** 1109–1122.

Dabovic, B., Lavassseur, R., Zambuto, L., Chen, Y., Karsenty, G., and Rifkin, D.B. 2005. Osteopetrosis-like phenotype in latent TGF-β binding protein 3 deficient mice. *Bone* **37:** 25–31.

Delaney, A.M., Amling, M., Priemel, M., Howe, C., Baron, R., and Canalis, E. 2000. Osteopenia and decreased bone formation in osteonectin-deficient mice. *J. Clin. Invest.* **105:** 915–923.

Delaney, A.M., Kalajzic, I., Bradshaw, A.D., Sage, E.H., and Canalis, E. 2003. Osteonectin-null mutation compromises osteoblast formation, maturation, and survival. *Endocrinology* **144:** 2588–2596.

Ducy, P., Desbois, C., Boyce, B., Pinero, G., Story, B., Dunstan, C., Smith, E., Bonadio, J., Goldstein, S., Gundberg, C., Bradley, A., and Karsenty, G. 1996. Increased bone formation in osteocalcin-deficient mice. *Nature* **382:** 448–452.

Engsig, M.T., Chen, Q.-J., Vu, T.H., Pedersen, A.-C., Therkidsen, B., Lund, L.R., Henriksen, K., Lenhard, T., Foged, N.T., Werb, Z., and Delaisse, J.-M. 2000. Matrix metalloproteinase 9 and vascular endothelial growth factor are essential for osteoclast recruitment into developing long bones. *J. Cell Biol.* **151:** 879–889.

Flannery, C.R. 2006. MMPs and ADAMTSs: Functional studies. *Front. Biosci.* **11:** 544–569.

Franzén, A., Hultenby, K., Reinholt, F.P., Onnerfjord, P., and Heingård D. 2008. Altered osteoclast development and function in osteopontin deficient mice. *J. Orthop. Res.* **26:** 721–728.

Ge, G. and Greenspan, D.S. 2006. Developmental roles of the BMP1/TLD metalloproteinases. *Birth Defects Res. Part C* **78:** 47–68.

Gregory, K.E., Ono, R.N., Charbonneau, N.L., Kuo, C.L., Keene, D.R., Bachinger, H.P., and Sakai, L.Y. 2005. The prodomain of BMP-7 targets the BMP-7 complex to the extracellular matrix. *J. Biol. Chem.* **280:** 27970–27980.

Gunderberg, C.M. and Nishimoto, S.K. 2006. Vitamin K dependent proteins of bone and cartilage. In *Dynamics of bone and cartilage metabolism*, 2nd ed. (ed. M.J. Seibel et al.), pp. 55–70. Academic, San Diego.

Hardingham, T. 2006. Proteoglycans and glycosaminoglycans. In *Dynamics of bone and cartilage metabolism*, 2nd ed. (ed. M.J. Seibel et al.), pp. 85–98. Academic, San Diego.

Heinegard, D., Lorenzo, P., and Saxne, T. 2006. Non-collagenous proteins; glycoproteins and related proteins. In *Dynamics of bone and cartilage metabolism*, 2nd ed. (ed. M.J. Seibel et al.), pp. 71–84. Academic, San Diego.

Heino, J. 2007. The collagen family members as cell adhesion proteins. *BioEssays* **29:** 1001–1010.

Holmbeck, K., Bianco, P., Caterina, J., Yamada, S., Kromer, M., Kuznetsov, S.A., Mankani, M., Robey, P.G., Poole, A.R., Pidoux, I., et al. 1999. MT1-MMP-deficient mice develop dwarfism, osteopenia, arthritis, and connective tissue disease due to inadequate collagen turnover. *Cell* **99:** 81–92.

Holmbeck, K., Bianco, P., Chrysovergis, K., Yamada, S., and Birkedal-Hansen, H. 2003. MT1-MMP-dependent, apoptotic remodeling of unmineralized cartilage: A critical process in skeletal growth. *J. Cell Biol.* **163:** 661–671.

Holmbeck, K., Bianco, P., Pidoux, I., Inoue, S., Billinghurst, R.C., Wu, W., Chrysovergis, K., Yamada, S., Birkedal-Hansen, H., and Poole, A.R. 2004. The metalloproteinase MT1-MMP is required for normal development and maintenance of osteocyte processes in bone. *J. Cell Sci.* **118:** 147–156.

Inada, M., Wang, Y., Byrne, M.H., Rahman, M.U., Miyaura, C., Lopez-Otin, C., and Krane, S.M. 2004. Critical roles for collagenase-3 (Mmp13) in development of growth plate cartilage and in endochondral ossification. *Proc. Natl. Acad. Sci.* **101:** 17192–17197.

Inoue, K., Mikuni-Takagaki, Y., Oikawa, K., Itoh, T., Inada, M., Noguchi, T., Park, J.-S., Onodera, T., Krane, S.M., Noda, M., and Itohara, S. 2006. A crucial role for matrix metalloproteinase 2 in osteocytic canalicular formation and bone metabolism. *J. Biol. Chem.* **281:** 33814–33824.

Kluppel, M., Wight, T.N., Chan, C., Hinek, A., and Wrana, J.L. 2005. Maintenance of chondroitin sulfation balance by chondroitin-4-sulfotransferase 1 is required for chondrocyte development and growth factor signaling during cartilage morphogenesis. *Development* **132:** 3989–4003.

Koziel, L., Kunath, M., Kelly, O.G., and Vortkamp, A. 2004. Ext1-dependent heparin sulfate regulates the range of Ihh signaling during endochondral ossification. *Dev. Cell* **6:** 801–813.

Kwan, K.M., Pang, M.K.M., Zhou, S., Cowan, S.K., Kong, R.Y.C., Pfordte, T., Olsen, B.R., Silllence, D.O., Tam, P.P.L., and Cheah, S.E. 1997. Abnormal compartmentalization of cartilage matrix components in mice lacking collagen X: Implications for function. *J. Cell Biol.* **136:** 459–471.

Labrador, J.P., Azcoitia, V., Tuckermann, J., Lin, C., Olaso, E., Manes, S., Bruckner, K., Goergen, J.-L., Lemke, G., Yancopoulos, G., et al. 2001. The collagen receptor DDR2 regulates proliferation and its elimination leads to dwarfism. *EMBO Rep.* **2:** 446–452.

Lee, N.K., Sowa, H., Hinoi, E., Ferron, M., Ahn, J.D., Confavreux, C., Dacquin, R., Mee, P.J., McKee, M.D., Jung, D.Y., et al. 2007. Endocrine regulation of energy metabolism by the skeleton. *Cell* **130:** 456–469.

Li, S.W., Prockop, D.J., Helminen, H., Fassler, R., Lapvetelainen, T., Kiraly, K., Peltarri, A., Arokoski, J., Lui, H., Arita, M., and Khillan, S. 1995a. Transgenic mice with targeted inactivation of the Col2α1 gene for collagen II develop a skeleton with membranes and periosteal bone but no endochondral bone. *Genes Dev.* **9:** 2821–2830.

Li, Y., Lacerda, D.A., Warman, M.L., Beier, D.R., Yoshioka, H., Ninomiya, Y., Oxford, J.T., Morris, N.P., Andrikopoulos, K., Ramirez, F., et al. 1995b. A fibrillar collagen gene, *Col11a1*, is essential for skeletal morphogenesis. *Cell* **80:** 423–430.

Little, C.B., Meeker, C.T., Hembry, R.M., Sims, N.A., Lawlor, K.E., Golub, S.B., Last, K., and Fosang, A.J. 2005b. Matrix metalloproteinases are not essential for aggrecan turnover during normal skeletal growth and development. *Mol. Cell. Biol.* **25:** 3388–3399.

Little, C.B., Mittaz, L., Belluccio, D., Rogerson, F.M., Campbell, I.K., Meeker, C.T., Bateman, J.F., Pritchard, M.A., and Fosang, A.J. 2005a. ADAMTS-1-knockout mice do not exhibit abnormalities in aggrecan turnover in vitro or in vivo. *Arthritis Rheum.* **52:** 1461–1472.

Luo, G., Ducy, P., McKee, M.D., Pinero, G.J., Loyer, E., Behringer, R.R., and Karsenty, G. 1997. Spontaneous calcification of arteries and cartilage in mice lacking matrix GLA protein. *Nature* **386:** 78–81.

McHugh, K.P., Hodivals-Dilke, K., Zheng, M.H., Namba, N., Lam, J., Novack, D., Feng,

X., Ross, F.P., Hynes, R.O., and Teitelbaum, S.L. 2000. Mice lacking β3 integrins are osteosclerotic because of dysfunctional osteoclasts. *J. Clin. Invest.* **105:** 433–440.

Mecham, R.P. 1998. Overview of extracellular matrix. *Curr. Protoc. Cell Biol.* **Ch. 10:** 10.1.1–10.1.14.

Morello, R., Bertin, T.K., Chen, Y., Hicks, J., Tonachini, L., Monticone, M., Castagnola, P., Rauch, F., Glorieux, F.H., Vranka, J., et al. 2006. CRTAP is required for prolyl 3-hydroxylation and mutations cause recessive osteogenesis imperfecta. *Cell* **127:** 291–304.

Morris, N.P., Keene, D.R., and Horton, W.A. 2002. Morphology and chemical composition of connective tissue: Cartilage. In *Connective tissue and its heritable disorders,* 2nd ed. (ed. P.M. Royce and B. Steinmann), pp. 41–65. Wiley-Liss, New York.

Mosig, R.A., Dowling, O., DiFeo, A., Ramirez, M.C.M., Parker, I.C., Abe, E., Diouri, J., Aqeel, A.A., Wylie, J.D., Oblander, S.A., et al. 2007. Loss of MMP-2 disrupts skeletal and craniofacial development and results in decreased bone mineralization, joint erosion and defects in osteoblast and osteoclast growth. *Hum. Mol. Genet.* **16:** 1113–1123.

Murshed, M., Schinke, T., McKee, M.D., and Karsenty, G. 2004. Extracellular matrix mineralization is regulated locally; different roles of two Gla-containing proteins. *J. Cell Biol.* **165:** 625–630.

Murshed, M., Harmey, D., Millan, J.L., McKee, M.D., and Karsenty, G. 2005. Unique coexpression in osteoblasts of broadly expressed genes accounts for the spatial restriction of ECM mineralization to bone. *Genes Dev.* **19:** 1093–1104.

Olsen, B.R. 1999. Life without perlecan has its problems. *J. Cell Biol.* **147:** 909–911.

Ra, H.J. and Parks, W.C. 2007. Control of matrix metalloproteinase catalytic activity. *Matrix Biol.* **26:** 587–596.

Ramirez, F., Carta, L., Lee-Arteaga, S., Liu, C., Nistala, K., and Smaldone, S. 2008. Fibrillin-rich microfibrils; structural and instructive determinants of mammalian development and organ function. *Connect. Tissue Res.* **49:** 1–6.

Rifkin, D.B. 2005. Latent transforming growth factor-β binding proteins: Orchestrators of TGF-β bioaivalability. *J. Biol. Chem.* **280:** 7409–7412.

Schenk, R.K., Hofstetter, W., and Felix, R. 2002. Morphology and chemical composition of connective tissue: Bone. In *Connective tissue and its heritable disorders,* 2nd ed. (ed. P.M. Royce and B. Steinmann), pp. 67–120. Wiley-Liss, New York.

Steiglitz, B.M., Kreider, J.M., Frankenburg, E.P., Pappano, W.N., Hoffman, G.G., Meganck, J.F., Liang, X., Hook, M., Birk, D.E., Goldstein, S.A., et al. 2008. Procollagen C proteinase enhancer 1 genes are important determinants of the mechanical properties and geometry of bone and the ultrastructure of connective tissues. *Mol. Cell. Biol.* **26:** 238–249.

Stickens, D., Behonick, D.J., Ortega, N., Heyer, B., Hartenstein, B., Yu, Y., Fosang, A.J., Schorpp-Kistner, M., Angel, P., and Werb, Z. 2004. Altered endochondral bone development in matrix metalloproteinase 13-deficient mice. *Development* **131:** 5883–5895.

Stickens, D., Zak, B.M., Rougler, N., Esko, J.D., and Werb, Z. 2005. Mice deficient in Ext2 lack heparin sulfate and develop exostoses. *Development* **132:** 5055–5068.

Takeda, S., Elefterious, F., Levesseur, R., Liu, X., Zhao, L., Parker, K.L., Armstrong, D., Ducy, P., and Karsenty, G. 2002. Leptin regulates bone formation via the sympathetic nervous system. *Cell* **111:** 305–317.

von der Mark, K. 2006. Structure, biosynthesis and gene regulation of collagens in cartilage and bone. In *Dynamics of bone and cartilage metabolism,* 2nd ed. (ed: M.J. Seibel et al.), pp. 3–40. Academic, San Diego.

Vu, T.H., Shipley, J.M., Bergers, G., Berger, J.E., Helms, J.A., Hanahan, D., Shapiro, S.D.,

Senior, R.M., and Werb, Z. 1998. MMP-9/gelatinase B is a key regulator of growth plate angiogenesis and apoptosis of hypertrophic chondrocytes. *Cell* **93:** 411–422.

Wai, A.W., Ng, L.J., Watanabe, H., Yamada, Y., Tam, P.P., and Cheah, K.S. 1998. Disrupted expression of matrix genes in the growth plate of the mouse cartilage matrix deficiency (cmd) mutant. *Dev. Genet.* **22:** 349–358.

Wang, Z., Jutterman, R., and Soloway, P.D. 2000. TIMP-2 is required for efficient activation of pro MMP-2 in vivo. *J. Biol. Chem.* **275:** 26411–26415.

Watanabe, H., Kimata, K., Line, S., Strong, D., Gao, L.-Y., Kozak, C.A., and Yamada, Y. 1994. Mouse cartilage matrix deficiency (cmd) caused by a 7bp deletion in the aggrecan gene. *Nat. Genet.* **7:** 154–157.

Xu, T., Bianco, P., Fisher, L.W., Longenecker, G., Smith, E., Goldstein, S., Bonadio, J., Boskey, A., Heegaard, A.-M., Sommer, B., et al. 1998. Targeted disruption of the biglycan gene leads to an osteoporosis-like phenotype in mice. *Nat. Genet.* **20:** 78–82.

Zhao, W., Byrne, M.H., Wang, Y., and Krane, S.M. 2000. Osteocyte and osteoblast apoptosis and excessive bone deposition accompany failure of collagenase cleavage of collagen. *J. Clin. Invest.* **106:** 941–949.

Index